Gauge Theories of Strong and Electroweak Interactions

Gauge Theories of Strong and Electroweak Interactions

PETER BECHER
and
MANFRED BÖHM
Würzburg University
and
HANS JOOS
Deutsches Elektronen-Synchrotron, Hamburg and *Hamburg University*

Translated by

VALERIE H. COTTRELL

A Wiley–Interscience Publication

JOHN WILEY & SONS
Chichester · New York · Brisbane · Toronto · Singapore

This edition is published by permission of Verlag B G Teubner, Stuttgart, and is the sole authorized English translation of the original German edition.

***Library of Congress Cataloging in Publication Data*:**

Becher, Peter.
Gauge theories of strong and electroweak interactions.

Translation of: Eichtheorien der starken und elektroschwachen Wechselwirkung.
'Wiley–Interscience.'
Includes bibliographical references and index.
1. Gauge fields (Physics) 2. Nuclear reactions.
3. Quantum chromodynamics. I. Böhm, Manfred, II. Joos, Hans. III. Title.
QC793.3.F5B4313 1983 530.1'4 83-6456

ISBN 0 471 10429 9

***British Library Cataloguing in Publication Data*:**

Becher, Peter
Gauge theories of strong and electroweak interactions.
1. Nuclear interactions 2. Gauge fields (Physics)
I. Title II. Böhm, Manfred III. Joos, Hans
539.7'54 QC794

ISBN 0 471 10429 9

Filmset and printed in Northern Ireland
at The Universities Press (Belfast) Ltd,
and bound at the Pitman Press, Bath, Avon

Contents

Preface

Our understanding of the basic elements of which matter is constituted has made considerable progress over the last decade. In theory, this led to the concept of gauge theories to describe the various interactions. The aim of the present book is to summarize this area of work for a wider range of interested parties in the realm of physics. It is in fact the elaboration of a manuscript containing lectures given to young students of elementary particle physics at the Autumn School for high energy physics in Maria Laach arranged by Professor J. K. Bienlein. The book is, therefore, mainly suited to more advanced students who want to become familiar with this area of work. Naturally, we hope that anyone with a general interest in gauge theories will learn something from it too.

As the detailed list of contents shows, the area covered is very extensive. It is a field of rapid development and no final conclusions are as yet in sight. The question then is whether a textbook can be justified at this stage. In the light of these considerations, the present book should be regarded as an experiment, surrounded by many loose ends. Over the course of time, many aspects will be seen in a different light, so that details which now seem important and which do in fact make the book difficult to read may eventually be omitted entirely. We tried to reduce the problem somewhat by giving relatively detailed summaries in the individual subsections, which could form a good guideline in themselves. On a first reading, technically complicated parts may be skipped over without detracting too much from a general understanding.

A serious problem was presented by the number of references selected from the gigantic number of original works. Apart from standard works, we have tried to include all review articles and papers which helped us originally towards a better understanding of the subject. Collective references are given for each chapter and these are placed before the actual full reference details in the bibliography section at the back, making for easier identification.

Conventions used are given on page 298.

This book owes much of its existence to the support offered by Deutsches Elektronen-Synchrotron DESY and its cooperation with the German universities. Many colleagues, from DESY and Würzburg University in particular, have assisted us both in word and deed: our thanks go to Dr A. Ali, Professor H. D. Dahmen, Professor H. Fraas, Dr W. Hollik, Dr M.

Krammer, W. Langguth, Professor G. Mack, Dr G. Münster, and Professor K. Symanzik in particular.

Finally we would like to thank Dr Spuhler from Teubner publications for his collaboration.

Hamburg/Würzburg
January 1981

P. Becher, M. Böhm, H. Joos

Preface to the Second Edition

Since the appearance of the first German edition, the ideas on the physics of elementary particles formulated with help of quantum gauge field theories have not changed very much. However, there has been some progress in details. The appendix to the second edition (p. 299) is a short report of these. The original text was changed only very little. Reference to new literature is marked by an N and collected in an appendix (p. 300). The experimental data as well as the conventions of the parton model have been changed according to [Pa 82]. Misprints in the first edition have been corrected.

Hamburg/Würzburg
January 1983

P. Becher, M. Böhm, H. Joos

CHAPTER 1

Phenomenological Basis of Gauge Theories of Strong, Electromagnetic, and Weak Interactions

The aim of elementary particle physics is to discover the fundamental law which will establish the dynamics of matter. Everything else remains a type of verbal painting, a 'super-review of particle properties' which is of interest to no-one for long. As this law combines the theory of relativity with quantum mechanics, it should take the form of a unified field theory for the strong, electromagnetic, weak, and ultimately gravitational interactions. This claim was made for elementary particle physics by W. Heisenberg at the Spring Conference of the German Physical Society in Munich in 1975 [He 76]. He tried to illustrate his point by presenting a speculative outline of such a theory [He 67].

Has research in the field of elementary particle physics brought us any nearer to this objective? Any answer to this question must take into earnest consideration the widespread hope that a theory, based on principles similar to those which have been so successful in quantum electrodynamics, can be built for the other interactions. The aim of this book is to give some foundation to these hopes, to describe the aforementioned dynamical principles and to discuss first applications of these theories.

In its purest form, quantum electrodynamics describes the interaction of electrons and positrons with photons. One special feature of this quantum theory of the electromagnetic and the Dirac-electron field is that the interaction of the field is uniquely defined by the principle of *minimal gauge-invariant coupling of the electric charge.* This important principle, first encountered in quantum mechanics in the form of the substitution rule $\partial_\mu \to \partial_\mu - ieA_\mu$, will be discussed in detail later on. It can be generalized to the interaction of complicated charge structures. One talks then of Yang–Mills field theories or non-Abelian gauge theories. One such Yang–Mills theory is quantum chromodynamics, which describes the interaction of hadrons by the gauge-invariant coupling of gluon fields to quark fields. A non-Abelian gauge theory can also describe the phenomena of the weak interaction when weak and electromagnetic fields are combined into a unified field theory, e.g. the Glashow–Salam–Weinberg theory. Thus, local gauge invariance may turn out to be an important general principle of the dynamics of matter.

The gauge theories for the interactions of elementary particles are based on simple models which supply a phenomenological explanation of experimental results. Among them the simple quark model is of special importance. The various aspects of the quark concept are the following:

(a) Quarks as hypothetical constituents of the strongly interacting particles explain the meson and baryon spectrum.
(b) Quark and lepton fields form the currents which are the sources of the universal electromagnetic and weak interaction.
(c) Quarks are seen as point like scattering centres—partons—in the deep inelastic lepton scattering.

The first section of this book deals with the phenomenological quark model of the hadron spectrum, the Fermi model of the weak interaction, and the simple parton model, plus a short introduction to the most simple concepts of group and field theory, by way of preparation for the formulation of gauge theories. [EP]

1.1 The hadron spectrum in the quark model

The quark model [Ko 69] was proposed by G. Zweig [Zw 64] and M. Gell-Mann [Ge 64] in 1964. In spite of an intensive search, quarks have not yet been found as free particles [Mo 79]. But pointlike scattering centres—partons—with quark properties have been seen in the deep inelastic lepton scattering. This apparent contradiction will, within the realm of quantum chromodynamics, form the basis of a discussion of the nature of quarks and especially their confinement in hadrons. First of all, a naive explanation of the properties of hadrons as determined by their quark composition is given.

1.1.1 Quantum numbers and wavefunctions of hadrons in the quark model

(a) *Quantum numbers*

It is well known that the many meson and baryon resonances observed can be classified by means of (conserved) *quantum numbers* [Lo 78, Ro 71a]. Currently, the flavour quantum numbers isospin I, I_3, strangeness S, charm C, baryon number B, hypercharge $Y=S+B-C/3$, electric charge $Q=I_3+Y/2+2C/3$ and the geometric quantum numbers spin j, j_3, parity π, and charge conjugation parity π_c are known. Figs 1.1 and 1.2 give a survey of the quantum numbers of the observed hadrons [Pa 82].

The remarkable experimental fact, that only certain values of these quantum numbers occur, can be explained with help of the quark model by

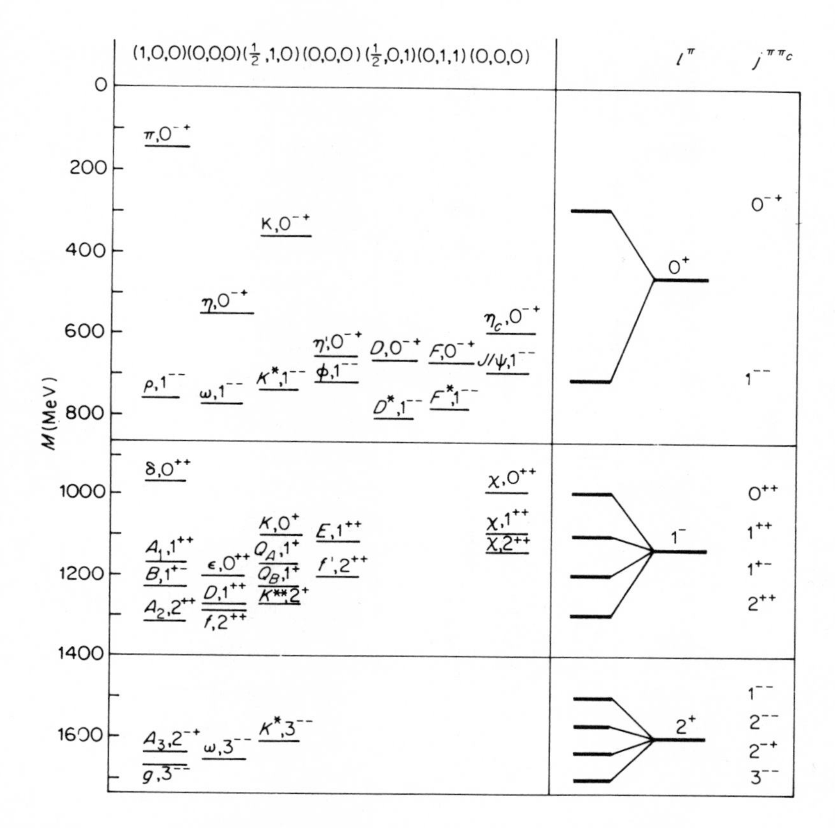

Figure 1.1 The meson spectrum in the quark model with four flavour degrees of freedom. The spectrum found experimentally is shown on the left. The number triplets in the first line stand for the quantum numbers (I, Y, C), $M-(150\text{ MeV})\cdot n_s-(1200\text{ MeV})\cdot n_c$ is written as mass, $n_s\equiv$number of s and $\bar{s}$ quarks, $n_c\equiv$number of c and $\bar{c}$ quarks in the meson. The low-lying multiplets expected in theory are shown on the right

the following hypothesis: mesons consist of one quark and one antiquark; baryons consist of three quarks.

In correspondence with the four quantum numbers I, I_3, S, and C, four quarks u, d, s, and c with flavour quantum numbers according to Table 1.1 are introduced. All quarks have baryon number 1/3 and spin 1/2.

The discovery of the Υ (9.46) particle [He 77] indicates that there is another family of hadrons with a new flavour quantum number, bottom, and thus another quark b. For this reason, the number of flavour degrees of freedom N_F will be left open in the following.

The quantum numbers and wavefunctions of hadrons are formed from the quark degrees of freedom, according to the rules for two- and three-particle bound states of non-relativistic quantum mechanics. The composition of

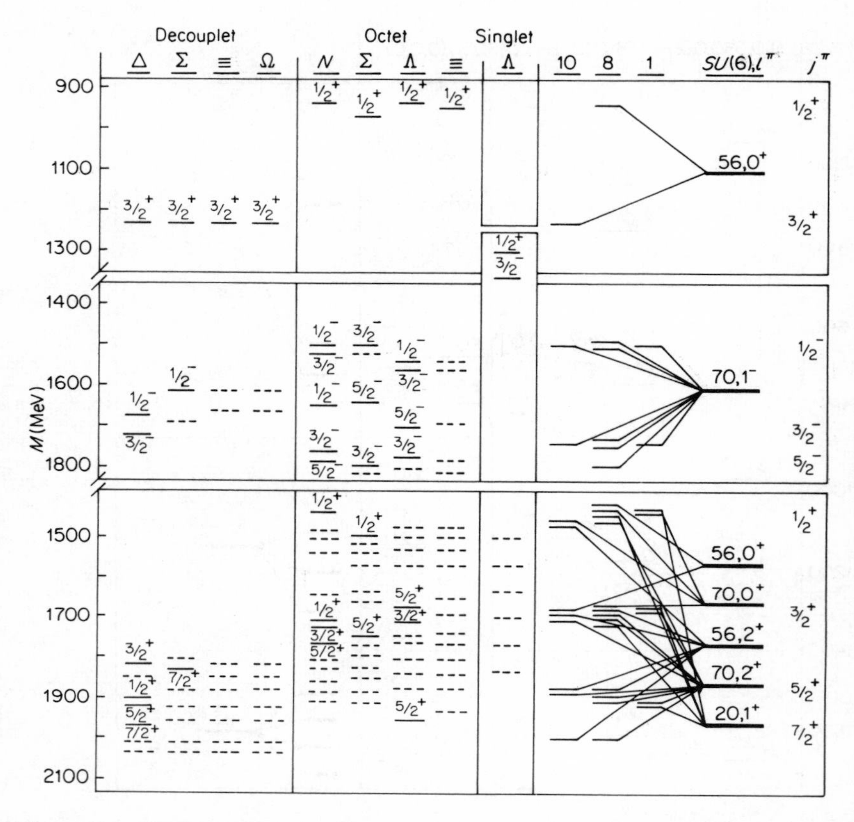

Figure 1.2 The baryon spectrum in the quark model with three flavour degrees of freedom. The spectrum observed experimentally is shown on the left. $M-(150\text{ MeV})\cdot n_s$ is written as mass, $n_s \equiv$ number of s quarks in the baryon. On the right are shown the low-lying particle multiplets predicted, arranged in order of total angular momentum values j^π. The particles still missing from the quark model are marked with dashed lines

Table 1.1 Quantum numbers of the quarks

Quark	Quantum number					
	I	I_3	S	C	Y	Q
u	1/2	+1/2	0	0	+1/3	+2/3
d	1/2	−1/2	0	0	+1/3	−1/3
s	0	0	−1	0	−2/3	−1/3
c	0	0	0	1	0	+2/3

wavefunctions from flavour, spin, and orbital parts is sketched in the following [Li 78, Cl 79].

(b) *Wavefunctions of* mesons *as quark–antiquark systems*

Construction of the flavour part $|I, I_3, S, C\rangle$ *from the quark degrees of freedom*

$$
\begin{aligned}
&|1,1,0,0\rangle = -\bar{d}u, \quad |1,0,0,0\rangle = (\bar{u}u - \bar{d}d)/\sqrt{2}, \quad |1,-1,0,0\rangle = \bar{u}d,\\
&|0,0,0,0\rangle = (\bar{u}u + \bar{d}d)/\sqrt{2};\\
&|\tfrac{1}{2},\tfrac{1}{2},1,0\rangle = \bar{s}u, \quad |\tfrac{1}{2},-\tfrac{1}{2},1,0\rangle = \bar{s}d,\\
&|\tfrac{1}{2},-\tfrac{1}{2},-1,0\rangle = \bar{u}s, \quad |\tfrac{1}{2},+\tfrac{1}{2},-1,0\rangle = -\bar{d}s,\\
&|0,0,0,0\rangle = \bar{s}s;\\
&|\tfrac{1}{2},\tfrac{1}{2},0,1\rangle = -\bar{d}c, \quad |\tfrac{1}{2},-\tfrac{1}{2},0,1\rangle = \bar{u}c,\\
&|\tfrac{1}{2},-\tfrac{1}{2},0,-1\rangle = \bar{c}d, \quad |\tfrac{1}{2},\tfrac{1}{2},0,-1\rangle = \bar{c}u;\\
&|0,0,1,1\rangle = \bar{s}c, \quad |0,0,-1,-1\rangle = \bar{c}s,\\
&|0,0,0,0\rangle = \bar{c}c;
\end{aligned}
\tag{1.1.1}
$$

The structure of multiplets of two quarks (u, d), three quarks (u, d, s), four quarks (u, d, s, c), and possibly more can be read off directly.

There are several particles with quantum numbers $|0, 0, 0, 0\rangle$ which can therefore mix, for example, pseudoscalar mesons show roughly SU(3) mixing:

$$|\eta(549)\rangle \simeq -(\bar{u}u + \bar{d}d - 2\bar{s}s)/\sqrt{6}, \quad |\eta'(958)\rangle \cong (\bar{u}u + \bar{d}d + \bar{s}s)/\sqrt{3}.$$

Spin part: $|s, s_3\rangle$. Quark and antiquark spins have the possible orientations $\uparrow$ and $\downarrow$. They can be combined to form a spin triplet and a spin singlet with the following wavefunctions:

$$
\begin{aligned}
&|1,1\rangle = \uparrow\uparrow, \quad |1,0\rangle = \frac{1}{\sqrt{2}}(\uparrow\downarrow + \downarrow\uparrow), \quad |1,-1\rangle = \downarrow\downarrow\\
&|0,0\rangle = \frac{1}{\sqrt{2}}(\uparrow\downarrow - \downarrow\uparrow);
\end{aligned}
\tag{1.1.2}
$$

Orbital part: $|l, m, n\rangle$. After separating off the centre of mass motion, the relative motion between quark and antiquark is described by a Schrödinger wavefunction of the relative coordinate r, which is made up of a spherical function $Y_{l,m}(\hat{r})$ and a radial part $f_{l,n}(r)$

$$|l, m, n\rangle = Y_{l,m}(r) f_{l,n}(r). \tag{1.1.3}$$

Spin and orbital angular momentum are added to meson spin j. This gives $j = l+1,\ l,\ l-1$ for the triplet and $j = l$ for the singlet states. The parity π and charge conjunction parity π_c of this fermion–antifermion system have the values $\pi=(-1)^{l+1}$ and $\pi_c=(-1)^{l+s}$. Fig. 1.1 shows the structure of the meson spectrum as it results from this composition of degrees of freedom.

(c) Baryons *as three-quark systems*

When three 'identical' particles are involved, the permutation symmetry of the wavefunction gives an important quantum number. In addition to the well-known symmetrical (*Sy*) and the antisymmetrical (*An*) there are the mixed-symmetrical wavefunctions ($\overline{Mi}$, $\underline{Mi}$), all of which can be constructed from the unsymmetrical wavefunctions as follows [Jo 70]:

$$|\text{Sy}\rangle=\frac{1}{\sqrt{6}}(|abc\rangle+|bca\rangle+|cab\rangle+|acb\rangle+|cba\rangle+|bac\rangle),$$

$$|\text{An}\rangle=\frac{1}{\sqrt{6}}(|abc\rangle+|bca\rangle+|cab\rangle-|acb\rangle-|cba\rangle-|bac\rangle). \tag{1.1.4}$$

For $a\neq b$, there exist two 2-dimensional, mixed symmetrical representations $Mi_{\pm}$:

$$|\overline{Mi}\rangle_{\pm}=\frac{1}{\sqrt{6}}((|abc\rangle\pm|bac\rangle)+\varepsilon(|bca\rangle\pm|acb\rangle)+\varepsilon^*(|cab\rangle\pm|cba\rangle)),$$

$$|\underline{Mi}\rangle_{\pm}=\frac{1}{\sqrt{6}}((|abc\rangle\pm|bac\rangle)+\varepsilon^*(|bca\rangle\pm|acb\rangle)+\varepsilon(|cab\rangle\pm|cba\rangle)), \tag{1.1.4}$$

where $\varepsilon:=\exp(2\pi i/3)=(-1+i\sqrt{3})/2$; in '$Mi$', the cyclic permutation (123) leads to multiplication with the phase ε, ε^* and the transposition (12) causes an exchange of both components. Mi_- is zero for $a=b$.

Flavour part. The number of states in the flavour multiplets can be obtained by decomposing the quark product states into symmetry types. Table 1.2 shows the result for N_F flavour degrees of freedom.

The flavour wavefunctions of a baryon can be constructed explicitly by using its quark content and (1.1.4) for its symmetry type; for instance $|uud\rangle$

Table 1.2 Number of states in the flavour multiplets

	N_F	$N_F=2$	$N_F=3$	$N_F=4$
Sy	$\binom{N_F+2}{3}$	4	10	20
Mi 2×	$2\binom{N_F+1}{3}$	2	8	20
An	$\binom{N_F}{3}$	0	1	4

is the quark content and

$$|\bar{P}\rangle = (|uud\rangle + \varepsilon\, |udu\rangle + \varepsilon^*\, |duu\rangle)/\sqrt{3}$$
$$|\underline{P}\rangle = (|uud\rangle + \varepsilon^*\, |udu\rangle + \varepsilon\, |duu\rangle)/\sqrt{3}$$

the flavour part of the proton wavefunction with mixed symmetry.

Spin part. The spins of the three quarks can be combined to form the total spin $(3/2)_{Sy}$ and $(1/2)_{Mi}$ by the same method. As the spin has only two possible orientations, there is no antisymmetrical combination:

$$|\tfrac{3}{2}, +\tfrac{3}{2}\rangle = \uparrow\uparrow\uparrow, \quad |\tfrac{3}{2}, +\tfrac{1}{2}\rangle = \frac{1}{\sqrt{3}}(\uparrow\uparrow\downarrow + \uparrow\downarrow\uparrow + \downarrow\uparrow\uparrow),$$

$$|\tfrac{3}{2}, -\tfrac{3}{2}\rangle = \downarrow\downarrow\downarrow, \quad |\tfrac{3}{2}, -\tfrac{1}{2}\rangle = \frac{1}{\sqrt{3}}(\downarrow\downarrow\uparrow + \downarrow\uparrow\downarrow + \uparrow\downarrow\downarrow);$$

$$\overline{|\tfrac{1}{2}, +\tfrac{1}{2}\rangle} = \frac{1}{\sqrt{3}}(\uparrow\uparrow\downarrow + \varepsilon\uparrow\downarrow\uparrow + \varepsilon^*\downarrow\uparrow\uparrow),$$

$$\overline{|\tfrac{1}{2}, -\tfrac{1}{2}\rangle} = \frac{1}{\sqrt{3}}(\downarrow\downarrow\uparrow + \varepsilon\downarrow\uparrow\downarrow + \varepsilon^*\uparrow\downarrow\downarrow); \tag{1.1.5}$$

$$\underline{|\tfrac{1}{2}, \pm\tfrac{1}{2}\rangle} = \overline{|\tfrac{1}{2}, \pm\tfrac{1}{2}\rangle} \quad \text{complex conjugate}$$

Orbital part. The relative motion of the three quarks is a function of two relative coordinates $\mathbf{z}_1 = (2\mathbf{x}_3 - \mathbf{x}_1 - \mathbf{x}_2)/\sqrt{6}$, $\mathbf{z}_2 = (\mathbf{x}_2 - \mathbf{x}_1)/\sqrt{2}$. Thus, there are many different combinations of internal angular momenta leading to one total orbital momentum l. In addition, Schrödinger wavefunctions

$$|l, m, n, \eta\rangle = F_l(\mathbf{z}_1, \mathbf{z}_2) \tag{1.1.6}$$

of all three symmetry types can be systematically constructed using the mixed-symmetrical, complex, relative coordinates $\mathbf{z}_1 = (\mathbf{z}_1 + i\mathbf{z}_2)/\sqrt{2}$, $\mathbf{z}^* = (\mathbf{z}_1 - i\mathbf{z}_2)/\sqrt{2}$. The structural details of Schrödinger wavefunctions are determined by the quark–quark interaction potentials (cf. [Gr 76, Bö 80]).

Composition of the baryon wavefunctions. When flavour, spin, and orbital parts are combined to give a total wavefunction, its symmetry character is founded upon the following composition rules

$$\begin{aligned} &Sy \otimes Sy = Sy, \quad An \otimes An = Sy, \quad Mi \otimes Mi = Sy \otimes Mi \otimes An, \\ &Sy \otimes An = An, \quad Sy \otimes Mi = Mi, \quad An \otimes Mi = Mi. \end{aligned} \tag{1.1.7}$$

In this way, the flavour and spin parts (Table 1.2 or Eqn. (1.1.5)) combine to give the mutliplets with defined symmetry as shown in Table 1.3. If the interaction shows only a slight flavour and spin dependence, it is advantageous to amalgamate multiplets of equal symmetry to form supermultiplets of the higher symmetry group $SU(2N_F)$ [Cl 79].

Table 1.3 (Flavour, spin)-multiplets with definite symmetry

	$N_F=2$	$N_F=3$	$N_F=4$
Sy	$20=(4,\frac{3}{2})+(2,\frac{1}{2})$	$56=(10,\frac{3}{2})+(8,\frac{1}{2})$	$120=(20,\frac{3}{2})+(20,\frac{1}{2})$
Mi 2×	$20=(4,\frac{1}{2})+(2,\frac{3}{2})$ $+(2,\frac{1}{2})$	$70=(10,\frac{1}{2})+(8,\frac{3}{2})$ $+(8,\frac{1}{2})+(1,\frac{1}{2})$	$168=(20,\frac{3}{2})+(20,\frac{1}{2})$ $+(20,\frac{1}{2})+(4,\frac{1}{2})$
An	$4=(2,\frac{1}{2})$	$20=(8,\frac{1}{2})+(1,\frac{3}{2})$	$56=(20,\frac{1}{2})+(4,\frac{3}{2})$

The final stage consists of the combination with the orbital wavefunction to give states with definite total angular momentum. If the total wavefunction is to be symmetrical under permutations, then the classification shown in Fig. 1.2 follows. This is a good description of the experimental spectrum for baryon resonances with strangeness zero. The many blanks which occur in the case of resonances with strangeness -2 or -3 are due to the impossibility of carying out phase shift analyses for this case. The first baryons with charm were discovered recently [Ab 79]. The mass of $\Lambda_c = udc$ is (2.285 ± 0.006)GeV, i.e. at the expected level [De 75].

The fact that a symmetrical total wavefunction must be chosen to explain the experimental baryon spectrum is a problem for the simple, phenomenological quark model: as quarks with spin 1/2 obey Fermi statistics, the total wavefunction must be antisymmetric. Accordingly, the orbital part of the wavefunction of the Δ^{++} resonance with $I=3/2$, $I_3=+3/2$ and $j=3/2$ must be antisymmetric, since the flavour and spin components are necessarily symmetric. On the other hand, the Δ^{++} is the ground state of all particles made up of three u quarks. It is possible to show that for relatively general potentials the orbital wavefunction of the ground state has no nodes and this is in contradiction to the antisymmetry required. The solution of this problem of quark statistics by using a quark model with colour was an important discovery on the way to quantum chromodynamics.

1.1.2 Quark model with colour

The antisymmetrical total wavefunction of baryons with symmetrical flavour, spin, and orbital parts, required by the Pauli principle, can be achieved by extending the quark degrees of freedom. This solution to the problem of quark statistics was first proposed by O. W. Greenberg [Gr 64] and M. Y. Han and Y. Nambu [Ha 65]; the present formulation originates in that of M. Gell-Mann [Ge 72a]. According to him, quarks have an additional degree of freedom, colour, which can assume three values (red, green, and blue) and in which the baryon wavefunction is antisymmetrical.

Thus the constituents of hadrons are quarks in three different colours:

$$q_{f,c} \quad \begin{pmatrix} u_r & u_g & u_b \\ d_r & d_g & d_b \\ s_r & s_g & s_b \\ c_r & c_g & c_b \\ \vdots & \vdots & \vdots \end{pmatrix} \downarrow N_F \text{ flavour degrees of freedom}$$

$\rightarrow$ 3 colour-degrees of freedom

Hadrons are constructed according to the rule 'hadrons are colourless'. This means that the meson and baryon wavefunctions have the following flavour and colour content:

$$\text{meson} = \sum_{c,c'} \bar{q}_{f,c} q_{f',c'} \, \delta_{cc'}/\sqrt{3}, \tag{1.1.8}$$

$$\text{baryon} = \sum_{c,c',c'} q_{f,c} q_{f',c'} q_{f'',c''} \varepsilon_{cc'c''}/\sqrt{6}; \tag{1.1.9}$$

here, $\varepsilon_{cc'c''}$ is the totally antisymmetric tensor of rank 3 with $\varepsilon_{123} = +1$.

The complete wavefunctions for the ρ^+ mson and the Δ^{++} resonance are given explicitly by way of illustration. A Gaussian wavefunction is a suitable ansatz for many dynamical problems. Therefore the radial wavefunction was chsoen of this form:

$$\rho^+ = -\frac{1}{\sqrt{3}}(\bar{d}_r u_r + \bar{d}_g u_g + \bar{d}_b u_b) \cdot |\uparrow\uparrow\rangle \cdot (\alpha_M/\pi)^{3/4} \exp(-\alpha_M \mathbf{r}^2/2),$$

$$\Delta^{++} = \frac{1}{\sqrt{6}}(u_r u_g u_b + u_b u_r u_g + u_g u_b u_r - u_g u_r u_b - u_b u_g u_r - u_r u_b u_g) \tag{1.1.10}$$
$$\times |\uparrow\uparrow\uparrow\rangle \cdot (\alpha_B/\pi)^{3/2} \exp[-\alpha_B(\mathbf{z}_1^2 + \mathbf{z}_2^2)/2].$$

The colour degree of freedom, which was introduced because of the statistics problem, is dynamically relevant. According to the colour rule, only colourless combinations of quarks form bound states, i.e. hadrons with conventional masses. This is a first look at the relationship between colour and quark confinement.

The introduction of colour was found to be of decisive importance for the formulation of the theory of strong interactions. Therefore, further phenomenological information on the colour degree of freedom is essential. Leaving the statistics problems aside, the classical examples [Fr 73] are the size of the total cross-section for e^+e^- annihilation into hadrons (see Section 1.4.4) and the decay rate for $\pi^0 \rightarrow 2\gamma$ (see Section 1.3.2). The structure of the colour part of the hadron wavefunctions indicates that there is an exact symmetry connected with colour. This aspect will be considered further in Section 1.2.

1.1.3 The concept of quark dynamics—quarkonia

(a) In Section 1.1.1 the quantum numbers of mesons and baryons and ansätze for their wavefunctions were discussed by means of a simple quark model. Dynamical questions were merely touched upon when assessing the significance of colour on the strong interaction. In this section, some further concepts of quark dynamics will be discussed within the framework of a non-relativistic potential model using the meson and baryon spectrum as a basis. This does, in fact, give a realistic picture for bound heavy quark–antiquark systems. Experimental examples for such systems are charmonium [Wi 79] ($J/\psi, \psi', \ldots$), bottomonium ($\Upsilon, \Upsilon', \ldots$) and possibly others, which can be considered under the general heading *quarkonia* [Kr 79]. Fig. 1.3 shows the experimental spectrum of the ψ and Υ family, their most significant decays and the assignments of quantum numbers.

(b) Before discussing a potential model for quarkonia, we would like to point out that according to current ideas the strong interaction is caused in a certain analogy to the electromagnetic interaction by the exchange of zero mass vector particles. These 'gluons' couple to the colour degree of freedom of the quarks. The consistent field theoretical formulation of these ideas leads to a gauge theory, quantum chromodynamics (QCD), which is discussed in Chapter 2. At this point, the main objective is to see what can be learned about the dynamics of quarks and gluons from a potential model for quarkonia. For systems of heavy quarks it can be expected that the quarks move non-relativistically in the bound state. Accordingly, the quark

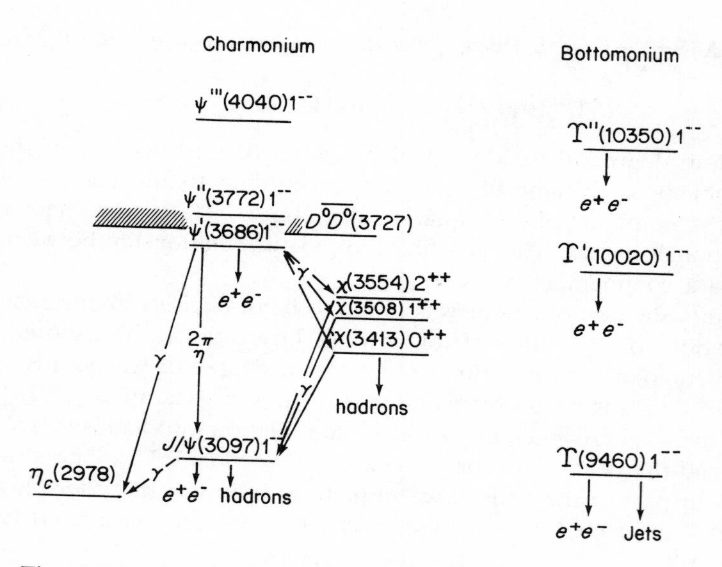

Figure 1.3 The experimental spectrum of the ψ and Υ family

dynamics can be described by a Hamilton operator:

$$H = \sum_i \left(m_i + \frac{\mathbf{p}_i^2}{2m_i} \right) + \sum_{i>j} (V_C(r_{ij}) + V_G(r_{ij})). \tag{1.1.11}$$

The quark masses m_i make an additive contribution to the hadron mass. In Figs 1.1 and 1.2 this is used by subtracting the mass differences of the various quarks from the hadron mass, i.e.

$$m_u \simeq m_d, \quad m_s \simeq m_u + 150\,\text{MeV}, \quad m_c \simeq m_u + 1200\,\text{MeV}, \quad m_b \simeq m_u + 4600\,\text{MeV}. \tag{1.1.12}$$

Obviously, this compensates the main effects of the mass splittings in the flavour multiplets. This means that the interaction potentials V_C and V_G of the quarks are approximately flavour independent.

In a potential model, quark confinement is forced by introducing an infinitely rising potential. Influenced by QCD (see Section 2.6), a linear ansatz is commonly accepted:

$$V_C(r) = \kappa r. \tag{1.1.13}$$

An attractive short-range Coulomb potential, which is produced in QCD by 1-gluon exchange, is added accordingly:

$$V_G(r) = -\frac{\alpha_s}{r} \begin{pmatrix} 4/3 \\ 2/3 \end{pmatrix} \begin{matrix} \text{quark–antiquark in the meson} \\ \text{quark–quark in the baryon} \end{matrix} \tag{1.1.14}$$

The factors 4/3 for mesons and 2/3 for baryons result from the colour component ((1.1.8) or (1.1.9) respectively) of the wavefunction (see Section 2.7.4 [Ap 75, Ap 75a, De 75]).

The hadron masses and wavefunctions can be calculated by solving the Schrödinger equation for H. Fig. 1.4 gives a comparison between the results thus obtained for the charmonium system [Ei 75] and experiment. The model parameters were determined from the masses of J/ψ (3100) and ψ' (3686) and the leptonic width of J/ψ, $\Gamma(J/\psi \to e^+e^-) = 4.7$ keV. Typical values for these parameters are

$$\alpha_s = 0.3\text{–}0.4, \quad \kappa = 0.8\text{–}1.0\,\text{GeV/fm}, \quad m_c = 1.2\text{–}1.6\,\text{GeV}. \tag{1.1.15}$$

The agreement between the general structure of the experimental spectrum and the calculated masses is striking. However, the spectrum can be well reproduced up to masses of 4 GeV with other potentials too. Therefore, these results alone cannot be regarded as an experimental confirmation of the potentials ((1.1.13), (1.1.14)) motivated by QCD. As non-relativistic approximations are the basis of the quarkonium model, a short remark on the quark velocity in the bound state would be appropriate. The expectation value of the velocity in the charmonium ground state is calculated as $\langle v \rangle \simeq 0.4c$. This roughly justifies the non-relativistic calculations. However,

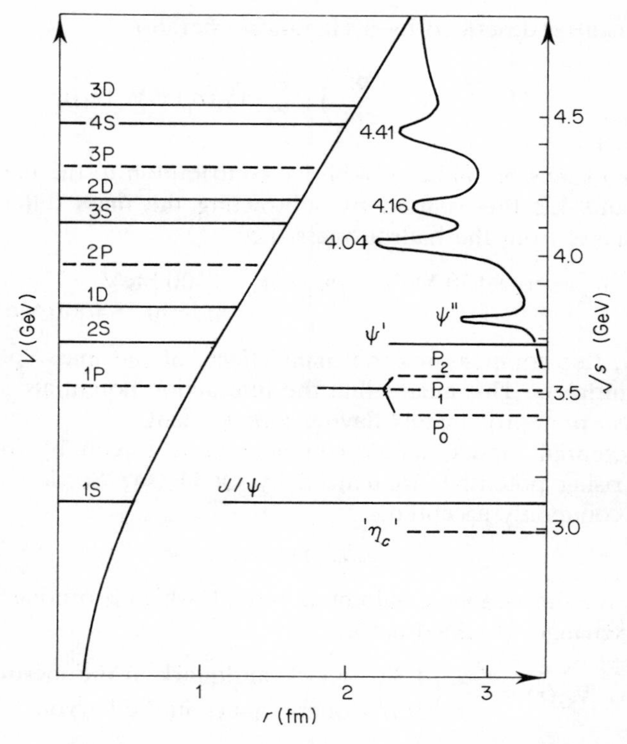

Figure 1.4 The standard charmonium potential ((1.1.13), (1.1.14)), the spectrum, and the charmonium states found experimentally. Parameters: $\alpha_s = 0.41$, $\kappa = 0.8665$ GeV/fm and $m_c = 1.6$ GeV [Kr 79]

we expect that relativistic effects will become significant for heavy quark bound states, too (see Section 2.7.4).

1.2 Quantum fields and currents

In our introductory remarks on the quark model, we tried to manage with the terminology of non-relativistic quantum mechanics. This scope is too narrow both for a phenomenological description of relativistic processes and also for the formulation of fundamental theories. General theories must satisfy the approved physical principles of causality, relativistic invariance, and quantum theory and include the known charge conservation theorems too. The description of elementary particle reactions by quantized, relativistic fields, which interact locally, satisfies all these requirements. The deep connection between the structure of conserved charges and the symmetry groups of the fields plays an important part in this. Therefore, some of the relevant group and field theory concepts are collected in the present chapter [GT, FT].

1.2.1 Flavour and colour symmetry groups

As we have seen, the hadrons can be arranged into multiplets which can be characterized by charge quantum numbers. These multiplet structures can formally be described by symmetry groups [Ro 71b, Li 78]. The most well-known examples of this are the isospin with its associated group $SU(2)_I$ and the $SU(3)_F$ flavour symmetry.

(a) As these examples show, the generalized charge operators (I_1, I_2, I_3 in the case of the isospin) correspond to the generators Q_a, $a = 1, \ldots, N$, of the *Lie algebra* of a symmetry group G of dimension N (dimension = number of parameters). In a suitable basis, the algebraic properties of the generators can be fully specified by their commutation relations:

$$[Q_a, Q_b] = \mathrm{i} f_{abc} Q_c, \quad a, b, c = 1, \ldots, N, \tag{1.2.1}$$

with totally antisymmetric *structure constants* f_{abc}.

For isospin, they are given by the Levi-Cività tensor ε_{abc} ($a, b, c = 1, 2, 3$, $\varepsilon_{123} = +1$); for the $SU(3)$ group, the following structure constants are—up to permutations of the indices—different from zero:

$$\begin{aligned} &f_{123} = +1, \quad f_{147} = -f_{156} = f_{246} = f_{257} = f_{345} = -f_{367} = \tfrac{1}{2}; \\ &f_{458} = f_{678} = \tfrac{1}{2}\sqrt{3}. \end{aligned} \tag{1.2.2}$$

If $f_{abc} = 0$, the group is called Abelian.

In the Lie algebra of the group G, a maximum number r of commuting charges H_i, $i = 1, \ldots, r$, can be found. r is called the *rank* of the group. The states of a multiplet are—up to degeneracies—characterized by the eigenvalues of H_i. The remaining charges which are linearly independent of H_i can be combined to form ladder operators E_α, $\alpha = 1, \ldots, N - r$, which transform the different multiplet states into one another. These notions can be illustrated by using the $SU(3)_F$ flavour group as an example (Fig. 1.5).

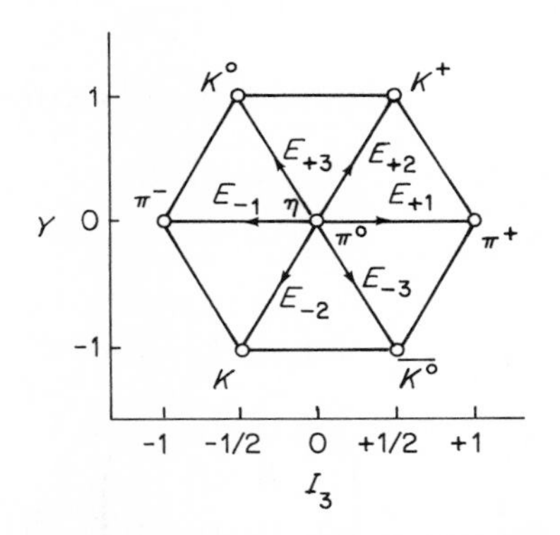

Figure 1.5 Graphical representation of the ladder operators of the pseudoscalar meson octet

$SU(3)_F$: the rank of the $SU(3)$ group is $r=2$

$$\begin{aligned}
&H_1 = I_3 \quad \text{third isospin component}\\
&H_2 = 2Y/\sqrt{3}: \quad \text{hypercharge } Y\\
&\left.\begin{aligned}
E_{\pm 1} &= Q_1 \pm \mathrm{i} Q_2 = I_1 \pm \mathrm{i} I_2\\
E_{\pm 2} &= Q_4 \pm \mathrm{i} Q_5\\
E_{\pm 3} &= Q_6 \pm \mathrm{i} Q_7
\end{aligned}\right\}\ \text{ladder operators}
\end{aligned}$$

(b) The algebra of charges of a symmetry group $SU(n)$ has a *representation* by $n \times n$ matrices, the fundamental representation of $SU(n)$.

In quark models with N_F flavour degrees of freedom, this is the representation of the flavour charges in the state space of the quarks. Examples:

$N_F=2$: $SU(2)_I \equiv$ isospin group, $I_a = \tau_a/2$, $a=1, 2, 3$,

$$\tau_1 = \begin{pmatrix} 0 & 1 \\ 1 & 0 \end{pmatrix}, \quad \tau_2 = \begin{pmatrix} 0 & -\mathrm{i} \\ \mathrm{i} & 0 \end{pmatrix}, \quad \tau_3 = \begin{pmatrix} 1 & 0 \\ 0 & -1 \end{pmatrix}; \tag{1.2.3}$$

$N_F=3$: $SU(3)_F$, $Q_a = \lambda_a/2$, $a=1, \ldots, 8$,

$$\lambda_1 = \begin{pmatrix} 0 & 1 & 0 \\ 1 & 0 & 0 \\ 0 & 0 & 0 \end{pmatrix}, \quad \lambda_2 = \begin{pmatrix} 0 & -\mathrm{i} & 0 \\ \mathrm{i} & 0 & 0 \\ 0 & 0 & 0 \end{pmatrix}, \quad \lambda_3 = \begin{pmatrix} 1 & 0 & 0 \\ 0 & -1 & 0 \\ 0 & 0 & 0 \end{pmatrix},$$

$$\lambda_4 = \begin{pmatrix} 0 & 0 & 1 \\ 0 & 0 & 0 \\ 1 & 0 & 0 \end{pmatrix}, \quad \lambda_5 = \begin{pmatrix} 0 & 0 & -\mathrm{i} \\ 0 & 0 & 0 \\ \mathrm{i} & 0 & 0 \end{pmatrix}, \tag{1.2.4}$$

$$\lambda_6 = \begin{pmatrix} 0 & 0 & 0 \\ 0 & 0 & 1 \\ 0 & 1 & 0 \end{pmatrix}, \quad \lambda_7 = \begin{pmatrix} 0 & 0 & 0 \\ 0 & 0 & -\mathrm{i} \\ 0 & \mathrm{i} & 0 \end{pmatrix}, \quad \lambda_8 = \frac{1}{\sqrt{3}} \begin{pmatrix} 1 & 0 & 0 \\ 0 & 1 & 0 \\ 0 & 0 & -2 \end{pmatrix};$$

The Gell-Mann matrices λ_a have the following properties:

$$\mathrm{Tr}\, \lambda_a = 0, \quad \mathrm{Tr}\, \lambda_a \lambda_b = 2\, \delta_{ab}. \tag{1.2.5}$$

$N_F=4$: $SU(4)_F$, $Q_a = \underline{\lambda}_a/2$, $a=1, \ldots, 15$;

$$\underline{\lambda}_a = \begin{pmatrix} \lambda_a & 0 \\ 0 & 0 \end{pmatrix} \quad \text{for } a=1, \ldots, 8;$$

$$\underline{\lambda}_9 = \begin{pmatrix} 0 & 0 & 0 & 1 \\ 0 & 0 & 0 & 0 \\ 0 & 0 & 0 & 0 \\ 1 & 0 & 0 & 0 \end{pmatrix}, \quad \underline{\lambda}_{10} = \begin{pmatrix} 0 & 0 & 0 & -\mathrm{i} \\ 0 & 0 & 0 & 0 \\ 0 & 0 & 0 & 0 \\ \mathrm{i} & 0 & 0 & 0 \end{pmatrix}, \quad \underline{\lambda}_{11} = \begin{pmatrix} 0 & 0 & 0 & 0 \\ 0 & 0 & 0 & 1 \\ 0 & 0 & 0 & 0 \\ 0 & 1 & 0 & 0 \end{pmatrix},$$

$$\underset{\sim}{\lambda}_{12}=\begin{pmatrix}0&0&0&0\\0&0&0&-i\\0&0&0&0\\0&i&0&0\end{pmatrix},\quad \underset{\sim}{\lambda}_{13}=\begin{pmatrix}0&0&0&0\\0&0&0&0\\0&0&0&1\\0&0&1&0\end{pmatrix},\quad \underset{\sim}{\lambda}_{14}=\begin{pmatrix}0&0&0&0\\0&0&0&0\\0&0&0&-i\\0&0&i&0\end{pmatrix},$$

$$\underset{\sim}{\lambda}_{15}=\frac{1}{\sqrt{6}}\begin{pmatrix}1&0&0&0\\0&1&0&0\\0&0&1&0\\0&0&0&-3\end{pmatrix}; \tag{1.2.6}$$

$N_F=5, 6, \ldots$: analogous.

The structure constants f_{abc} can be used to define matrices T_a according to $(T_a)_{bc}=-\mathrm{i}f_{abc}$, which satisfy the commutation relations (1.2.1) because of the Jacobi identity for commutators. This representation of the Lie algebra which is of the dimension of the group is called the *adjoint representation.*

Polynomials C in the charges Q_a, which commute with all Q_a, are called *Casimir operators.* In irreducible representations, according to Schur's lemma, the C's are multiples of the unit matrix.

Examples:

$SU(2)$: $C_2=\sum_{a=1}^{3} I_a^2$

$$\text{fundamental representation:}\quad \sum_{a=1}^{3}(\tau_a/2)^2=\tfrac{3}{4}\underset{\sim}{1},$$

$SU(3)$: $C_2=\sum_{a=1}^{8} Q_a^2$

$$\text{fundamental representation:}\quad \tfrac{1}{4}\sum_{a,j}(\lambda_a)_{ij}(\lambda_a)_{jk}=\tfrac{4}{3}\,\delta_{ik}=:C_2^F\,\delta_{ik}, \tag{1.2.7}$$

$$\text{adjoint representation:}\quad \sum_a\sum_b f_{abc}f_{abd}=3\,\delta_{cd}=:C_2\,\delta_{cd}; \tag{1.2.8}$$

(c) The exponential series of infinitesimal transformations

$$g(\theta^a)=\exp(-\mathrm{i}\theta^aQ_a)=\underset{\sim}{1}+\frac{-\mathrm{i}}{1!}\,\theta^aQ_a+\frac{(-\mathrm{i})^2}{2!}\,\theta^aQ_a\cdot\theta^bQ_b+\ldots, \tag{1.2.9}$$

the *finite transformations*, form a group:

$$gg'=\exp(-\mathrm{i}\theta^aQ_a)\exp(-\mathrm{i}\theta'^bQ_b)=\exp(-\mathrm{i}\theta''^cQ_c),\quad \theta''^c=\theta''^c(\theta^a,\theta'^b),$$
$$g^{-1}=\exp(\mathrm{i}\theta^aQ_a),\quad g(0)=\underset{\sim}{1}. \tag{1.2.10}$$

Example: Fundamental representation of $SU(2)$ (which represents spatial rotations in spinor space):

$$g(\theta^a) = \exp\left(-i \sum_{a=1}^{3} \theta^a \frac{\tau_a}{2}\right)$$

with

$$\boldsymbol{\theta} = (\theta^1, \theta^2, \theta^3) = \theta \mathbf{e}, \quad \mathbf{e}^2 = 1$$

or

$$\theta^1 = \theta \sin \vartheta \cos \phi, \quad \theta^2 = \theta \sin \vartheta \sin \phi, \quad \theta^3 = \theta \cos \vartheta$$
$$0 \leqslant \theta < 2\pi, \quad 0 \leqslant \phi < 2\pi, \quad 0 \leqslant \vartheta \leqslant \pi.$$

Then

$$g = \cos \tfrac{1}{2}\theta \cdot \underline{1} - i \sin \tfrac{1}{2}\theta \cdot (\mathbf{e} \cdot \boldsymbol{\tau})$$
$$= \begin{pmatrix} \cos \frac{1}{2}\theta - i \sin \frac{1}{2}\theta \cdot e_3 & -i \sin \frac{1}{2}\theta \cdot (e_1 - ie_2) \\ -i \sin \frac{1}{2}\theta \cdot (e_1 + ie_2) & \cos \frac{1}{2}\theta + i \sin \frac{1}{2}\theta \cdot e_3 \end{pmatrix} \qquad (1.2.11)$$

is the most general 2-dimensional unitary matrix of determinant +1. It describes rotation around axis $\mathbf{e}$ by angle θ.

A Lie group becomes a manifold by the analytical dependence of the group elements on the N parameters θ^a. Further mathematical investigation [He 62] shows that for many groups (e.g. the compact, semi-simple ones) this manifold has a metric, which is invariant under right and left translations $g \to g \cdot f$ or $g \to f \cdot g$. Later, in Section 2.6.3, it will be important that a measure $d\mu(g)$ on the group is connected with this metric:

$$d\mu(g) = I(\theta^a) \prod_{a=1}^{N} d\theta^a. \qquad (1.2.12)$$

This measure is right- or left-invariant too:

$$d\mu(g) = d\mu(f \cdot g) = d\mu(g \cdot f) = d\mu(g^{-1}), \quad f \in G. \qquad (1.2.13)$$

This means that under transformation (1.2.10) with the corresponding Jacobi determinant, the expression (1.2.12) remains invariant. For $SU(2)$ with the group parameters of our example this measure has the form:

$$d\mu(g) = \frac{1}{4\pi^2} \sin^2 \tfrac{1}{2}\theta \sin \vartheta \, d\theta \, d\vartheta \, d\phi, \quad \int_{SU(2)} d\mu(g) = 1. \qquad (1.2.14)$$

(d) The group theoretical concepts just described are widely used in elementary particle physics. Thus, the discussion of the flavour multiplets of hadrons in Section 1,1.1 corresponds to the arrangement of the particles in representations of the *flavour group* $SU(N_F)$. Electric charge Q, strangeness S, charm C, hypercharge Y, isospin I, etc., are closely connected with the flavour group generators. For $N_F = 4$:

$$Q = Q_3 - \frac{1}{\sqrt{3}} Q_8 - \sqrt{\tfrac{2}{3}} Q_{15} + \tfrac{1}{6}\underline{1}, \quad S = \frac{2}{\sqrt{3}} Q_8 - \frac{1}{\sqrt{6}} Q_{15} - \tfrac{1}{4}\underline{1},$$
$$C = -(\sqrt{\tfrac{3}{2}}) Q_{15} + \tfrac{1}{4}\underline{1}, \quad Y = \frac{2}{\sqrt{3}} Q_8, \quad I_3 = Q_3. \tag{1.2.15}$$

Because the unit operator $\underline{1}$ does not belong to the $SU(4)$ generators, Q, S, and C are not directly charges of $SU(4)$.

The flavour multiplets show a strong mass splitting. Therefore the groups $SU(3)_F$, $SU(4)_F, \ldots$ are only approximate 'symmetry groups' at most. Fig. 1.1 shows that the masses of mesons with charm are much higher than those of the 'old' mesons. This applies equally to bottom and presumably to other possible flavour degrees of freedom. Experimentally, one observes a decoupling of the dynamics of mesons with great mass differences (Zweig rule, see Section 2.7.4). This explains why the hadron dynamics without charm can be well described within the framework of a quark model with three quark-flavour charges.

The structure of the electromagnetic and weak interaction is described group theoretically by means of the *weak isospin* $SU(2)_W$ and the *weak hypercharge* $U(1)$ (see Section 1.3).

(e) The generators of the $SU(3)$ *colour group* Q_a^C, $a = 1, \ldots, 8$, the colour charges, discriminate the observed hadrons from quarks and similar exotic states. The colour charges of hadrons are zero:

$$Q_a^C |\text{had}\rangle = 0 \tag{1.2.16}$$

This is the group theoretical formalization of the rule 'hadrons are colourless'. If the fundamental representation of $SU(3)$ is ascribed to the colour degrees of freedom of the quarks, then the colour part of the hadron wavefunction, that is the scalar product (1.1.8) or the determinant (1.1.9), satisfies the invariance condition (1.2.16).

It is the simplicity of this scheme which establishes the structure of the colour group as $SU(3)$. The experimental evidence for (1.2.16) is obviously not direct. So far, attempts to arrange hadrons into non-trivial colour multiplets have met with no success [St 76].

Operators which commute with all symmetry transformations are represented in irreducible representations by multiples of the unit matrix. This makes them suitable for the characterization of multiplets. Reference has been made in this context to the importance of Casimir operators, which describe quantum numbers such as total spin etc. (Eqns (1.2.7) and (1.2.8)). Finite group transformations which commute with the whole group, form the centre of the group. The centre of the $SU(3)$ colour group is

$$\mathfrak{z} = \{\underline{1},\ z = e^{-2\pi i/3}\underline{1},\ z^2 = e^{-2\pi i/3}\underline{1}\}. \tag{1.2.17}$$

States, for which $z \to 1$, $e^{2\pi i/3}$, $e^{-2\pi i/3}$, have zero, positive, and negative triality, respectively. Quarks have positive, antiquarks negative triality. Hadrons and colour-octet representations (gluons!) have zero triality. z is a multiplicative quantum number.

1.2.2 Elements of relativistic quantum field theory

(a) We begin our introduction to relativistic quantum mechanics by describing a basis for the *quantum-mechanical states of a free particle*:

$$\left|\begin{matrix} M & j \\ p & j_3 \end{matrix}\right\rangle,$$

which obeys relativistic kinematics [Ca 71]. These free one-particle states are eigenstates of the relativistic energy–momentum operator P_μ and thus transform under space–time translations $x^\mu \to x^\mu + a^\mu$, according to:

$$U(a)\left|\begin{matrix} M & j \\ p & j_3 \end{matrix}\right\rangle = e^{ipa}\left|\begin{matrix} M & j \\ p & j_3 \end{matrix}\right\rangle, \quad p = (p^\mu) = (p^0 = +\sqrt{(M^2 + \mathbf{p}^2)} = E_p, \mathbf{p}) \tag{1.2.18}$$

(see p. 3?? for conventions used). Homogeneous Lorentz transformations $x^\mu = \Lambda^\mu_\nu x^\nu$, the generators of which are combined to form the relativistic angular momentum operator $M_{\mu\nu}$, transform the one-particle state into one with four-momentum Λp and with a changed orientation of the third component j_3 of the particle spin j, which is given by the Wigner rotation $R(\Lambda, p)$:

$$U(\Lambda)\left|\begin{matrix} M & j \\ p & j_3 \end{matrix}\right\rangle = D^{(j)}_{j_3' j_3}(R(\Lambda, p))\left|\begin{matrix} M & j \\ \Lambda p & j_3' \end{matrix}\right\rangle. \tag{1.2.18'}$$

The $U(\Lambda)$, $U(a)$ generate an irreducible representation of the group of homogeneous Lorentz transformations and space–time translations (Poincaré group) [Wi 39].

(b) One- and multiparticle states, which are symmetrical (for bosons) or antisymmetrical (for fermions) under permutation of identical particles, are described by *creation operators* $a^\dagger_{j_3}(p)$ and *annihilation operators* $a_{j_3}(p)$. These are characterized by their commutation relations

$$\begin{aligned} [a_{j_3}(p), a^\dagger_{j_3'}(p')]_\pm &= 2E_p\,\delta(\mathbf{p} - \mathbf{p}')\,\delta_{j_3 j_3'}, \\ [a^\dagger_{j_3}(p), a^\dagger_{j_3'}(p')]_\pm &= [a_{j_3}(p), a_{j_3'}(p')]_\pm = 0, \end{aligned} \tag{1.2.19}$$

$+$ for fermions, $-$ for bosons, and by the fact that the annihilation operators $a_{j_3}(p)$ annihilate the relativistic invariant vacuum $|0\rangle$:

$$a_{j_3}(p)\,|0\rangle = 0. \tag{1.2.20}$$

The one-particle states (1.2.18) are created by $a^\dagger_{j_3}(p)$ from the vacuum:

$$\left|\begin{matrix} M & j \\ p & j_3 \end{matrix}\right\rangle = a^\dagger_{j_3}(p)\,|0\rangle, \quad \left\langle\begin{matrix} M & j \\ p & j_3 \end{matrix}\right|\left.\begin{matrix} M & j \\ p' & j_3' \end{matrix}\right\rangle = 2E_p\,\delta(\mathbf{p}-\mathbf{p}')\,\delta_{j_3 j_3'}, \tag{1.2.21}$$

and free multiparticle states from repeated application of $a^\dagger$. Creation and annihilation operators of the associated antiparticles are denoted by $b^\dagger_{j_3}(p)$ and $b_{j_3}(p)$, respectively.

(c) A complementary description of free particles which emphasizes their causal propagation, uses *free quantized fields.* These are defined by field equations and commutation relations. For scalar charged fields these are the Klein–Gordon equation

$$(\partial_\mu \partial^\mu + M^2)\psi(x) = 0 \tag{1.2.22}$$

and the c- number commutation relations:

$$\begin{aligned} &[\psi(x), \psi(x')]_- = [\psi^\dagger(x), \psi^\dagger(x')]_- = 0, \\ &[\psi(x), \psi^\dagger(x')]_- = \mathrm{i}\Delta(x-x', M), \end{aligned} \tag{1.2.23}$$

where

$$\begin{aligned} \Delta(x, M) &= -\mathrm{i}(2\pi)^{-3} \int \mathrm{d}^4 p\, \delta(p^2 - M^2)\varepsilon(p^0)\mathrm{e}^{-\mathrm{i}px} \\ &= \frac{-1}{2\pi}\, \varepsilon(x^0)\left(\delta(x^2) - \frac{M^2}{2}\,\theta(x^2)\,\frac{J_1(M\sqrt{x^2})}{M\sqrt{x^2}}\right). \end{aligned} \tag{1.2.24}$$

The commutativity of the field operators for space like distances is an expression of causality.

Free fields and creation and annihilation operators of particles with arbitrary spin are connected by a Fourier transformation:

$$\psi_\alpha(x) = (2\pi)^{-3/2} \sum_{j_3} \int \frac{\mathrm{d}^3 p}{2E_p}\,[u_\alpha(p, j_3) a_{j_3}(p) \mathrm{e}^{-\mathrm{i}px} + v_\alpha(p, j_3) b^\dagger_{j_3}(p)\mathrm{e}^{+\mathrm{i}px}], \quad p^0 = E_p. \tag{1.2.25}$$

The spinors $u_\alpha(p, j_3)$, $v_\beta(p, j_3)$ are determined such that $\psi_\alpha(x)$ transforms locally under Lorentz transformations: $\psi_\alpha(x) \to S_{\alpha\alpha'}(\Lambda)\psi_{\alpha'}(\Lambda^{-1}x)$ ($u(p) = v(p) \equiv 1$ for scalar particles). Eqn. (1.2.25) incorporates the wave–particle dualism: *particles are field quanta.*

Spin-$\frac{1}{2}$ particles are described by the Dirac field $\psi_\alpha(x)$ which satisfies the Dirac equation:

$$(\mathrm{i}\gamma^\mu{}_{\alpha\beta}\, \partial_\mu - M\,\delta_{\alpha\beta})\psi_\beta(x) = 0 \tag{1.2.26}$$

and the commutation relations:

$$\begin{aligned} &[\psi_\alpha(x), \bar{\psi}_{\alpha'}(x')]_+ = -\mathrm{i}S_{\alpha\alpha'}(x-x', M), \\ &[\psi_\alpha(x), \psi_{\alpha'}(x')]_+ = [\bar{\psi}_\alpha(x), \bar{\psi}_{\alpha'}(x')]_+ = 0, \end{aligned} \tag{1.2.27}$$

$$S_{\alpha\beta}(x, M) = -(\mathrm{i}\gamma^\mu{}_{\alpha\beta}\, \partial_\mu + M\,\delta_{\alpha\beta})\,\Delta(x, M) \tag{1.2.28}$$

See Table 1.4 for other formulae.

Table 1.4 Formulae for free spin-$\frac{1}{2}$ and zero mass spin-1 fields

Spin-$\frac{1}{2}$ fields

Dirac matrices:

$$[\gamma_\mu, \gamma_\nu]_+ = 2g_{\mu\nu} \cdot 1$$

$$\gamma_0 = \begin{pmatrix} 0 & 1 \\ 1 & 0 \end{pmatrix}, \quad \gamma_i = \begin{pmatrix} 0 & -\sigma_i \\ \sigma_i & 0 \end{pmatrix}, \quad \sigma_i \equiv \text{Pauli matrices}$$

$$\gamma_5 := -i\gamma_0\gamma_1\gamma_2\gamma_3 = \begin{pmatrix} -1 & 0 \\ 0 & 1 \end{pmatrix}, \quad [\gamma_5, \gamma_\mu]_+ = 0,$$

$$\sigma_{\mu\nu} := \frac{i}{2}[\gamma_\mu, \gamma_\nu]_-, \quad a^\mu\gamma_\mu \equiv a \cdot \gamma \equiv \not{a}, \quad \not{a}\not{a} = a^2 \cdot 1$$

Dirac spinors:

$$(\not{p} - M)u(p, j_3) = 0, \quad \bar{u} = u^\dagger\gamma_0,$$
$$(\not{p} + M)v(p, j_3) = 0, \quad \bar{v} = v^\dagger\gamma_0;$$

Normalization:

$$\sum_\alpha \bar{u}_\alpha(p, j_3')u_\alpha(p, j_3) = 2M\,\delta_{j_3'j_3},$$
$$\sum_\alpha \bar{v}_\alpha(p, j_3')v_\alpha(p, j_3) = -2M\,\delta_{j_3'j_3};$$

Polarization sum:

$$\sum_{j_3} u_\alpha(p, j_3)\bar{u}_\beta(p, j_3) = (\gamma^\mu p_\mu + M)_{\alpha\beta},$$
$$\sum_{j_3} v_\alpha(p, j_3)\bar{v}_\beta(p, j_3) = (\gamma^\mu p_\mu - M)_{\alpha\beta}.$$

Spin-1 fields

Polarization vectors $\varepsilon_\mu(k, \lambda)$, $\lambda = 1, 2$, $k^2 = 0$, $k^0 > 0$ for the gauge $\partial^\mu A_\mu = 0$ and $n^\mu A_\mu = 0$, $n^\mu \equiv$ external four-vector

$$k^\mu\varepsilon_\mu(k, \lambda) = 0, \quad n^\mu\varepsilon_\mu(k, \lambda) = 0, \quad \varepsilon^*_\mu(k, \lambda)\varepsilon^\mu(k, \lambda') = -\delta_{\lambda\lambda'};$$

$$\sum_{\lambda=1}^{2} \varepsilon^*_\mu(k, \lambda)\varepsilon_\nu(k, \lambda) = -g_{\mu\nu} - \frac{k_\mu k_\nu}{(k \cdot n)^2}n^2 + \frac{n_\mu k_\nu + n_\nu k_\mu}{k \cdot n}$$

(d) Relativistic field equations can be obtained from an invariant Lagrangian density $\mathscr{L}(\psi, \partial_\mu\psi)$ with help of the Hamilton principle

$$\delta \int \mathscr{L}\, d^\mu x = 0 \tag{1.2.29}$$

as Euler–Lagrange equations

$$\frac{\partial\mathscr{L}}{\partial\psi} - \partial_\mu \frac{\partial\mathscr{L}}{\partial\, \partial_\mu\psi} = 0. \tag{1.2.30}$$

Use of the *Lagrangian formalism* is advantageous for the discussion of the relationship between symmetries and conservation laws (see Section 1.2.3); it is of utmost importance for the formulation of theories with interaction and their quantization. This can be illustrated by quantum electrodynamics (QED), which describes the interaction of photons, electrons, and positrons [Bj 66, QED]. The QED Lagrangian density is

$$\mathscr{L} = -\tfrac{1}{4}F_{\mu\nu}F^{\mu\nu} + \bar{\psi}(\mathrm{i}\gamma^{\mu}\,\partial_{\mu} - M)\psi + eA_{\mu}\bar{\psi}\gamma^{\mu}\psi \tag{1.2.31}$$

with the electromagnetic (em) field-strength tensor

$$F_{\mu\nu} = \partial_{\mu}A_{\nu} - \partial_{\nu}A_{\mu}. \tag{1.2.32}$$

Variation with respect to the four-potential A_{μ} gives the Maxwell equation:

$$\partial^{\mu}F_{\mu\nu}(x) = -e\bar{\psi}(x)\gamma_{\nu}\psi(x) = +ej_{\nu}^{\mathrm{em}}(x) \tag{1.2.33}$$

(The homogeneous Maxwell equations $\partial_{\mu}\varepsilon^{\mu\nu\rho\sigma}F_{\rho\sigma} = 0$ follow from (1.2.32).) The Dirac equation for the electron field

$$(\mathrm{i}\gamma^{\mu}\,\partial_{\mu} - M)\psi(x) = -e\gamma^{\mu}A_{\mu}\psi(x) \tag{1.2.34}$$

is obtained by variation with respect to $\bar{\psi}_{\alpha}(x)$. The elementary charge $e>0$ or Sommerfeld's fine structure constant $\alpha = e^2/4\pi \approx 1/137$, determine the strength of the coupling between the em field and the Dirac field. $j_{\mu}^{\mathrm{em}}(x)$ denotes the electromagnetic current normalized to the elementary charge. For the negatively charged electron, this is equal to the negative 'particle current' $j_{\mu}^{\mathrm{em}}(x) = -\bar{\psi}(x)\gamma_{\mu}\psi(x)$. For $e = 0$, the variation of the first term in $\mathscr{L}$ gives the free Maxwell equations, that of the second term gives the free Dirac equation (1.2.26). The em current $-\bar{\psi}\gamma_{\mu}\psi$ as the source of $F_{\mu\nu}(x)$ in (1.2.33) and the potential term in the Dirac equation result from the variation of the trilinear expression, i.e. the interaction term:

$$\mathscr{L}_{\mathrm{I}} = eA_{\mu}\bar{\psi}\gamma^{\mu}\psi. \tag{1.2.35}$$

(e) The systematic quantization of the non-linear field equations ((1.2.33) and (1.2.34)) is considered in Section 2.3.2. For small values of the coupling constant α the result can be expressed by *Feynman graphs*. These describe intuitively the interaction in terms of scattering processes between the particles (quanta) of the free fields. For example, the reaction $e^+e^- \to \mu^+\mu^-$ is described in the lowest order of the em interaction by the graph shown in Fig. 1.6. The in- and outgoing particles of the reactions are represented by external lines. The inner lines describe the field propagation between the interaction points (vertices). In this sense, the Feynman graph in Fig. 1.6 describes the above reaction as the annihilation of an e^+e^- pair into a photon at point 1, the propagation of the photon from point 1 to point 2 and the creation of a $\mu^+\mu^-$ pair from the photon at point 2.

(f) In accordance with the *Feynman rules*, these graphs represent analytical expressions for *S-matrix elements*. The rules described here are for QED, i.e. the interaction of electrons, muons, and photons.

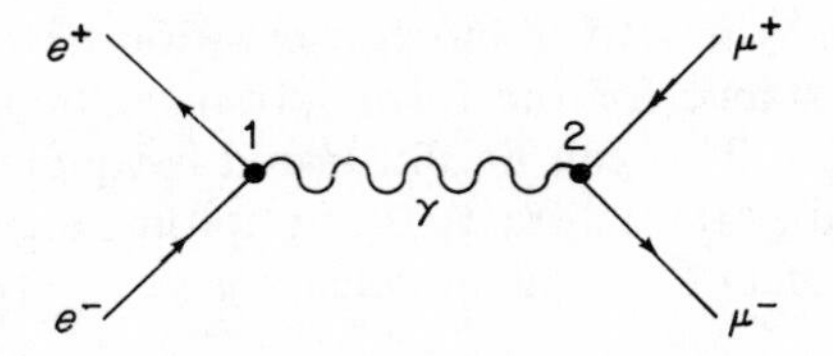

Figure 1.6 Feynman graph for the 1-photon approximation of the reaction $e^+e^- \to \mu^+\mu^-$

(1) The analytical expressions for the *external lines* are given by the plane wave solutions of the free field equations. Thus, particles with momentum p and spin j_3 of the reaction, which are represented by external lines, contribute wavefunctions $u(p, j_3), \ldots, \varepsilon_\mu(k, j_3)$ as factors:
incoming:

	ℓ^-: →•	ℓ^+: ←•	γ: ∿•
$(2\pi)^{-3/2}\times$	$u(p, j_3)$	$\bar{v}(p, j_3)$	$\varepsilon_\mu(k, j_3)$

outgoing: (1.2.36)

	ℓ^-: •→	ℓ^+: •←	γ: •∿
$(2\pi)^{-3/2}\times$	$\bar{u}(p, j_3)$	$v(p, j_3)$	$\varepsilon_\mu(k, j_3)$

(2) The interaction between photons and electrons or muons, corresponding to the Lagrange term $\mathscr{L}_\mathrm{I} = eA^\mu\bar{\psi}\gamma_\mu\psi$ (Eqn. (1.2.35)) represents a *vertex* with a charge line passing through and momentum conservation.

p_2, p_1, k:
$$= \mathrm{i}(2\pi)^4 e\gamma_\mu\,\delta^{(4)}(p_1 + p_2 - k) \tag{1.2.37}$$

(3) The propagation of the interacting fields is described by the causal Green functions of the free field equations, the so-called Feynman propagators Δ_F, S_F. For the Klein–Gordon equation:

$$\Delta_F(x, M) = -\mathrm{i}(\theta(x^0)\,\Delta_+(x, M) + \theta(-x^0)\,\Delta_-(x, M)) = \frac{1}{(2\pi)^4}\int \mathrm{d}^4p\,\frac{\mathrm{e}^{-\mathrm{i}px}}{p^2 - M^2 + \mathrm{i}\varepsilon}, \tag{1.2.38}$$

$$(\partial^\mu\,\partial_\mu + M^2)\,\Delta_F(x - x', M) = -\delta^{(4)}(x - x') \tag{1.2.38$'$}$$

with

$$\Delta_\pm(x, M) = (2\pi)^{-3}\int \frac{\mathrm{d}^3p}{2E_p}\,\mathrm{e}^{\mp\mathrm{i}px} = \frac{\pm\varepsilon(x^0)\,\delta(x^2)}{4\pi\mathrm{i}} + \frac{M^2\varepsilon(x^2)K_1(M\sqrt{(-x^2 \pm \mathrm{i}\varepsilon x^0)})}{4\pi^2 M\sqrt{(-x^2 \pm \mathrm{i}\varepsilon x^0)}}$$

and for the Dirac equation:

$$S_F(x, M) = (\mathrm{i}\gamma^\mu\,\partial_\mu + M)\,\Delta_F(x, M). \tag{1.2.39}$$

In Feynman graphs, the propagation is symbolized by an *internal line*; for this, the Fourier transform of iS_F, $i\Delta_F$ etc. is inserted as a factor in the analytical expression:

$$\ell: \quad \bullet\!\longrightarrow\!\bullet \qquad \frac{i}{(2\pi)^4}\frac{1}{\gamma p - M + i\varepsilon}$$
$$\gamma: \quad \bullet\!\sim\!\sim\!\sim\!\bullet \qquad \frac{i}{(2\pi)^4}\frac{-g_{\mu\nu}}{p^2 + i\varepsilon} \tag{1.2.40}$$

(4) The composition of the elements of the graph, i.e. the factors of the analytical expression, follows the structure of the reaction. The arrows indicate the direction of charge flow. The momentum flux from the incoming to the outgoing particles can be freely chosen, provided the conservation of momentum is observed. Integration over internal momenta has to be performed. The relative signs of Feynman graphs to be added up are determined by the Pauli principle:

(i) any commutation of two outgoing fermion lines gives a factor -1;
(ii) closed fermion lines give a factor -1 and require summation over fermion indices;
(iii) graph contributions have to be divided by the symmetry factor, which is defined by the number of possibilities of mapping the graph on itself by permutation of lines and/or vertices; e.g. a factor 1/2 thus results for any closed boson line.

(g) If the invariant S-matrix element: $1 + i\mathcal{M}\delta(p-q)$ for the reaction of two particles with momenta p_1 and p_2, forming n particles with momenta $q_1, \ldots, q_n$ has been calculated in this way, the differential cross-section can be determined in accordance with the formula

$$d\sigma = \frac{\pi^2}{\sqrt{[(p_1p_2)^2 - M_1^2M_2^2]}}|\mathcal{M}|^2\,\delta^{(4)}\left(p_1 + p_2 - \sum_{i=1}^{n} q_i\right)\frac{d^3q_1}{2E_{q_1}}\cdots\frac{d^3q_n}{2E_{q_n}}. \tag{1.2.41}$$

The decay probability of a particle with momentum $p = (E, \mathbf{p})$ into a final state of n particles with momenta $q_1, \ldots, q_n$ is given by

$$d\Gamma = \frac{1}{2\pi}\frac{|\mathcal{M}|^2}{2E}\,\delta^{(4)}\left(p - \sum_{i=1}^{n} q_i\right)\frac{d^3q_1}{2E_{q_1}}\cdots\frac{d^3q_n}{2E_{q_n}}. \tag{1.2.42}$$

(h) As an example, the e^+e^- *annihilation into muon pairs* is calculated. It is given in lowest order, the so-called 1-photon exchange approximation, by the Feynman graph in Fig. 1.6. The kinematic quantities of the reaction are labelled as follows:

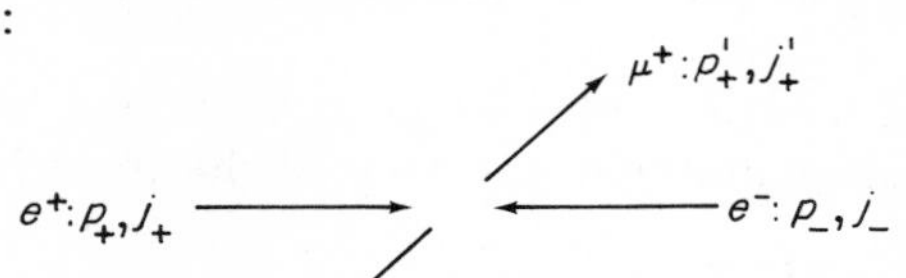

$\mu^+: p'_+, j'_+$

$e^+: p_+, j_+$ ⟶ ⟵ $e^-: p_-, j_-$

$\mu^-: p'_-, j'_-$

Centre of mass system:

$p_\pm = (E, 0, 0, \pm p)$ 4-momentum of the incoming $e^\pm$

$p'_\pm = (E, \pm p' \sin\vartheta \cos\phi, \pm p' \sin\vartheta \sin\phi, \pm p' \cos\vartheta)$

4-momentum of the outgoing $\mu^\pm$

$$s = (p_+ + p_-)^2 = (p'_+ + p'_-)^2 = 2m_e^2 + 2p_+p_- = 2m_\mu^2 + 2p'_+p'_- = 4E^2$$
$$t = (p_+ - p'_+)^2 = (p_- - p'_-)^2 = m_e^2 + m_\mu^2 - 2p_+p'_+; \quad p_+p'_+ = p_-p'_- = E^2 - pp' \cos\varphi$$
$$u = (p_+ - p'_-)^2 = (p_- - p'_+)^2 = m_e^2 + m_\mu^2 - 2p_+p'_-; \quad p_+p'_- = p_-p'_+ = E^2 + pp' \cos\varphi$$

Mandelstam variables s, t, and u satisfy the condition $s + t + u = 2m_e^2 + 2m_\mu^2$.

According to the Feynman rules, the associated invariant matrix element is given by

$$\mathcal{M} = \frac{\alpha}{\pi}\, \bar{u}(p'_-, j'_-)\gamma_\mu v(p'_+, j'_+) \frac{-g^{\mu\nu}}{s}\, \bar{v}(p_+, j_+)\gamma_\nu u(p_-, j_-). \tag{1.2.43}$$

Using the electromagnetic current

$$j_\nu^{\text{em}}(x) = -\bar{\psi}_e(x)\gamma_\nu\psi_e(x) - \bar{\psi}_\mu(x)\gamma_\nu\psi_\mu(x)$$

(compare Eqns (1.2.33) and (1.2.25)), which yields the current matrix element

$$\langle 0|\, j_\mu(0)\, |e^+; e^-\rangle = (2\pi)^{-3}\bar{v}(p_+, j_+)\gamma_\mu u(p_-, j_-), \tag{1.2.44}$$

this expression can also be written as

$$\mathcal{M} = \frac{\alpha}{\pi}(2\pi)^6\langle \mu^+; \mu^-|\, j_\mu(0)\, |0\rangle \frac{-g^{\mu\nu}}{s} \langle 0|\, j_\nu(0)\, |e^+; e^-\rangle. \tag{1.2.45}$$

The differential cross-section for unpolarized incoming and outgoing particles is as follows in the centre of mass system:

$$\begin{aligned}
\frac{d\sigma}{d\Omega} &= \frac{\alpha^2}{s^2}\frac{p'}{16E^2\cdot p}\frac{1}{4}\Big[\sum_{j_+j_-} \bar{v}(p_+, j_+)\gamma_\mu u(p_-, j_-)\cdot \bar{u}(p_-, j_-)\gamma_\nu v(p_+, j_+)\Big] \\
&\quad \times \Big[\sum_{j'_+j'_-} \bar{u}(p'_-, j'_-)\gamma^\mu v(p'_+, j'_+)\cdot \bar{v}(p'_+, j'_+)\gamma^\nu u(p'_-, j'_-)\Big] \\
&= \frac{\alpha^2}{s^2}\frac{p'}{16E^2\cdot p}\tfrac{1}{4}\operatorname{Tr}[(\not{p}_+ - m_e)\gamma_\mu(\not{p}_- + m_e)\gamma_\nu] \\
&\qquad \cdot \operatorname{Tr}[(\not{p}'_- + m_\mu)\gamma^\mu(\not{p}'_+ - m_\mu)\gamma^\nu] \\
&= \frac{\alpha^2}{s^2}\frac{p'}{16E^2\cdot p}\tfrac{1}{4}L^{(e)}_{\mu\nu}L^{(\mu)\mu\nu}, \quad d\Omega = \sin\vartheta\, d\vartheta\, d\phi.
\end{aligned} \tag{1.2.46}$$

Table 1.4 was used to work out the spin summations. The leptonic tensors $L^{(e)}_{\mu\nu}$ and $L^{(\mu)}_{\mu\nu}$ are calculated with the help of the trace formulae

$$\begin{aligned}
\operatorname{Tr}\, \gamma_\mu\gamma_\nu &= 4g_{\mu\nu} \\
\operatorname{Tr}\, \gamma_\mu\gamma_\nu\gamma_\rho\gamma_\sigma &= 4(g_{\mu\nu}g_{\rho\sigma} - g_{\mu\rho}g_{\nu\sigma} + g_{\mu\sigma}g_{\nu\rho}).
\end{aligned} \tag{1.2.47}$$

It follows that:

$$L^{(e)}_{\mu\nu} = \sum_{j_+ j_-} \bar{v}(p_+, j_+)\gamma_\mu u(p_-, j_-) \cdot \bar{u}(p_-, j_-)\gamma_\nu v(p_+, j_+)$$
$$= \mathrm{Tr}(\not{p}_+ - m_e)\gamma_\mu(\not{p}_- + m_e)\gamma_\nu$$
$$= 4(p_{+\mu}p_{-\nu} + p_{-\mu}p_{+\nu} - \tfrac{1}{2}sg_{\mu\nu}). \qquad (1.2.48)$$

With the above-defined kinematic variables in the centre of mass system one gets:

$$\frac{d\sigma}{d\Omega} = \frac{\alpha^2}{s^2}\frac{p'}{E^2 \cdot p}[E^4 + p^2p'^2\cos^2\vartheta + E^2(m_e^2 + m_\mu^2)]. \qquad (1.2.49)$$

For storage ring energies, $m_e \ll E$, this can be simplified to

$$\frac{d\sigma}{d\Omega} = \frac{\alpha^2}{4s}\frac{p'}{E}\left(1 + \cos^2\vartheta + \frac{m_\mu^2}{E^2}\sin^2\vartheta\right). \qquad (1.2.50)$$

Integration then gives the total cross-section:

$$\sigma = \frac{4\pi}{3}\frac{\alpha^2}{s}\frac{p'}{E}\left(1 + \frac{2m_\mu^2}{s}\right). \qquad (1.2.51)$$

If the μ mass can be neglected, Eqns (1.2.50) and (1.2.51) can be simplified to give:

$$\frac{d\sigma}{d\Omega} = \frac{\alpha^2}{4s}(1 + \cos^2\vartheta) \qquad (1.2.52)$$

and

$$\sigma = \frac{4\pi\alpha^2}{3s}. \qquad (1.2.53)$$

1.2.3 Currents and charges

After using the QED example to show how cross-sections can be calculated from a given Lagrangian density with the help of the appropriate Feynman rules, the next step is to adapt the group theoretical description of symmetries of Section 1.2.1 to Lagrangian field theories.

(a) Lagrangian densities which contain a complex field in the form $\psi^\dagger(x)\Gamma\psi(x)$ (QED falls into this class too) are invariant under transformations of the phase of the field, i.e. *Abelian gauge transformations of the first kind*:

$$\psi_\alpha(x) \to e^{i\theta}\psi_\alpha(x), \quad \psi_\alpha^\dagger(x) \to \psi_\alpha^\dagger(x)e^{-i\theta}. \qquad (1.2.54)$$

For infinitesmal transformations $\delta\theta$:

$$\psi_\alpha(x) \to (1 + i\,\delta\theta)\psi_\alpha(x) = \psi_\alpha(x) + \delta\psi_\alpha(x),$$
$$\psi_\alpha^\dagger(x) \to \psi_\alpha^\dagger(x)(1 - i\,\delta\theta) = \psi_\alpha^\dagger(x) + \delta\psi_\alpha^\dagger(x).$$

that is,

$$\delta\psi_\alpha(x) = \mathrm{i}\,\delta\theta\psi_\alpha(x), \quad \delta\psi^\dagger_\alpha(x) = \psi^\dagger_\alpha(x)(-\mathrm{i}\,\delta\theta) \tag{1.2.55}$$

The Lagrangian density $\mathscr{L}(\psi, \psi^\dagger, \partial_\mu\psi, \partial_\mu\psi^\dagger)$ does not change under these transformations:

$$\begin{aligned}\mathscr{L}(\psi, \psi^\dagger, \partial_\mu\psi, \partial_\mu\psi^\dagger) &\to \mathscr{L}(\psi + \delta\psi, \psi^\dagger + \delta\psi^\dagger, \partial_\mu\psi + \partial_\mu\,\delta\psi, \partial_\mu\psi^\dagger + \partial_\mu\,\delta\psi^\dagger)\\ &= \mathscr{L}(\psi, \psi^\dagger, \partial_\mu\psi, \partial_\mu\psi^\dagger) + \delta\mathscr{L},\end{aligned} \tag{1.2.56}$$

i.e. $\delta\mathscr{L} = 0$. The Noether theorem guarantees the existence of a conserved current for any invariance of a Lagrangian density [La 69]. The construction of these currents from the fields ψ and $\psi^\dagger$ and the fact that its divergence vanishes can be seen from the following calculation which makes use of the equations of motion (1.2.30):

$$\begin{aligned}0 = \delta\mathscr{L} &= \delta\psi^\dagger \frac{\partial\mathscr{L}}{\partial\psi^\dagger} \pm \frac{\partial\mathscr{L}}{\partial\psi}\,\delta\psi + \delta(\partial_\mu\psi^\dagger)\frac{\partial\mathscr{L}}{\partial(\partial_\mu\psi^\dagger)} \pm \frac{\partial\mathscr{L}}{\partial(\partial_\mu\psi)}\,\delta(\partial_\mu\psi)\\ &= \delta\psi^\dagger\,\partial_\mu \frac{\partial\mathscr{L}}{\partial(\partial_\mu\psi^\dagger)} \pm \partial_\mu \frac{\partial\mathscr{L}}{\partial(\partial_\mu\psi)}\,\delta\psi + \delta(\partial_\mu\psi^\dagger)\frac{\partial\mathscr{L}}{\partial(\partial_\mu\psi^\dagger)} \pm \frac{\partial\mathscr{L}}{\partial(\partial_\mu\psi)}\,\delta(\partial_\mu\psi)\\ &= \partial_\mu\left(\delta\psi^\dagger \frac{\partial\mathscr{L}}{\partial(\partial_\mu\psi^\dagger)} \pm \frac{\partial\mathscr{L}}{\partial(\partial_\mu\psi)}\,\delta\psi\right)\\ &= \partial_\mu\left[-\mathrm{i}\left(\psi^\dagger \frac{\partial\mathscr{L}}{\partial(\partial_\mu\psi^\dagger)} \mp \frac{\partial\mathscr{L}}{\partial(\partial_\mu\psi)}\,\psi\right)\right]\delta\theta = \partial_\mu j^\mu\,\delta\theta\end{aligned}$$

with

$$j^\mu(x) = -\mathrm{i}\left(\psi^\dagger \frac{\partial\mathscr{L}}{\partial(\partial_\mu\psi^\dagger)} \mp \frac{\partial\mathscr{L}}{\partial(\partial_\mu\psi)}\,\psi\right) \quad \text{for} \begin{cases}\text{bosons}\\ \text{fermions}\end{cases}. \tag{1.2.57}$$

In QED, the electromagnetic current which occurs in the Maxwell equation (1.2.33) results from Eqns (1.2.57) and (1.2.31):

$$j^{\mathrm{em}}_\mu(x) = -\bar{\psi}(x)\gamma_\mu\psi(x). \tag{1.2.58}$$

The invariance of QED under phase transformations thus means the conservation of the electric charge measured in units of the elementary charge e:

$$Q = \int j_0(t, \mathbf{x})\,\mathrm{d}^3x, \quad \frac{\mathrm{d}Q}{\mathrm{d}t} = 0. \tag{1.2.59}$$

The phase transformations of the fields $\psi, \psi^\dagger$ form the group $U(1)$.

(b) In elementary particle physics, a generalization to *non-Abelian symmetry groups* such as flavour $SU(N_\mathrm{F})$ and colour $SU(3)_\mathrm{C}$ mentioned in Section 1.2.1 is necessary. Quarks are therefore described by Dirac fields, which carry flavour and colour indices:

$\psi_{\alpha,f,c}(x)$	α	Dirac index
	$f=u, d, s, c, \ldots$	flavour index
	$c=r, g, b$	colour index

Alternatively, $u_\alpha(x)$, $d(x), \ldots$.

If the quarks were free particles, the fields would obey the Dirac equation

$$(\mathrm{i}\gamma^\mu_{\alpha\alpha'}\,\delta_{ff'}\,\delta_{cc'}\,\partial_\mu - M_f\,\delta_{\alpha\alpha'}\,\delta_{ff'}\,\delta_{cc'})\psi_{\alpha',f',c'}(x)=0 \tag{1.2.60}$$

written with all indices, or

$$(\mathrm{i}\not\partial - M)\psi(x)=0$$

in matrix notation, for short.

The field equations (1.2.60) and the commutation relations of the quark fields are invariant under colour transformations:

$$\psi_{\alpha,f,c}(x)\to(\exp(-\tfrac{1}{2}\mathrm{i}\theta^a\lambda_a))_{cc'}\psi_{\alpha,f,c'}(x), \tag{1.2.61}$$

or infinitesimally:

$$\delta\psi_{\alpha,f,c}(x)=-\tfrac{1}{2}\mathrm{i}(\lambda_a)_{cc'}\psi_{\alpha,f,c'}(x)\;\delta\theta^a. \tag{1.2.62}$$

where λ_α are Gell-Mann matrices.

If the quark masses are flavour independent, there is flavour symmetry too, i.e. invariance under flavour transformations:

$$\psi_{\alpha,f,c}(x)\to(\exp(-\tfrac{1}{2}\mathrm{i}\eta^a\underline{\lambda}_a)_{ff'}\psi_{\alpha,f',c}(x),$$

where λ_a is the fundamental representation of $SU(N_F)$. Infinitesimally,

$$\delta\psi_{\alpha,f,c}(x)=-\tfrac{1}{2}\mathrm{i}(\underline{\lambda}_a)_{ff'}\psi_{\alpha,f',c}(x)\;\mathrm{d}\eta^a. \tag{1.2.63}$$

As already shown, conserved colour and flavour currents are linked with these invariances:

$$j_{\mu,a}(x)=\mathrm{i}\left(\frac{\partial\mathcal{L}}{\partial(\partial_\mu\psi)}\frac{\lambda_a}{2}\psi+\psi^\dagger\frac{\lambda_a}{2}\frac{\partial\mathcal{L}}{\partial(\partial_\mu\psi^\dagger)}\right). \tag{1.2.64}$$

In a pure quark field theory, they have the form:

$$j^{\mathrm{C}}_{\mu,a}(x)=\bar{\psi}_{f,c}(x)\left(\frac{\lambda_\alpha}{2}\right)_{cc'}\gamma_\mu\psi_{f,c'}(x), \tag{1.2.65}$$

$$j^{\mathrm{F}}_{\mu,a}(x)=\bar{\psi}_{f,c}(x)\left(\frac{\lambda_\alpha}{2}\right)_{ff'}\gamma_\mu\psi_{f',c}(x). \tag{1.2.66}$$

Eqn. (1.2.64) agrees with definition (1.2.57) if $\lambda_a/2$ is replaced by -1 according to the negative charge of the electron.

As a consequence of the commutation relations of the quark field operators, the generalization of Eqn. (1.2.27):

$$\begin{aligned} [\psi_{\alpha,f,c}(x), \bar{\psi}_{\alpha',f',c'}(x')]_+ &= -\mathrm{i}S_{\alpha\alpha'}(x-x', M)\,\delta_{ff'}\,\delta_{cc'}, \\ [\psi_{\alpha,f,c}(x), \psi_{\alpha',f',c'}(x')]_+ &= 0. \end{aligned} \tag{1.2.67}$$

reflects the composition of $\psi(x)$ from creation and annihilation operators. The currents (1.2.58) and (1.2.66) have a direct physical interpretation, which will be illustrated by the following two examples.

The current $j_\mu^{\mathrm{F,em}}(x)$, which contains as flavour matrix the electric charges of the quarks:

$$j_\mu^{\mathrm{F,em}}(x) = \bar{\psi}(x) Q \gamma_\mu \psi(x), \quad \text{with } Q = \mathrm{diag}\,(2/3, -1/3, -1/3, 2/3, \ldots) \tag{1.2.68}$$

describes not only the charges, magnetic moments, and radiative transitions of hadrons (e.g. $\omega \to \pi^0 + \gamma$, $\Delta^+ \to P + \gamma$) but also the creation or annihilation of quark–antiquark pairs (e.g. $e^+e^- \to J/\psi(c\bar{c})$). In a similar way, the isospin current:

$$j_\mu^{\mathrm{F},(+)}(x) = \bar{\psi}(x)\tau^+\gamma_\mu\psi(x) \quad \text{with } \tau^+ = \tfrac{1}{2}\lambda_1 + \mathrm{i}\tfrac{1}{2}\lambda_2$$

converts a d quark in a hadron into a u quark or a $\bar{u}$ quark into a $\bar{d}$ quark (e.g. $N \to Pe\nu$, $\pi^- \to \pi^0 e\nu$) or creates a $u\bar{d}$ pair or annihilates a $d\bar{u}$ pair. Our examples have already anticipated how important current operators are to the phenomenological descriptions of the electromagnetic and weak interaction. The next section discusses this in more detail.

(c) The *operators for the conserved colour and flavour charges*

$$\begin{aligned} Q_a^{\mathrm{C}} &= \int \mathrm{d}^3x\, j_{0,a}^{\mathrm{C}}(t, \mathbf{x}) \\ Q_a^{\mathrm{F}} &= \int \mathrm{d}^3x\, j_{0,a}^{\mathrm{F}}(t, \mathbf{x}) \end{aligned} \tag{1.2.69}$$

satisfy the commutation relations (1.2.1):

$$\begin{aligned} [Q_a^{\mathrm{C}}, Q_b^{\mathrm{C}}] &= \mathrm{i}f_{abc}Q_c^{\mathrm{C}} \\ [Q_a^{\mathrm{F}}, Q_b^{\mathrm{F}}] &= \mathrm{i}f_{abc}Q_c^{\mathrm{F}} \end{aligned} \tag{1.2.70}$$

where f_{abc} are the structure constants of $SU(N_{\mathrm{F}})$ according to Eqn. (1.2.1), because of the field commutation relations (1.2.67). They generate the symmetry transformations (1.2.62) and (1.2.63)

$$\delta\theta^a[Q_a^{\mathrm{C}}, \psi_{f,c}(x)] = -\tfrac{1}{2}\mathrm{i}(\lambda_a)_{cc'}\psi_{f,c'}(x)\,\delta\theta^a$$

and

$$\delta\eta^a[Q_a^{\mathrm{F}}, \psi_{f,c}(x)] = -\tfrac{1}{2}\mathrm{i}(\lambda_a)_{ff'}\psi_{f',c}(x)\,\delta\eta^a \tag{1.2.71}$$

respectively. With this, the symmetries considered above can be written as unitary transformations on the quantified fields. Besides the above-mentioned direct physical interpretation of currents and charges, this shows their importance in connection with symmetries in quantum field theories.

If a symmetry, such as the flavour symmetry is only approximately valid, the associated currents are not conserved and the corresponding charges are time-dependent. Despite this, the equal time commutation relations according to (1.2.70) are satisfied. The investigation of such structures in the framework of current algebra gave interesting results (see Eqns (1.3.36) and (1.3.37)) [Ad 68].

One general remark may now serve to end this section on the close connection between symmetry laws and field theory. In field theory, dynamics is described by means of the interation of local quantities—fields. Global changes of the fields, as for example of their phase with help of gauge transformations of the first kind (Eqn. (1.2.54)) contradicts the spirit of this physical principle of only local action. This objection suggests that global symmetry transformations as described in Eqns (1.2.54), (1.2.61), etc. should be replaced by transformations in which the fields are transformed locally independent. For phase transformations (1.2.54), this would mean that θ becomes space–time dependent: $\theta = \theta(x)$. It was the elaboration of this idea of local symmetry which led to the concept of gauge theories [We 29].

1.3 Phenomenology of the electromagnetic and weak interactions

Electromagnetic and weak processes are successfully described phenomenologically by the one-photon exchange model and the Fermi model, respectively [FM]. As the example of e^+e^- annihilation into μ pairs in Section 1.2.2 shows, S-matrix elements can be built up from the matrix elements of the electromagnetic current (see Eqn. (1.2.45)). In a similar way, the weak interaction of elementary particles is based on current matrix elements in the Fermi model. This section describes the structure of electromagnetic and weak currents and compares it with experimental results. Although the agreement is satisfactory, there are principle objections to the use of these models, e.g. because they fail to give a unitary S-matrix. These problems are overcome in the case of the electromagnetic interaction by the higher-order approximations of QED. The Fermi model can be improved by introducing heavy vector bosons, but electromagnetic and weak interactions must be combined in the form of a gauge theory before a consistent theory can be formulated (Chapter 3).

1.3.1 The electromagnetic and weak interactions of leptons

(a) In the *Fermi model*, the interaction density $\mathscr{L}_\text{I}$ of charged and neutral weak processes is of the current–current type:

$$\mathscr{L}_\text{I}^\text{weak}(x) = \frac{4G_F}{\sqrt{2}}(j_\mu^{(+)}(x)j^{(-)\mu}(x) + j_\mu^{(n)}(x)j^{(n)\mu}(x)). \qquad (1.3.1)$$

The charged current $j_\mu^{(\pm)}(x)(j_\mu^{(-)} = (j_\mu^{(+)})^\dagger)$ and the neutral current $j_\mu^{(n)}(x)$ are made of a leptonic and a hadronic component:

$$j_\mu^{(\pm)}(x) = l_\mu^{(\pm)}(x) + h_\mu^{(\pm)}(x), \tag{1.3.2}$$

$$j_\mu^{(n)}(x) = l_\mu^{(n)}(x) + h_\mu^{(n)}(x). \tag{1.3.3}$$

In contrast to the electric charge, the Fermi coupling constant G_F has non-vanishing dimension even in the natural system of units ($\hbar = c = 1$):

$$G_F = 1.02/10^5 M_p^2 = (294\ \mathrm{GeV})^{-2}. \tag{1.3.4}$$

So far, three two-particle families of leptons are known: the electron e and the electron neutrino ν_e, the muon μ and the muon neutrino ν_μ and the tau τ and the tau neutrino ν_τ. A lepton number conserved in any process can be ascribed to every one of these families. This rules out, for example, the radiative decay $\mu^- \to e^- + \gamma$. Experimentally, the branching ratio is $B(\mu \to e\gamma) < 3.6 \times 10^{-9}$.

(b) A remarkable property of the em and weak interaction is its universality: all leptons make a similarly structured contribution to $\mathscr{L}_\mathrm{I}^{\mathrm{em}}$ and $\mathscr{L}_\mathrm{I}^{\mathrm{weak}}$, i.e. their *currents* are built up in the same way from the Dirac fields of the leptons $e(x)$, $\nu_e(x)$, $\mu(x)$, $\nu_\mu(x)$, $\tau(x)$, and $\nu_\tau(x)$.

$$l_\mu^{\mathrm{em}}(x) = -\bar{e}(x)\gamma_\mu e(x) - \bar{\mu}(x)\gamma_\mu \mu(x) - \bar{\tau}(x)\gamma_\mu \tau(x), \tag{1.3.5}$$

$$\begin{aligned} l_\mu^{(+)}(x) &= \bar{\nu}_e(x)\gamma_{\mu L} e(x) + \bar{\nu}_\mu(x)\gamma_{\mu L}\mu(x) + \bar{\nu}_\tau(x)\gamma_{\mu L}\tau(x), \\ l_\mu^{(-)}(x) &= \bar{e}(x)\gamma_{\mu L}\nu_e(x) + \bar{\mu}(x)\gamma_{\mu L}\nu_\mu(x) + \bar{\tau}(x)\gamma_{\mu L}\nu_\tau(x), \end{aligned} \tag{1.3.6}$$

$$\begin{aligned} l_\mu^{(0)}(x) = \tfrac{1}{2}(&\bar{\nu}_e(x)\gamma_{\mu L}\nu_e(x) - \bar{e}(x)\gamma_{\mu L} e(x) \\ &+ \bar{\nu}_\mu(x)\gamma_{\mu L}\nu_\mu(x) - \bar{\mu}(x)\gamma_{\mu L}\mu(x) \\ &+ \bar{\nu}_\tau(x)\gamma_{\mu L}\nu_\tau(x) - \bar{\tau}(x)\gamma_{\mu L}\tau(x)), \end{aligned} \tag{1.3.7}$$

$$l_\mu^{(n)}(x) = l_\mu^{(0)}(x) - \sin^2\theta_\mathrm{W} l_\mu^{\mathrm{em}}(x), \tag{1.3.8}$$

$$\gamma_{\mu L} := \gamma_\mu \frac{1-\gamma_5}{2}, \quad \gamma_{\mu R} := \gamma_\mu \frac{1+\gamma_5}{2}.$$

In contrast to the em current which is a pure vector (V) current, the currents $l_\mu^{(\pm)}$ and $l_\mu^{(0)}$ consist of equally strong vector and axial vector (A) components, they are left-handed ($L = (V-A)/2$) currents. This *ansatz* reflects the experimental fact that parity is violated maximally in charged, weak processes. In neutral, weak processes, the parity violation depends on the charge of the particles, because the neutral current contains an admixture of the em current. The strength of this admixture can be parametrized using the Weinberg angle θ_W (experimentally, $\sin^2\theta_\mathrm{W} = 0.228 \pm 0.010$ [Pa 82]). In principle, scalar ($g_s \bar{\psi}\psi \cdot \bar{\psi}\psi$), pseudoscalar ($g_p \bar{\psi}\gamma_5\psi \cdot \bar{\psi}\gamma_5\psi$) as well as tensorial and derivative couplings could contribute to the 4-fermion coupling, in addition to the current–current interaction. However, these terms are not necessary for achieving a reasonable level of agreement with the present experimental data.

(c) The charges:

$$
\begin{aligned}
I_W^{(\pm)}(t) &= \int d^3x l_0^{(\pm)}(t, \mathbf{x}) \\
I_W^{(0)}(t) &= \int d^3x l_0^{(0)}(t, \mathbf{x})
\end{aligned}
\tag{1.3.9}
$$

of the currents $l_\mu^{(\pm)}(x)$ and $l_\mu^{(0)}(x)$ have the commutators:

$$
\begin{aligned}
[I_W^{(+)}(t), I_W^{(-)}(t)]_- &= 2I_W^{(0)}(t), \\
[I_W^{(0)}(t), I_W^{(\pm)}(t)]_- &= \pm I_W^{(\pm)}(t)
\end{aligned}
\tag{1.3.10}
$$

due to the equal-time commutation relations of the Dirac fields as in Eqn. (1.2.67). This is the Lie algebra of an $SU(2)$ group, *the weak isospin group.* Because of this structure it makes sense to arrange the left-handed leptons (e.g. $e_L(x) = (1-\gamma_5)e(x)/2$ and $\nu_{eL}(x) = (1-\gamma_5)\nu_e(x)/2)$) into weak isospin doublets and the right-handed ones into singlets [Gl 61].

(d) The *weak hypercharge*

$$
\tfrac{1}{2}Y_W(t) = Q - I_W^{(0)}(t) = \int d^3x (l_0^{em}(t, \mathbf{x}) - l_0^{(0)}(t, \mathbf{x}))
\tag{1.3.11}
$$

commutes with all components of the weak isospin with which it generates the group $SU(2)_W \times U(1)$. This group plays an important role for the construction of a gauge theory of the unified electroweak interaction, thus leading to a theoretical explanation of universality (Section 2.2.2b).

(e) After the definition of the Fermi model for leptons, it can now be compared with *experimental data* by means of a few examples. The $(V-A)$ structure of the current (1.3.6) can be tested in the muon decay process $\mu^- \rightarrow \nu_\mu + e^- + \bar{\nu}_e$ (see Fig. 1.7). For the decay matrix element $\mathcal{M}$, the lowest

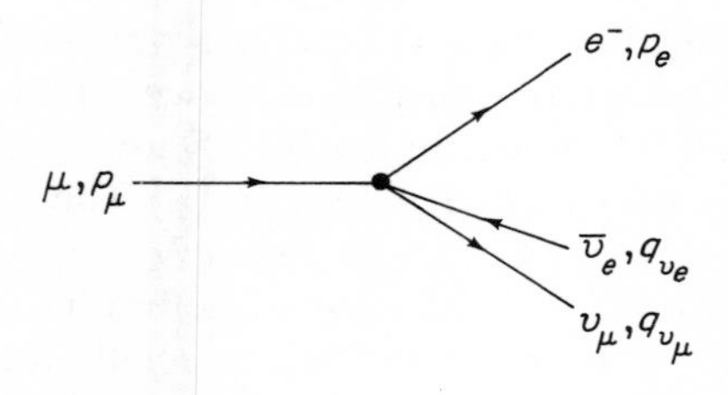

Figure 1.7 Kinematics of μ decay

order perturbation theory in the Fermi constant is:

$$\mathcal{M} = (2\pi)^4 \langle e^- \bar{\nu}_e \nu_\mu \, | \mathcal{L}_I^{\text{weak}}(0) | \, \mu^- \rangle$$
$$= \frac{4G_F}{\sqrt{2}} (2\pi)^{-2} (\bar{u}(p_e) \gamma_{\mu L} v(q_{\nu_e}))(\bar{u}(q_{\nu_\mu}) \gamma^{\mu L} u(p_\mu)).$$

The second line follows from the first by representing the fields and currents by creation and annihilation operators in accordance with Eqn. (1.2.25), and using the relations (1.2.19)–(1.2.21) systematically. The transition probability per unit time is obtained from $\mathcal{M}$ using Eqn. (1.2.42).

Neglecting the electron mass, the following electron spectrum ($x = E_e/M_\mu$) can be calculated for the decay of polarized muons at rest ($\mathcal{P}$ = degree of muon polarization, $\vartheta \equiv$ angle between μ polarization and electron momentum) [Ma 69]:

$$\mathrm{d}\Gamma = \frac{G_F^2 M_\mu^5}{192\pi^3} 12x^2 \{1 - x + \tfrac{2}{3}\rho(\tfrac{4}{3}x - 1)$$
$$+ \mathcal{P} \cos\vartheta [\alpha(1-x) + \beta(\tfrac{4}{3}x - 1)]\} \frac{\mathrm{d}\Omega_e}{4\pi} \, \mathrm{d}x. \tag{1.3.12}$$

The Michel parameters ρ, α, and β are equal to 3/4, −1/3, and −3/4, respectively, for pure (V–A) coupling. Deviations from this can be interpreted as an admixture of other couplings. However, these are small because of the experimental results: $\rho = 0.752 \pm 0.003$, $\alpha = 0.324 \pm 0.004$, $\beta = 0.755 \pm 0.009$ [Pa 82].

Integration of $\mathrm{d}\Gamma$ over the electron momentum gives the width Γ or lifetime τ of the muon:

$$\Gamma = \tau^{-1} = \int \mathrm{d}\Gamma = \frac{G_F^2 M_\mu^5}{192\pi^3}. \tag{1.3.13}$$

The value given for the Fermi constant G_F in Eqn. (1.3.4) follows from $\tau_{\text{exp}} = 2.19 \times 10^{-6}$ s.

The universality and the (V–A) structure of the coupling can be tested using the leptonic tau decays ($\tau^- \to \nu_\tau + e^- + \bar{\nu}_e$ and $\tau^- \to \nu_\tau + \mu^- + \bar{\nu}_\mu$). The Michel parameter ρ is 0.66 ± 0.13 [Fl 79] and the branching ratio $B_{\mu e} = \Gamma(\tau^- \to \mu^- \bar{\nu}_\mu \nu_\tau)/\Gamma(\tau^- \to e^- \bar{\nu}_e \nu_\tau) = 0.99 \pm 0.20$, whereas theoretically $\Gamma(\tau \to e\bar{\nu}_e \nu_\tau) = 6.2 \times 10^{11}\ \mathrm{s}^{-1}$, $\Gamma(\tau \to \mu^- \bar{\nu}_\mu \nu_\tau) = 6.0 \times 10^{11}\ \mathrm{s}^{-1}$, i.e. $B_{\mu e} = 0.98$.

The interaction density $\mathcal{L}_I$ (1.3.1) describes not only the weak decays of leptons, but also their scattering processes. Since only electrons are available as a target, the reactions shown in Table 1.5 are possible. Here, $\sigma_0 = G_F^2 s/\pi = G_F^2 2M_e E_\nu/\pi = 1.68 \times 10^{-41}(E_\nu/\mathrm{GeV})\ \mathrm{cm}^2$. The experimental data $\sigma(\nu_\mu e^- \to \nu_\mu e^-)/E_\nu = (1.6 \pm 0.4) \times 10^{-42}\ \mathrm{cm}^2/\mathrm{GeV}$ and $\sigma(\bar{\nu}_\mu e^- \to \bar{\nu}_\mu e^-)/E_\nu = (1.3 \pm 0.6) \times 10^{-42}\ \mathrm{cm}^2/\mathrm{GeV}$ can be used to determine the Weinberg angle. The result is $\sin^2\theta_W = 0.23$ (see [Wi 79a] especially for a discussion of the experimental error).

Table 1.5 Neutrino–electron scattering

Reaction	Contributions from	Total cross-section
$\nu_\mu e^- \to \nu_\mu e^-$	$l_\mu^{(n)}$	$(\frac{1}{4}-\sin^2\theta_W+\frac{4}{3}\sin^4\theta_W)\sigma_0$
$\bar{\nu}_\mu e^- \to \bar{\nu}_\mu e^-$	$l_\mu^{(n)}$	$\frac{1}{3}(\frac{1}{4}-\sin^2\theta_W+4\sin^4\theta_W)\sigma_0$
$\nu_\mu e^- \to \nu_e \mu^-$	$l_\mu^{(\pm)}$	σ_0
$\bar{\nu}_e e^- \to \bar{\nu}_\mu \mu^-$	$l_\mu^{(\pm)}$	$\frac{1}{3}\sigma_0$
$\nu_e e^- \to \nu_e e^-$	$l_\mu^{(\pm)}$ and $l_\mu^{(n)}$	$(\frac{1}{4}+\sin^2\theta_W+\frac{4}{3}\sin^4\theta_W)\sigma_0$
$\bar{\nu}_e e^- \to \bar{\nu}_e e^-$	$l_\mu^{(\pm)}$ and $l_\mu^{(n)}$	$\frac{1}{3}(\frac{1}{4}+\sin^2\theta_W+4\sin^4\theta_W)\sigma_0$

(f) Although experiments in the currently accessible energy range are compatible with the Fermi model, the increase of the cross section σ_0 indicates that the pointlike current–current coupling fails at higher energies. On the one hand, one calculates the S-matrix element of the reaction $\nu_\mu + e^- \to \nu_e + \mu^-$:

$$\mathcal{M} = \frac{4G_F}{\sqrt{2}}(2\pi)^{-2}(\bar{u}(p_\mu)\gamma^{\mu L}u(q_{\nu_\mu})(\bar{u}(q_{\nu_e})\gamma_{\mu L}u(p_e)) \qquad (1.3.14)$$

and herefrom the differential cross section:

$$\frac{d\sigma(\nu_\mu^- \to \nu_e\mu^-)}{d\Omega} = \frac{G_F^2}{4\pi^2}\frac{(s-M_\mu^2)^2}{s} \simeq \frac{G_F^2}{4\pi^2}s \quad \text{for } s \gg M_\mu^2. \qquad (1.3.15)$$

On the other hand, the unitarity of the S-matrix for s-wave scattering requires the inequality, $d\sigma/d\Omega \leqslant 1/4p_{\text{CMS}}^2 = 1/s$. For $\sqrt{s} \simeq 500$ GeV, Eqn. (1.3.15) violates this unitarity bound. By introducing *vector bosons* of sufficiently high mass, the current–current coupling can be 'softened' such that this difficulty is overcome, at least for neutrino–lepton scattering [Le 60]. The charged (neutral) currents are thereby coupled to a charged (neutral) vector boson $W^\pm(Z^0)$, their exchange mediates the interaction. Thus, the reaction mechanism is very similar to the 1-photon exchange for em processes.

Neutrino–electron scattering will now be used to show how the W exchange can improve the high-energy behaviour of this reaction. For this, the following Feynman diagram must be calculated:

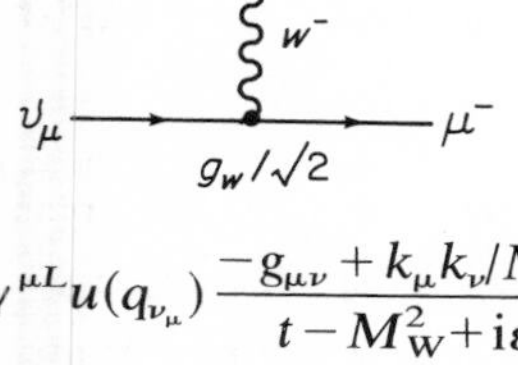

$$\mathcal{M} = \frac{g_W^2}{2}(2\pi)^{-2}(\bar{u}(p_\mu)\gamma^{\mu L}u(q_{\nu_\mu})\frac{-g_{\mu\nu}+k_\mu k_\nu/M_W^2}{t-M_W^2+i\varepsilon}(\bar{u}(q_{\nu_e})\gamma^{\nu L}u(p_e)), \qquad (1.3.16)$$

$$t = k^2 = (p_e - q_{\nu_e})^2, \quad 0 \leqslant |t| \leqslant s.$$

In comparison with Eqn. (1.3.15) one gains a factor t^{-1}* which means that the total cross section behaves like g_W^4/s for high energies. For energies $s \ll M_W^2$, the matrix element (1.3.16) goes over into that of the Fermi theory (1.3.14) up to corrections of the order of $(\sqrt{s})/M_W$, if the vector boson coupling constant is identified with the Fermi constant according to

$$\frac{4G_F}{\sqrt{2}} = \frac{g_W^2}{2M_W^2} \quad \text{or} \quad g_W = \frac{M_W}{124\,\text{GeV}}. \tag{1.3.17}$$

Similarly, one has to substitute for neutral current processes:

$$\frac{4G_F}{\sqrt{2}} j_\mu^{(n)} j^{(n)\mu} \to g_Z^2 \frac{-g_{\mu\nu} + (k_\mu k_\nu / M_Z^2)}{k^2 - M_Z^2 + i\varepsilon} j^{(n)\mu} j^{(n)\nu} \tag{1.3.18}$$

For $s \approx M_W^2, M_Z^2$, the modification of the Fermi theory by intermediate bosons has drastic effects. However, at lower energies this hypothesis can only be tested by a careful analysis of the energy dependence of the cross-sections. So far, no such propagator effect has been found. A comparison of the Fermi model with the experimental data gives a lower bound of about 59 GeV/c^2 for the mass of the Z boson [Da 82N]. From the fact that W and Z bosons are massive particles and thus have longitudinal polarization states in contrast to the zero mass photon, it follows that the cross-sections for W pair production diverge too much at high energies if there is no unification with the em interaction. This makes the vector boson hypothesis only an initial, but nevertheless important, stage in the development of a theory of the electroweak interaction.

1.3.2 The electromagnetic and weak interactions of hadrons

(a) In the simple quark model it was successful to trace back the *hadron currents* to those of their constituents, i.e. to quark currents. Expressed in quark fields, the em hadronic current is given by (compare Eqn. (1.2.68)):

$$h_\mu^{\text{em}}(x) = \bar\psi_{f,c}(x)\gamma_\mu Q_{ff'}\,\delta_{cc'}\psi_{f',c'}(x) \quad \text{with } Q_{ff'} = \text{diag}(\tfrac{2}{3}, -\tfrac{1}{3}, -\tfrac{1}{3}, \tfrac{2}{3}, \ldots).$$

Experiments on weak interaction show that the $SU(2)_W \times U(1)$ structure

* Because of the Dirac equation for the spinors u, v the $k_\mu k_\nu$ terms in the nominator of the W-propagator give contributions proportional to the lepton masses:

$$\bar u(q_{\nu_e})\not{k}(1-\gamma_5)u(p_e) = \bar u(q_{\nu_e})(\not{p}_e - \not{q}_{\nu_e})(1-\gamma_5)u(p_e)$$
$$= \bar u(q_{\nu_e})(1+\gamma_5)\not{p}_e u(p_e) = M_e \bar u(q_{\nu_e})(1+\gamma_5)u(p_e)).$$

and thus do not spoil the high-energy behaviour.

formulated for leptons also applies in hadronic weak processes.* Thus one extends the definition (1.3.9) of the generators of weak isospin with commutation relations (1.3.10) by the hadronic current component:

$$\begin{aligned} I_{\mathrm{W}}^{(\pm)}(t) &= \int \mathrm{d}^3x \,(l_0^{(\pm)}(t, \mathbf{x}) + h_0^{(\pm)}(t, \mathbf{x})), \\ I_{\mathrm{W}}^{(0)}(t) &= \int \mathrm{d}^3x \,(l_0^{(0)}(t, \mathbf{x}) + h_0^{(0)}(t, \mathbf{x})). \end{aligned} \tag{1.3.19}$$

As with leptons, the N_{F} left-handed quarks are arranged in doublets:*

$$\begin{pmatrix} u \\ d \end{pmatrix}_L, \begin{pmatrix} c \\ s \end{pmatrix}_L, \begin{pmatrix} t \\ b \end{pmatrix}_L, \ldots \begin{matrix} \leftarrow Q = 2/3 \\ \leftarrow Q = -1/3 \end{matrix}$$

the right-handed ones in singlets.

The charged current for the weak interaction of hadrons has the form [Ko 73a, Ja 79],

$$h_\mu^{(+)}(x) = (\bar{u}(x), \bar{c}(x), \bar{t}(x), \ldots, \gamma_{\mu L} U \begin{pmatrix} d(x) \\ s(x) \\ b(x) \\ \vdots \end{pmatrix} \tag{1.3.20}$$

with a $(N_{\mathrm{F}}/2) \times (N_{\mathrm{F}}/2)$ matrix U. In the corresponding construction for leptons, U had to be diagonal because of the separate conservation of lepton numbers and equal to the unit matrix because of universality (see Eqn. (1.3.6)). The non-electromagnetic component of the neutral current $h_\mu^{(0)}(x)$ is constructed as commutator of $I_{\mathrm{W}}^{(+)}$ and $h_\mu^{(-)}$ as the zeroth component of weak isospin (1.3.19):

$$2h_\mu^{(0)}(x) = (\bar{u}(x), \bar{c}(x), \bar{t}(x), \ldots ; \bar{d}(x), \bar{s}(x), \bar{b}(x), \ldots)\gamma_{\mu L} \begin{pmatrix} UU^\dagger & 0 \\ 0 & -U^\dagger U \end{pmatrix} \begin{pmatrix} u(x) \\ c(x) \\ t(x) \\ \vdots \\ d(x) \\ s(x) \\ b(x) \\ \vdots \end{pmatrix}.$$

If one generalizes the fact that no strangeness-changing neutral currents are observed, to the statement that there are no flavour-changing neutral currents, it follows that $UU^\dagger$ must be diagonal and in view of Eqn. (1.3.10) must be equal to the unit matrix (GIM mechanism, [Gl 70]): $UU^\dagger = U^\dagger U = \underline{1}$.

(b) The unitarity of the extended *Cabibbo matrix U* expresses the universality

* Colour indices are suppressed in this section, since all currents under discussion are colour singlets (see summation convention).

of the weak interaction. As a unitary $(N_F/2)\times(N_F/2)$ matrix, U depends on $N_F^2/4$ real parameters. Of these, N_F-1 phases can be standardized by changing the phases of the quark fields $u(x)\to\exp(i\eta_u)u(x),\ldots$ The $(N_F/2)\times(N_F/2-1)/2$ rotation angles, i.e. the Cabibbo angles θ_i and $(N_F/2-1)\times(N_F/2-2)/2$ phases δ_i [Ha 76] which, can be related to CP-violating weak processes [Ko 73a], remain.

For $N_F=4$, i.e. $\binom{u}{d}$, $\binom{c}{s}$ quarks, U is a 2×2 matrix. The phases of the matrix elements can be standardized to zero so that the unitary matrix which is then real can be parametrized by the Cabibbo angle θ_C alone:

$$U=\begin{pmatrix}\cos\theta_C & \sin\theta_C\\ -\sin\theta_C & \cos\theta_C\end{pmatrix}. \tag{1.3.21}$$

The charged current (1.3.20) is explicitly given by

$$\begin{aligned}h_\mu^{(+)}(x)&=\bar{u}(x)\gamma_{\mu L}(\cos\theta_C\cdot d(x)+\sin\theta_C\cdot s(x))\\&\quad+\bar{c}(x)\gamma_{\mu L}(-\sin\theta_C\cdot d(x)+\cos\theta_C\cdot s(x)).\end{aligned} \tag{1.3.22}$$

It includes the selection rules for weak, semi-leptonic decays of hadrons as summarized in Fig. 1.8. Especially, it contains no terms with $\Delta S=-\Delta Q$ or $|\Delta S|=2$. θ_C measures the strength of the $\Delta I=1/2$ transitions and has the value

$$\sin\theta_C\simeq0.21,\quad \cos\theta_C\simeq0.97,\quad \tan^2\theta_C=0.05. \tag{1.3.23}$$

The 3×3 matrix U for six quarks is parametrized in accordance with M. Kobayashi and K. Maskawa [Ko 73a]. If $c_i\equiv\cos\theta_i$, $s_i\equiv\sin\theta_i$, then

$$U=\begin{pmatrix}c_1 & s_1c_3 & s_1s_3\\ -s_1c_2 & c_1c_2c_3+s_2c_3e^{i\delta} & c_1c_2s_3-s_2c_3e^{i\delta}\\ -s_1s_2 & c_1s_2c_3-c_2s_3e^{i\delta} & c_1s_2c_3+c_2c_3e^{i\delta}\end{pmatrix}. \tag{1.3.24}$$

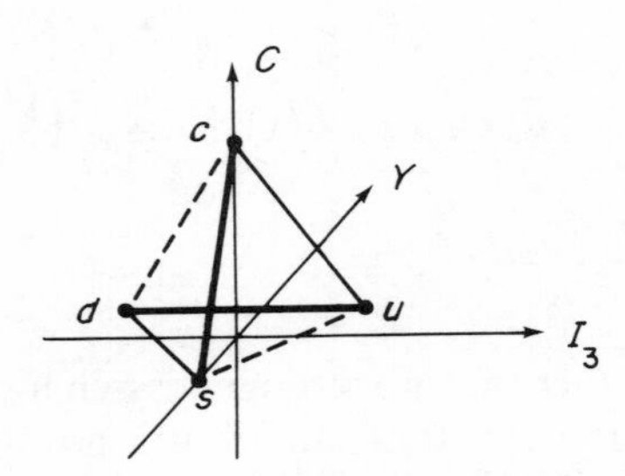

Figure 1.8 Cabibbo-allowed (—— $\sim\cos\theta_C$) and Cabbibo-suppressed (- - - $\sim\sin\theta_C$) transitions in the (I_3, Y, C) diagram

$$\begin{aligned}\Delta C=0:&\ \Delta S=0,\quad \Delta Q=1,\quad \Delta I=0;1\sim\cos\theta_C\\&\ \Delta S=\Delta Q=1,\quad \Delta I=\tfrac{1}{2}\sim\sin\theta_C\\ \Delta C=1:&\ \Delta S=0,\quad \Delta Q=1,\quad \Delta I=\tfrac{1}{2}\sim-\sin\theta_C\\&\ \Delta S=\Delta Q=1,\quad \Delta I=0;1\sim\cos\theta_C\end{aligned}$$

The angle θ_1 is identical to the Cabibbo angle θ_C (for $\theta_2 = \theta_3 = \delta = 0$); estimates for the rotation angles θ_2 and θ_3 and the phase δ exist [El 79]:

$$0.1 < |s_2| < 0.7, \quad |s_3| \simeq 0.28^{+0.21}_{-0.28}, \quad s_2 s_3 \sin\delta = O(10^{-3}).$$

(c) As in the case of leptons, the neutral quark current is made up of the zeroth component of the weak isospin current and an admixture of the em quark current which is parametrized by the same *Weinberg angle.* The explicit expression is thus:

$$\begin{aligned} h_\mu^{(n)}(x) &= h_\mu^{(0)}(x) - \sin^2\theta_W h_\mu^{em}(x) \\ &= \bar{u}(x)[(\tfrac{1}{2} - \tfrac{2}{3}\sin^2\theta_W)\gamma_{\mu L} - \tfrac{2}{3}\sin^2\theta_W \gamma_{\mu R}]u(x) \\ &\quad + \bar{c}(x)[(\tfrac{1}{2} - \tfrac{2}{3}\sin^2\theta_W)\gamma_{\mu L} - \tfrac{2}{3}\sin^2\theta_W \gamma_{\mu R}]c(x) + \dots \\ &\quad + \bar{d}(x)[(-\tfrac{1}{2} + \tfrac{1}{3}\sin^2\theta_W)\gamma_{\mu L} + \tfrac{1}{3}\sin^2\theta_W \gamma_{\mu R}]d(x) \\ &\quad + \bar{s}(x)(-\tfrac{1}{2} + \tfrac{1}{3}\sin^2\theta_W)\gamma_{\mu L} + \tfrac{1}{3}\sin^2\theta_W \gamma_{\mu R}]s(x) + \dots \end{aligned} \tag{1.3.25}$$

Like $h_\mu^{(0)}$, $h_\mu^{(n)}$ is flavour diagonal, i.e. neutral weak processes obey the selection rules

$$\Delta Q = \Delta S = \Delta C = \dots = 0. \tag{1.3.26}$$

The concepts of universality and weak isospin form the structural basis of the Fermi model of the weak interaction of leptons and hadrons. These are the quintessence of the phenomenological analysis of electroweak processes of leptons and hadrons of the first (e, ν_e; u, d) and second (μ, ν_μ; c, s) generation. At the moment, its extension to particles (τ, ν_τ; t, b) of the third generation is speculative. Measurements of the decays of mesons with 'open bottom' will show whether this extension of universality is actually justified.

(d) The object of this section is to show how concrete *results on physically possible processes* can be obtained with the help of the above currents. Semi-leptonic meson decays and deep inelastic lepton–hadron scattering are chosen for this purpose. These reactions are determined by the matrix elements $(2\pi)^4(4G_F/\sqrt{2})\langle \text{out}\,|l_\mu(0)h^\mu(0)|\,\text{in}\rangle$ in first-order perturbation theory. As the structure of the leptonic current is sufficiently clarified, this is a good basis for tests of the properties of the hadronic currents.

The matrix elements for the decays of charged, pseudoscalar mesons into leptons (e.g. $\pi^- \to \mu^-\bar{\nu}_\mu$, $\pi^- \to e^-\bar{\nu}_e$) and of the heavy lepton τ into mesons and its neutrino (e.g. $\tau \to \pi^-\nu_\tau$) are given by

$$\mathcal{M} = \frac{4G_F}{\sqrt{2}} 2\pi(\bar{u}(p_l)\gamma_{\mu L} v(q_{\nu_L}))\langle 0|\, h^{(+)\mu}(0)\,|\text{meson}, P\rangle. \tag{1.3.27}$$

Here, $(2\pi)^{-3}\bar{u}\gamma_{\mu L} v$ is the matrix element of the leptonic current $l_\mu^{(-)}$. The hadronic matrix element is expanded in covariants of the Lorentz group

$$(2\pi)^{3/2}\langle 0|\, h^{(+)\mu}(0)\,|PS; P\rangle = \mathrm{i}P^{\mu} f_{\mathrm{PS}} \frac{1}{2}\begin{pmatrix}\cos\theta_{\mathrm{C}}\\ \sin\theta_{\mathrm{C}}\end{pmatrix} \quad \text{for} \quad \begin{matrix}\Delta I = 0, 1\\ \Delta I = \frac{1}{2}\end{matrix}, \tag{1.3.28}$$

$$(2\pi)^{3/2}\langle 0|\, h^{(+)\mu}(0)\,|V; P, \varepsilon\rangle = \varepsilon^{\mu} f_{\mathrm{V}} \frac{1}{2}\begin{pmatrix}\cos\theta_{\mathrm{C}}\\ \sin\theta_{\mathrm{C}}\end{pmatrix} \quad \text{for} \quad \begin{matrix}\Delta I = 0, 1\\ \Delta I = \frac{1}{2}\end{matrix}, \tag{1.3.29}$$

$$(2\pi)^{3/2}\langle 0|\, h^{(+)\mu}(0)\,|A; P, \varepsilon\rangle = \varepsilon^{\mu} f_{\mathrm{A}} \frac{1}{2}\begin{pmatrix}\cos\theta_{\mathrm{C}}\\ \sin\theta_{\mathrm{C}}\end{pmatrix} \quad \text{for} \quad \begin{matrix}\Delta I = 0, 1\\ \Delta I = \frac{1}{2}\end{matrix}. \tag{1.3.30}$$

f_{PS}, f_{V}, and f_{A} are the decay constants of the pseudosclar, vector, and axial vector mesons. Decay widths can then be calculated using Eqn. (1.2.42):

$$\Gamma(PS \to l + \nu_l) = \frac{G_{\mathrm{F}}^2}{8\pi} f_{\mathrm{PS}}^2 \begin{pmatrix}\cos^2\theta_{\mathrm{C}}\\ \sin^2\theta_{\mathrm{C}}\end{pmatrix} M_{\mathrm{PS}} M_l^2 \left(1 - \frac{M_l^2}{M_{\mathrm{PS}}^2}\right)^2, \tag{1.3.31}$$

$$\Gamma(\tau \to PS + \nu_\tau) = \frac{G_{\mathrm{F}}^2}{8\pi} f_{\mathrm{PS}}^2 \begin{pmatrix}\cos^2\theta_{\mathrm{C}}\\ \sin^2\theta_{\mathrm{C}}\end{pmatrix} \frac{M_\tau^3}{4}\left[1 + O\left(\left(\frac{M_{\mathrm{PS}}}{M_\tau}\right)^2\right)\right], \tag{1.3.32}$$

$$\Gamma(\tau \to V(A) + \nu_\tau) = \frac{G_{\mathrm{F}}^2}{8\pi} f_{\mathrm{V(A)}}^2 \begin{pmatrix}\cos^2\theta_{\mathrm{C}}\\ \sin^2\theta_{\mathrm{C}}\end{pmatrix} \frac{M_\tau^3}{2M_{\mathrm{V(A)}}^2}\left[1 + O\left(\left(\frac{M_{\mathrm{V(A)}}}{M_\tau}\right)^4\right)\right]. \tag{1.3.33}$$

The ratio

$$B_{\mathrm{PS}} = \frac{\Gamma(PS \to e\nu_e)}{\Gamma(PS \to \mu\nu_\mu)} = \frac{M_e^2}{M_\mu^2}\left(\frac{M_{\mathrm{PS}}^2 - M_e^2}{M_{\mathrm{PS}}^2 - M_\mu^2}\right)^2 \tag{1.3.34}$$

is independent of the decay constant f_{PS} and proportional to the square of the quotient of the lepton masses. This is typical for the $(V–A)$ interaction. For example, a scalar coupling would give a one instead of M_e^2/M_μ^2. Experimentally $B_\pi = (1.267 \pm 0.023) \times 10^{-4}$ and $B_K = (2.43 \pm 0.14) \times 10^{-5}$ whereas $B_\pi = 1.23 \times 10^{-4}$ and $B_K = 2.55 \times 10^{-5}$ follow from Eqn. (1.3.34).

The decay constants $f_\pi = 138$ MeV, $f_K = 160$ MeV can be calculated from the experimental widths $\Gamma(\pi^- \to \mu^- + \bar{\nu}_\mu) = 3.84 \times 10^7\,\mathrm{s}^{-1}$ and $\Gamma(K^- \to \mu^- + \bar{\nu}_\mu) = 5.13 \times 10^7\,\mathrm{s}^{-1}$ using Eqns (1.3.31) and (1.3.23). With this, the τ decay widths $\Gamma(\tau^- \to \pi^- + \nu_\tau) = 2.5 \times 10^{11}\,\mathrm{s}^{-1}$ and $\Gamma(\tau^- \to K^- + \nu_\tau) = 1.9 \times 10^{10}\,\mathrm{s}^{-1}$ can be calculated from Eqn. (1.3.32). Experimentally: $\Gamma(\tau^- \to K^- + \ldots)/\Gamma(\tau^- \to \pi^- + \ldots) = 0.07 \pm 0.06$ and $\Gamma(\tau^- \to \pi^- + \ldots)/\Gamma(\tau^- \to l\bar{\nu}_l) = 0.50 \pm 0.06$ which is in agreement with the theoretical results [Fl 79].

The approximate equality of f_π and f_K ($f_K = 1.15 f_\pi$) points to a relatively weak flavour dependence of the decay constants f_{PS}. If $f_K \approx f_D \approx f_F$, the partial widths of mesons with charm can be predicted:

$$\Gamma(D^- \to \mu^- + \bar{\nu}_\mu) = 2.0 \times 10^8\,\mathrm{s}^{-1}, \quad \Gamma(F^- \to \mu^- + \bar{\nu}_\mu) = 3.7 \times 10^9\,\mathrm{s}^{-1}.$$

As an example of an em decay we choose the decay of vector mesons (ρ, ω, ϕ, J/ψ, ...) into lepton pairs. The decay width can be calculated from the em current matrix element which corresponds to that of Eqn. (1.3.29):

$$\Gamma(V \to l^+l^-) = \frac{4\pi}{3}\alpha^2 \frac{f_V^2}{M_V^3}.$$

f_V can be determined from the measured $\Gamma(V \to l^+l^-)$. Assuming the validity of the Weinberg sum rule [We 67a], then $f_A = f_V$. With this the decay widths $\Gamma(\tau^- \to \rho^-\nu_\tau) = 5.5 \times 10^{11}\,\mathrm{s}^{-1}$, $\Gamma(\tau^- \to A_1^- + \nu_\tau) = 1.5 \times 10^{11}\,\mathrm{s}^{-1}$ can be calculated from Eqns (1.3.32) and (1.3.33). The corresponding branching ratios agree with the experimental results.

The GIM mechanism forbids weak meson decays into lepton pairs which are induced only by the neutral current $h_\mu^{(n)}$ and involve a change in the flavour quantum number (see Eqn. (1.3.26)). Experimentally, $\Gamma(K_L^0 \to \mu^+\mu^-)/\Gamma_{K^0} = (9.1 \pm 1.8) \times 10^{-9}$ and $\Gamma(K_L^0 \to e^+e^-)/\Gamma_{K_L^0} < 2.0 \times 10^{-9}$. The order of magnitude of these branching ratios can be explained by higher-order effects [Ga 74].

The colour degree of freedom of the quarks does not lead to directly visible consequences in processes covered so far. This is different for the decay $\pi^0 \to 2\gamma$ which can be described by the Feynman graphs:

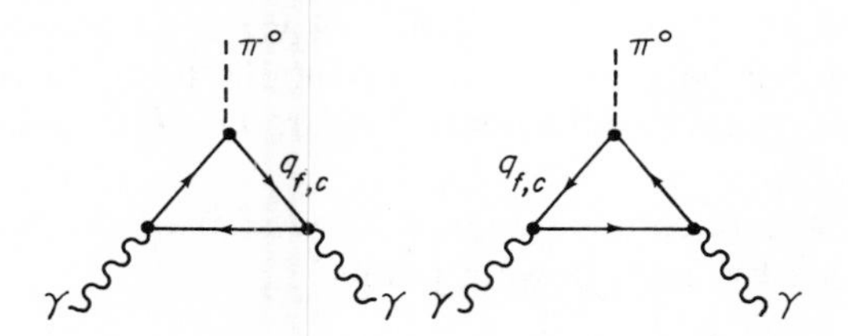

The decay width may be calculated herefrom using the PCAC hypothesis [Ad 70]:

$$\Gamma(\pi^0 \to 2\gamma) = \frac{\alpha^2}{32\pi^3} \frac{M_\pi^3}{f_\pi^2} \frac{N_c^2}{9}.$$

Experimentally, $\Gamma(\pi^0 \to 2\gamma) = (7.95 \pm 0.55)$ eV. For $f_\pi = 138$ MeV, the number of quark colours $N_C = 3$ follows. Three colours are also required to solve the problem of the baryon wavefunction symmetry.

The matrix elements of the β-decays of the pseudoscalar mesons $PS' \to PS + l + \bar{\nu}_l$ are given by

$$\mathcal{M} = \frac{4G_F}{\sqrt{2}} 2\pi(\bar{u}(p_e)\gamma_{\mu L} v(q_{\nu_e}))\langle PS; P|\, h^{(+)\mu}(0)\,|PS'; P'\rangle.$$

Only the vector part $V^{(+)\mu}(0)$ of the charged hadronic current, which is not divergence-free due to the flavour symmetry breaking, contributes. Lorentz invariance is used to decompose its matrix element into form factors:

$$(2\pi)^3\langle PS; P|\, V^{(+)\mu}(0)\,|PS'; P'\rangle = (P+P')^\mu f_+(q^2) + q^\mu f_-(q^2), \quad q^\mu = (P-P')^\mu. \tag{1.3.36}$$

For $q^2=0$, the form factor $f_+(0)$ of the divergence-free current component is identical to the matrix element of the corresponding charge. According to Eqn. (1.2.70), this is the corresponding flavour generator (CVC hypothesis). With this, $f_+(0)$ is calculable. In the language of quark wavefunctions (see Eqn. (1.1.1)) this can be shown for example for $\langle\pi^+|\,I^{(+)}\,|\pi^0\rangle$:

$$f_+(0)=-\underbrace{\langle u\bar{d}|}_{\langle\pi^+|}\,\underbrace{(|u\rangle\langle d|-|\bar{d}\rangle\langle\bar{u}|)}_{I^{(+)}}\,\underbrace{|u\bar{u}-d\bar{d}\rangle}_{|\pi^0\rangle}\frac{1}{\sqrt{2}}=\frac{1}{\sqrt{2}}(1+1)=\sqrt{2}. \tag{1.3.37}$$

Essentially, only $f_+(0)$ occurs in the decay widths. Table 1.6 shows that the widths thus calculated agree well with the experimental numbers. There, also predictions for the semileptonic decays of mesons with charm are given.

The GIM mechanism forbids neutral, strangeness-changing semileptonic decays in accordance with the $\Delta S=0$ selection rule. Experimentally,

$$\frac{\Gamma(K^+\to\pi^+e^+e^-)}{\Gamma(K^+\to\pi^0e^+\nu_e)}=(5.4\pm1.0)\times10^{-6},\quad \frac{\Gamma(K^+\to\pi^+\nu\bar{\nu})}{\Gamma(K^+\to\pi^0e^+\nu_e)}<3\cdot10^{-6}.$$

This ends our comparison of the Fermi model with experimental data by means of selected examples. A discussion of weak baryon decays based on the Fermi model can be found in the textbooks listed in reference [FM]. Section 1.4 covers tests of the neutral hadronic current in $\nu(\bar{\nu})$–nucleon scattering.

The results of the phenomenological discussion of the weak interaction can be summarized by the following points:

(1) The leptonic and semileptonic weak processes can be well described by an effective Hamilton operator of the current–current type. However, problems do occur with non-leptonic weak decays, e.g. $K\to2\pi$, $K\to3\pi$, $\Lambda\to N\pi$. Experimentally the $\Delta I=1/2$ rule is observed in these decays (octet

Table 1.6 Semileptonic decays of pseudoscalar mesons.

Decay	$f_+(0)$	$\Gamma_{\text{theor}}(\text{s}^{-1})$	$\Gamma_{\text{exp}}(\text{s}^{-1})$
$\pi^+\to\pi^0e^+\nu_e$	$\sqrt{2}$	3.94×10^{-1}	$(3.8\pm0.25)\times10^{-1}$
$K^+\to\pi^0e^+\nu_e$	$1/\sqrt{2}$	3.60×10^{6}	3.9×10^{6}
$\quad\pi^0\mu^+\nu_\mu$		2.34×10^{6}	2.6×10^{6}
$K^0\to\pi^-e^+\nu_e$	1	7.30×10^{6}	7.53×10^{6}
$\quad\pi^-\mu^+\nu_\mu$		4.80×10^{6}	5.3×10^{6}
$D^+\to K^0e^+\nu_e,\ \mu^+\nu_\mu$	1	1.10×10^{11}	
$D^+\to\pi^0e^+\nu_e,\ \mu^+\nu_\mu$	$1/\sqrt{2}$	5.0×10^{9}	
$D^0\to K^-e^+\nu_e,\ \mu^+\nu_\mu$	1	1.1×10^{11}	
$\quad\pi^-e^+\nu_e,\ \mu^+\nu_\mu$	1	1.0×10^{10}	
$F^+\to K^0e^+\nu_e,\ \mu^+\nu_\mu$	1	1.0×10^{10}	
$\quad\eta e^+\nu_e,\ \mu^+\nu_\mu$	$-2/\sqrt{6}$	2.0×10^{11}	

enhancement [Da 64]). Theoretically, this has not yet been clarified satisfactorily. It can be regarded as a problem of the strong interaction and must not throw doubt on the Fermi model (1.3.1). However, this seems to be a problem in which QCD and the theory of the electroweak interaction have yet to stand their test [Al 74, Ga 74].

(2) No dynamic deviations from the pointlike 4-fermion interaction were found up to the available energies.

(3) The currents of the weak interaction can be built up from the fields of the leptons and quarks.

(4) The weak interaction is universal for leptons and hadrons. This universality can be partially expressed by the algebraic structure of the charges (group $SU(2)_W \times U(1)$).

(5) The charges of the left-handed charged and neutral currents generate the weak isospin group $SU(2)_W$. The neutral current is composed of the left-handed weak isospin current and the electromagnetic current with the Weinberg angle as mixing parameter, leading to the $SU(2)_W \times U(1)$ symmetry (see Eqn. (1.3.10)).

(6) According to GIM, the hadronic, left-handed weak isospin current is built up from the Cabibbo-rotated quark fields (see Eqns (1.3.20) and (1.3.25)).

1.4 The quark-parton model [PM]

The hypothesis that mesons and baryons are made up of quarks was the basis of a successful description of the hadron spectrum. However, so far there has been no direct confirmation of this concept, because free quarks have not been observed experimentally. Nevertheless, the constituents of hadrons can be investigated by means of probes such as electrons, muons, and neutrinos which have no strong interaction. The discovery of scale invariance in deep inelastic lepton–nucleon scattering shows that the nucleons contain pointlike constituents. Tests on the quantum numbers of these constituents confirm that they are identical to quarks. Moreover, there are hints to the quanta of the strong interaction, the gluons.

1.4.1 Scaling in deep inelastic lepton scattering

The differential cross-section for deep inelastic lepton scattering, i.e. for the reaction $l+\mathcal{N} \rightarrow l'+X$ ($\mathcal{N}$, nucleon; X, arbitrary hadronic final state) can be calculated from the graph shown in Fig. 1.9. In the case of electron–nucleon scattering, this 1-photon exchange graph results in the following expression for the unpolarized cross-section:

$$\frac{d^2\sigma^{e\mathcal{N}}}{dE'\,d\Omega} = \frac{\alpha^2}{Q^4}\frac{E'}{M_{\mathcal{N}}E}\frac{1}{2}L^{\mu\nu}W^{e\mathcal{N}}_{\mu\nu} \tag{1.4.1}$$

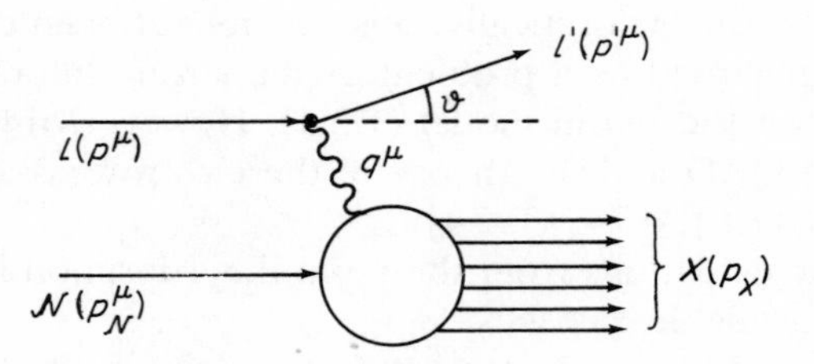

Figure 1.9 Kinematics of deep inelastic lepton–nucleon scattering. Kinematical variables in the laboratory system:

$$p^\mu = (E, \mathbf{p}), \quad p'^\mu = (E', \mathbf{p}'), \quad q^\mu = p^\mu - p'^\mu$$
$$p^\mu_{\mathcal{N}} = (M_{\mathcal{N}}, \mathbf{0})$$
$$Q^2 = -q^2 = 4EE' \sin^2 \tfrac{1}{2}\vartheta, \quad \nu = p_{\mathcal{N}} q = M_{\mathcal{N}}(E-E')$$
$$x = Q^2/2\nu, \quad 0 \leqslant x \leqslant 1, \quad y = \nu/p_{\mathcal{N}} p = (E-E')/E, 0 \leqslant y \leqslant 1$$

with the leptonic tensor (see Eqn. (1.2.48))

$$\begin{aligned} L^{\mu\nu} &= \sum_{j_3 j_3'} \bar{u}(p', j_3')\gamma^\mu u(p, j_3) \cdot \bar{u}(p, j_3)\gamma^\nu u(p', j_3') \\ &= 4(p'^\mu p^\nu + p^\mu p'^\nu - g^{\mu\nu} pp'). \end{aligned} \tag{1.4.2}$$

Lorentz-covariance can be used to parametrize the hadronic tensor

$$\begin{aligned} W^{e\mathcal{N}}_{\mu\nu} &= \tfrac{1}{2} \sum_{\text{spins}} \sum_X (2\pi)^3 \, \delta^{(4)}(p_{\mathcal{N}} - p_X - q)\langle \mathcal{N}|\, j^{\text{em}}_\mu(0)\, |X\rangle\langle X|\, j^{\text{em}}_\nu(0)\, |\mathcal{N}\rangle \\ &= (-g_{\mu\nu} + q_\mu q_\nu/q^2) W^{e\mathcal{N}}_1(\nu, Q^2) \\ &\quad + (p_{\mathcal{N}\mu} - q_\mu (p_{\mathcal{N}} \cdot q)/q^2)(p_{\mathcal{N}\nu} - q_\nu (p_{\mathcal{N}} \cdot q)/q^2) \frac{1}{M^2_{\mathcal{N}}} W^{e\mathcal{N}}_2(\nu, Q^2), \end{aligned} \tag{1.4.3}$$

by the structure functions $W^{e\mathcal{N}}_1$ and $W^{e\mathcal{N}}_2$, which depend only on the relativistic invariants ν and Q^2.

The cross-section for deep inelastic $e\mathcal{N}$ or $\nu(\bar{\nu})\mathcal{N}$ scattering via charged currents then takes the following form:

$$\frac{d^2\sigma}{dE' 2\pi \sin\vartheta \, d\vartheta} = \frac{M_{\mathcal{N}} E E'}{\pi} \frac{d^2\sigma}{dQ^2 \, d\nu} = \frac{E'}{2\pi M_{\mathcal{N}} E \cdot y} \frac{d^2\sigma}{dx \, dy} \tag{1.4.4}$$

$$= \begin{cases} \dfrac{4\alpha^2}{Q^4} \dfrac{E'^2}{M_{\mathcal{N}}} \left(W^{e\mathcal{N}}_1 \cdot 2\sin^2 \dfrac{\vartheta}{2} + W^{e\mathcal{N}}_2 \cdot \cos^2 \dfrac{\vartheta}{2} \right) & \text{for } e\mathcal{N}, \\ \dfrac{G_F^2}{2\pi^2} \dfrac{E'^2}{M_{\mathcal{N}}} \left(W^{\nu(\bar{\nu})\mathcal{N}}_1 \cdot 2\sin^2 \dfrac{\vartheta}{2} + W^{\nu(\bar{\nu})\mathcal{N}}_2 \cdot \cos^2 \dfrac{\vartheta}{2} \right. & \text{for } \nu(\bar{\nu})\mathcal{N} \text{ via charged} \\ \qquad \left. (\mp) W^{\nu(\bar{\nu})\mathcal{N}}_3 \dfrac{E+E'}{M_{\mathcal{N}}} \sin^2 \dfrac{\vartheta}{2} \right) & \quad \text{current} \end{cases}$$

In the case of $\nu(\bar{\nu})\mathcal{N}$ scattering, the parity-violating components in the weak current of leptons and hadrons give rise to a further structure function $W_3(\nu, Q^2)$.

Fig. 1.10 shows the experimental results for the structure functions $W_1^{e\mathcal{N}}(\nu, Q^2)$ and $W_2^{e\mathcal{N}}(\nu, Q^2)$. It gives the data as functions of the variables $x = Q^2/2\nu$ and Q^2. The fact that the measured values lie relatively close along one curve is called scaling [Bj 69a] and means that the structure functions depend only on the scaling variable x. Thus, the Bjorken limit $F_i(x)$ of the structure functions W_i can be defined as

$$\begin{aligned} \text{Bj-lim}\ W_1(\nu, Q^2) &= F_1(x), \\ \text{Bj-lim}\ \nu W_{2,3}(\nu, Q^2)/M_{\mathcal{N}}^2 &= F_{2,3}(x), \end{aligned} \qquad \text{Bj-lim} = \lim_{Q^2, \nu \to \infty; x \text{ fixed}} . \qquad (1.4.5)$$

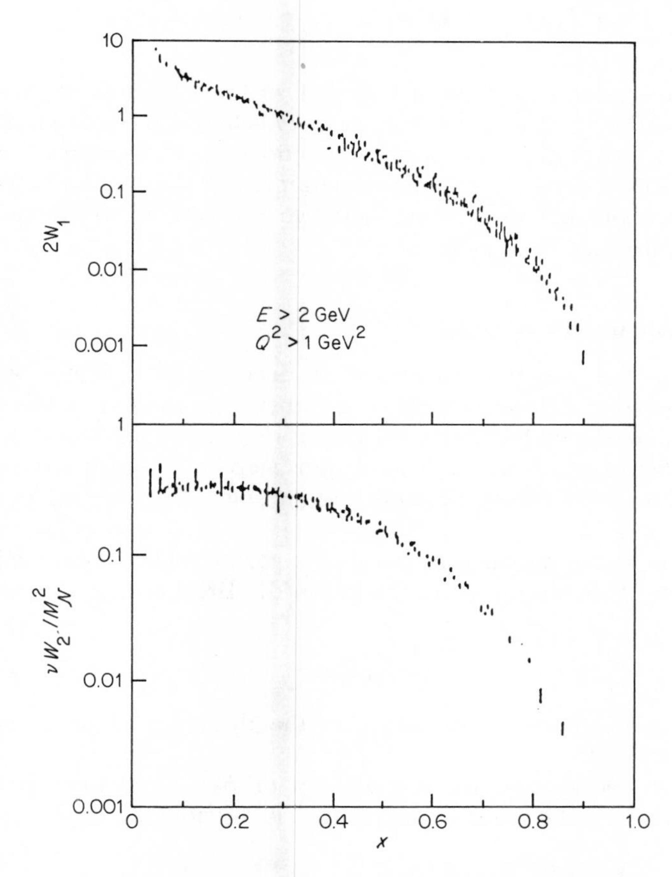

Figure 1.10 The structure functions $W_1^{e\mathcal{N}}(\nu, Q^2)$ and $W_2^{e\mathcal{N}}(\nu, Q^2)$ [Ta 75]

In the Bjorken limit, the cross-section (1.4.4) is then given by

$$\frac{d^2\sigma^{e\mathcal{N}}}{dx\,dy} = \frac{4\pi\alpha^2}{Q^2}\left(yF_1^{e\mathcal{N}} + \frac{1-y}{xy}F_2^{e\mathcal{N}}\right),$$
$$\frac{d^2\sigma^{\nu\mathcal{N}}}{dx\,dy} = \frac{G_F^2 M_{\mathcal{N}} E}{\pi}\left[(1-y)F_2^{\nu\mathcal{N}} + y^2 x F_1^{\nu\mathcal{N}} \mp y(1-\tfrac{1}{2}y)xF_3^{\nu\mathcal{N}}\right]$$

or by

$$\frac{d^2\sigma^{e\mathcal{N}}}{dQ^2\,d\nu} = \frac{4\pi\alpha^2}{Q^4}\frac{1}{s^2(u+s)}\left[2(u+s)^2 x F_1^{e\mathcal{N}}(x) - 2usF_2^{e\mathcal{N}}(x)\right] \tag{1.4.6}$$

when expressing it with help of

$$s = 2M_{\mathcal{N}}E, \quad u = -2M_{\mathcal{N}}E', \quad 2\nu = s+u, \quad x = Q^2/(s+u),$$
$$\sin^2\vartheta/2 = -Q^2 M_{\mathcal{N}}^2/(us) = -(u+s)xM_{\mathcal{N}}^2/(us)$$

by invariants.

The parton model, described in the next section attempts to provide a simple explanation of this scaling behaviour which is also found in neutrino scattering and gives information on the functions F_i. A detailed analysis of the data reveals a small systematic deviation from the scaling behaviour. This can be explained by gluonic radiative corrections within quantum chromodynamics (see Section 2.7).

1.4.2 The simple parton model

The fact that the structure functions of deep inelastic lepton–nucleon scattering depend almost exclusively on dimensionless quantity x shows that no structure determined by a mass parameter makes any significant dynamical contribution to this reaction. The obvious step is then to try to describe the Bjorken limit by means of lepton scattering off zero mass pointlike nucleon constituents.

This concept forms the basis of the simple parton model [Fe 69, Bj 69b]. In this model, $l\mathcal{N}$ scattering is considered in the Breit system, defined by:

$$q^\mu = (0, 0, 0, \surd Q^2),$$
$$p_{\mathcal{N}}^\mu = (p_{\mathcal{N}}^0, \mathbf{p}) = (\surd(M_{\mathcal{N}}^2 + p^2), 0, 0, p), \quad p = \nu/\surd Q^2, \quad \text{Bj-lim}\, p = \infty, \tag{1.4.7}$$

The following assumptions are made on the dynamics of deep inelastic scattering processes:

(i) The high-energy nucleon is made up of free, zero mass particles, partons, over which the nucleon momentum is distributed

$$\mathbf{p}_i = \xi_i \mathbf{p} + \mathbf{p}_i^{\mathrm{T}}, \quad \mathbf{p}\cdot\mathbf{p}_i^{\mathrm{T}} = 0, \quad \xi_i \geqslant 0, \quad \sum_i \xi_i = 1.$$

ξ_i is the part of the nucleon momentum which the ith parton has as longitudinal momentum.

The parton model is also based on the assumption that the transverse momenta $\mathbf{p}_i^T$ have a cut-off, i.e. remain finite especially in the Bjorken limit. Under these conditions, the parton energy can be written

$$p_i^0 = \sqrt{[\xi_i^2(\mathbf{p}^2 + M_{\mathcal{N}}^2) + (\mathbf{p}_i^{\mathrm{T}})^2 - \xi_i^2 M_{\mathcal{N}}^2]}$$
$$\approx \xi_i p_{\mathcal{N}}^0 + \frac{1}{2\xi_i p_{\mathcal{N}}^0}((\mathbf{p}_i^{\mathrm{T}})^2 - \xi_i^2 M_{\mathcal{N}}^2), \quad \text{for } \mathbf{p}^2 \gg M_{\mathcal{N}}^2. \tag{1.4.8}$$

and thus the following ensues for the 4-momentum of the parton i, ignoring the finite $\mathbf{p}^{\mathrm{T}}$:

$$p_i^\mu \approx (\xi_i p_{\mathcal{N}}^0, 0, 0, \xi_i p) = \xi_i p_{\mathcal{N}}^\mu. \tag{1.4.9}$$

(ii) The cross-section for deep inelastic lepton–nucleon scattering is the incoherent sum of the parton cross-sections weighed with the distribution function $f_i^{\mathcal{N}}(\xi_i)$. This is made up of the probability P_n ($\sum_n P_n = 1$) for n partons and the probability $r_i^{(n)}(\xi_i)$ ($\int_0^1 r_i^{(n)}(\xi_i)\, d\xi_i = 1$) that from these, parton i has the momentum fraction ξ_i: $f_i^{\mathcal{N}}(\xi_i) = \sum_n P_n r_i^{(n)}(\xi_i)$. The parton cross-sections are calculated as scattering cross-sections of elastically scattered pointlike particles in the 1-photon approximation according to Fig. 1.11. The calculation is similar in the crossed channel to that for e^+e^- annihilation into μ pairs (see Section 1.2.2) and gives

$$\left.\frac{d\sigma}{dQ^2}\right|_{\substack{\text{electron-}\\ \text{parton}}} = \frac{4\pi\alpha^2 Q_i^2}{Q^4} \cdot \frac{1}{2}\left(1 + \frac{u_i^2}{s_i^2}\right). \tag{1.4.10}$$

The Mandelstam condition between Q^2, s_i, and u_i is $-Q^2 + s_i + u_i = 0$ for vanishing parton masses. It is made explicit in Eqn. (1.4.10) by introducing a δ-function distribution for u_i and using this to transform to the kinematic

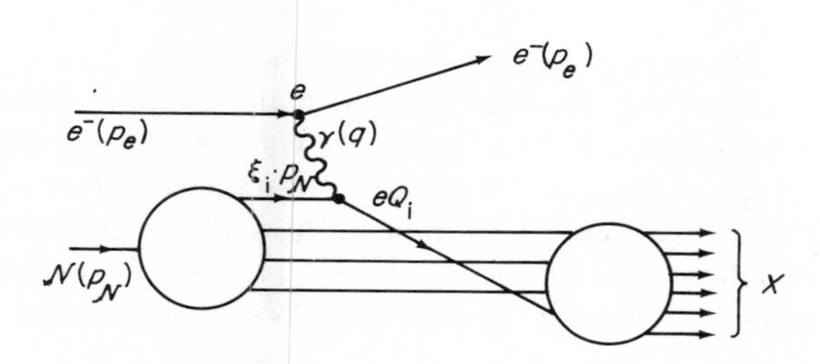

Figure 1.11 Lepton–nucleon scattering in the simple parton model. Kinematical variables:

$Q^2 = -(p_e - p_e')^2$

$s_i = (p_e + \xi_i p_{\mathcal{N}})^2 \simeq \xi_i \cdot 2p_e p_{\mathcal{N}} \simeq \xi_i s, \quad s = (p_e + p_{\mathcal{N}})^2$

$u_i = (p_e' - \xi_i p_{\mathcal{N}})^2 \simeq -\xi_i \cdot 2p_e' p_{\mathcal{N}} \simeq \xi_i u, \quad u = (p_e' - p_{\mathcal{N}})^2$

eQ_i = charge of the ith parton

variables s and u:

$$\frac{d^2\sigma}{dQ^2\,du_i} = \frac{4\pi\alpha^2 Q_i^2}{Q^4} \cdot \frac{1}{2}\left(1+\frac{u_i^2}{s_i^2}\right)\delta(-Q^2+s_i+u_i)$$
$$= 4\pi\alpha^2 \frac{Q_i^2}{Q^4} \cdot \frac{1}{2}\left(1+\frac{u^2}{s^2}\right)\delta(Q^2-\xi_i(u+s)).$$

The deep inelastic $e\mathcal{N}$ scattering cross-section can now be built up with the parton distribution functions and these parton cross-sections:

$$\frac{d^2\sigma}{dQ^2\,d\nu} = \sum_i \int_0^1 d\xi_i f_i^{\mathcal{N}}(\xi_i) \frac{d^2\sigma^{i\mathcal{N}}}{dQ^2\,du_i} 2\xi_i$$
$$= \frac{4\pi\alpha^2}{Q^4}\left(1+\frac{u^2}{s^2}\right)\sum_i \int_0^1 d\xi_i f_i^{\mathcal{N}}(\xi_i) Q_i^2 \xi_i\, \delta(Q^2-\xi_i(u+s))$$
$$= \frac{4\pi\alpha^2}{Q^4}\frac{1}{s+u}\frac{s^2+u^2}{s^2}\sum_i Q_i^2 x f_i^{\mathcal{N}}(x),$$
$$s+u = 2\nu. \tag{1.4.12}$$

This parton model result for the cross-section of deep inelastic electron–nucleon scattering shows scaling behaviour. According to the δ-function in (1.4.12): $x=\xi_i$. Thus, the scaling variable x measures the longitudinal part of the parton momentum. Comparison of Eqns. (1.4.12) and (1.4.8) gives the other most important results of the parton model, namely, the Callan–Gross relation [Ca 69]:

$$F_2^{e\mathcal{N}}(x) = 2xF_1^{e\mathcal{N}}(x), \tag{1.4.13}$$

and the connection between the structure functions $F_i(x)$ and the parton distributions $f_i(x)$:

$$F_1^{e\mathcal{N}}(x) = \tfrac{1}{2}\sum_i Q_i^2 f_i^{\mathcal{N}}(x). \tag{1.4.14}$$

(iii) In the quark-parton model, the partons, which contribute to deep inelastic lepton scattering, are identified with the nucleon constituents, the quarks. The fact that quarks have spin 1/2 was anticipated when deriving Eqn. (1.4.12). If the partons had spin zero, for example, then $F_1(x)\equiv 0$. The parton distribution functions $f_i^{\mathcal{N}}(x)$ indicate how the nucleon momentum is distributed throughout the quarks $u, d, s, c, \ldots$. The following quark distribution functions for proton P and neutron N are introduced using charge symmetry:

$$\begin{aligned}
f_u^P(x) &= f_d^N(x) =: u(x), & f_{\bar u}^P(x) &= f_{\bar d}^N(x) =: \bar u(x),\\
f_d^P(x) &= f_u^N(x) =: d(x), & f_{\bar d}^P(x) &= f_{\bar u}^N(x) =: \bar d(x),\\
f_s^P(x) &= f_s^N(x) =: s(x), & f_{\bar s}^P(x) &= f_{\bar s}^N(x) =: \bar s(x),\\
f_c^P(x) &= f_c^N(x) =: c(x), & f_{\bar c}^P(x) &= f_{\bar c}^N(x) =: \bar c(x).
\end{aligned} \tag{1.4.15}$$

The coupling constants Q_i are identical to the quark charges.

Obviously, scaling and the Callan–Gross relation are also obtained for deep inelastic $\nu(\bar{\nu})$ scattering in the parton model by using the neutral and charged quark currents (compare Eqns. (1.3.20) and (1.3.25)). The structure functions $F_1(x)$ and $F_3(x)$ are built up from the quark distributions $u(x), \ldots, \bar{c}(x)$ in analogy to (1.4.14). This is summarized in Table 1.7. For example,

$$2F_1^{eP}(x) = \tfrac{4}{9}u(x) + \tfrac{1}{9}d(x) + \tfrac{1}{9}s(x) + \cdots + \tfrac{4}{9}\bar{c}(x),$$
$$F_1^{\nu P} = \cos^2\theta_C d(x) + \sin^2\theta_C s(x) + \bar{u}(x).$$

The couplings ε^2 and Δ^2 of the neutral current were taken from Eqn. (1.3.25):

$$\varepsilon_Q^2 = \varepsilon_{Q,L}^2 + \varepsilon_{Q,R}^2, \quad \Delta_Q^2 = \varepsilon_{Q,L}^2 - \varepsilon_{Q,R}^2. \tag{1.4.16}$$

For quarks with charge $Q = \frac{2}{3}$ $(u, c, \ldots)$ and $Q = -\frac{1}{3}$ $(d, s, \ldots)$ it follows that:

$$\begin{aligned} \varepsilon_{2/3,L} &= \tfrac{1}{2} - \tfrac{2}{3}\sin^2\theta_W, & \varepsilon_{2/3,R} &= -\tfrac{2}{3}\sin^2\theta_W, \\ \varepsilon_{-1/3,L} &= -\tfrac{1}{2} + \tfrac{1}{3}\sin^2\theta_W, & \varepsilon_{-1/3,R} &= +\tfrac{1}{3}\sin^2\theta_W, \end{aligned} \tag{1.4.17}$$

respectively. The large mass of the hadrons with charm leads to scaling violation in the region of the charm threshold. Therefore, the couplings above and below the charm threshold are given separately in Table 1.7. Because of the GIM mechanism, the values above the charm threshold no longer depend on the Cabibbo angle. Relations between various structure functions, in particular sum rules and inequalities, can be derived from Table 1.7.

1.4.3 Applications of the simple parton model

In this section a comparison is made between the results of the quark-parton model and experiments with special emphasis on the question to what extent the partons carry the quark quantum numbers.

(a) *Parton spin*

The Callan–Gross relation (1.4.13) holds for partons with spin 1/2. It states that in the Bjorken limit, the ratio of the longitudinal and transverse cross-section should disappear:

$$\text{Bj-lim}\,\frac{\sigma_L}{\sigma_T} = \text{Bj-lim}\,\frac{W_2(1+\nu/2M_{\mathcal{N}}^2 x) - W_1}{W_1} = \frac{F_2/2x - F_1}{F_1} = 0.$$

The result of the CDHS νP experiment [Ab 83N] for $0.4 \leq x \leq 0.7$, $Q^2 = 38\ \text{GeV}^2$ is $\langle\sigma_L/\sigma_T\rangle = 0.006 \pm 0.012\ (\text{stat}) \pm 0.025\ (\text{syst})$.

Table 1.7 Coefficients for building up the structure functions for electromagnetic and weak lepton–nucleon scattering from the quark distribution functions

	u	d	s	c	$\bar{u}$	$\bar{d}$	$\bar{s}$	$\bar{c}$	
$2F_1^{eP}$	$\frac{4}{9}$	$\frac{1}{9}$	$\frac{1}{9}$	$\frac{4}{9}$	$\frac{4}{9}$	$\frac{1}{9}$	$\frac{1}{9}$	$\frac{4}{9}$	
$2F_1^{eN}$	$\frac{1}{9}$	$\frac{4}{9}$	$\frac{1}{9}$	$\frac{4}{9}$	$\frac{1}{9}$	$\frac{4}{9}$	$\frac{1}{9}$	$\frac{4}{9}$	
$F_1^{\nu P}$	0 (0)	$\cos^2\theta_C$ (1)	$\sin^2\theta_C$ (1)	— (0)	1 (1)	0 (0)	0 (0)	— (1)	charged current
$F_1^{\bar{\nu}N}$	0 (0)	1 (1)	0 (0)	— (1)	$\cos^2\theta_C$ (1)	0 (0)	$\sin^2\theta_C$ (1)	— (0)	charged current
$F_1^{\nu N}$	$\cos^2\theta_C$ (1)	0 (0)	$\sin^2\theta_C$ (1)	— (0)	0 (0)	1 (1)	0 (0)	— (1)	charged current
$F_1^{\bar{\nu}P}$	1 (1)	0 (0)	0 (0)	— (1)	0 (0)	$\cos^2\theta_C$ (1)	$\sin^2\theta_C$ (1)	— (0)	charged current
$-\frac{1}{2}F_3^{\nu P}$	0 (0)	$-\cos^2\theta_C$ (−1)	$-\sin^2\theta_C$ (−1)	— (0)	1 (1)	0 (0)	0 (0)	— (1)	charged current
$-\frac{1}{2}F_3^{\bar{\nu}N}$	0 (0)	−1 (−1)	0 (0)	— (−1)	$\cos^2\theta_C$ (1)	0 (0)	$\sin^2\theta_C$ (1)	— (0)	charged current
$-\frac{1}{2}F_3^{\nu N}$	$-\cos^2\theta_C$ (−1)	0 (0)	$-\sin^2\theta_C$ (−1)	— (0)	0 (0)	1 (1)	0 (0)	— (1)	charged current
$-\frac{1}{2}F_3^{\bar{\nu}P}$	−1 (−1)	0 (0)	0 (0)	— (−1)	0 (0)	$\cos^2\theta_C$ (1)	$\sin^2\theta_C$ (1)	— (0)	charged current
$F_1^{\nu P}=F_1^{\bar{\nu}N}$	$\varepsilon_{2/3}^2$	$\varepsilon_{-1/3}^2$	$\varepsilon_{-1/3}^2$	$\varepsilon_{2/3}^2$	$\varepsilon_{2/3}^2$	$\varepsilon_{-1/3}^2$	$\varepsilon_{-1/3}^2$	$\varepsilon_{2/3}^2$	Neutral current
$F_1^{\nu N}=F_1^{\bar{\nu}P}$	$\varepsilon_{-1/3}^2$	$\varepsilon_{2/3}^2$	$\varepsilon_{-1/3}^2$	$\varepsilon_{2/3}^2$	$\varepsilon_{-1/3}^2$	$\varepsilon_{2/3}^2$	$\varepsilon_{-1/3}^2$	$\varepsilon_{2/3}^2$	Neutral current
$-\frac{1}{2}F_3^{\nu P}=-\frac{1}{2}F_3^{\bar{\nu}N}$	$-\Delta_{2/3}^2$	$-\Delta_{-1/3}^2$	$-\Delta_{-1/3}^2$	$-\Delta_{2/3}^2$	$\Delta_{2/3}^2$	$\Delta_{-1/3}^2$	$\Delta_{-1/3}^2$	$\Delta_{2/3}^2$	Neutral current
$-\frac{1}{2}F_3^{\nu N}=-\frac{1}{2}F_3^{\bar{\nu}P}$	$-\Delta_{-1/3}^2$	$-\Delta_{2/3}^2$	$-\Delta_{-1/3}^2$	$-\Delta_{2/3}^2$	$\Delta_{-1/3}^2$	$\Delta_{2/3}^2$	$\Delta_{-1/3}^2$	$\Delta_{2/3}^2$	Neutral current

(b) *Sum rules for internal quantum numbers*

The electric charges of proton and neutron can be calculated as

$$1=\int_0^1 \mathrm{d}x[\tfrac{2}{3}(u(x)-\bar{u}(x))-\tfrac{1}{3}(d(x)-\bar{d}(x))] \tag{1.4.19}$$

and

$$0=\int_0^1 \mathrm{d}x[\tfrac{2}{3}(d(x)-\bar{d}(x))-\tfrac{1}{3}(u(x)-\bar{u}(x))], \tag{1.4.20}$$

respectively. As the nucleon has strangeness and charm zero, it follows:

$$0=\int_0^1 \mathrm{d}x(s(x)-\bar{s}(x))=\int_0^1 \mathrm{d}x(c(x)-\bar{c}(x)). \tag{1.4.21}$$

From these four equations sum rules for the structure functions can be derived. With help of Table 1.7 the difference between Eqns. (1.4.19) and

(1.4.20) gives the Adler sum rule [Ad 66]:

$$\int_0^1 \mathrm{d}x(F_{1,CC}^{\bar{\nu}P} - F_{1,CC}^{\nu P}) = 1, \tag{1.4.22}$$

where CC stands for weak charged current (see ref. [Cl 77] for the experimental situation). Similarly, the sum of Eqns. (1.4.19), (1.4.20) and (1.4.21) leads to the Gross–Llewellyn-Smith sum rule [Gr 69], which measures the difference betwen quarks and antiquarks in the proton:

$$\frac{1}{2}\int_0^1 \mathrm{d}x(F_{3,CC}^{\bar{\nu}P} + F_{3,CC}^{\nu P}) = 3. \tag{1.4.23}$$

The experimental value is 3.2 ± 0.6 [Pe 75]. It confirms that the proton consists of three valence quarks. In addition, the Llewellyn-Smith relation [Ll 70]

$$F_{3,CC}^{\nu P} - F_{3,CC}^{\bar{\nu}P} = 12(F_1^{eP} - F_1^{eN}), \tag{1.4.24}$$

which has not yet been tested, follows from Table 1.7.

(c) *Momentum of the partons*

The momentum distributions $f_i(x)$ of the partons are normalized to one. Therefore,

$$\begin{aligned}\sum_i \int_0^1 \mathrm{d}x x f_i(x) &= \int_0^1 \mathrm{d}x x(u(x) + \bar{u}(x) + d(x) + \ldots + \bar{s}(x)) \\ &= \int_0^1 \mathrm{d}x[\tfrac{9}{2}(F_2^{eP} + F_2^{eN}) - \tfrac{3}{4}(F_{2,CC}^{\nu P} + F_{2,CC}^{\nu N})] \\ &= 1 - \varepsilon \end{aligned} \tag{1.4.25}$$

(below the charm threshold), measures the fraction of the momentum which is transported by quarks and antiquarks in the nucleon. The experimental result [Pe 72] of $\varepsilon \approx \frac{1}{2}$ indicates that about half the momentum in the nucleon is carried, presumably by other flavour-neutral partons. According to quantum chromodynamics, one expects that these are the gluons, the quanta mediating the interaction of the quarks.

(d) *Inqualities*

The Nachtmann inequality [Na 72]

$$\frac{1}{4} \leqslant \frac{F_2^{eN}}{F_2^{eP}} \leqslant 4 \tag{1.4.26}$$

is derived from Table 1.7. It agrees with experiment [Bo 79]. The lower bound is reached when x is large.

The ratio:

$$\frac{F_2^{eN}+F_2^{eP}}{F_{2,CC}^{\nu N}+F_{2,CC}^{\nu P}}=\frac{\frac{5}{9}(u+\bar{u}+d+\bar{d})+\frac{2}{9}(s+\bar{s})+\frac{8}{9}(c+\bar{c})}{2(u+\bar{u}+d+\bar{d})}$$

$$=\frac{5}{18}\left(1+\frac{2}{5}\frac{s+\bar{s}+4(c+\bar{c})}{u+\bar{u}+d+\bar{d}}\right)\geqslant\frac{5}{18}$$

is equal to $\frac{5}{18}$ in the absence of the non-valence quarks $s, c, \ldots$.

Fig. 1.12 gives a comparison between $F_2^{\nu\mathcal{N}}$ and $\frac{18}{5}F_2^{e\mathcal{N}}$ [Pe 75]. The results agree for $x>0.2$ within experimental errors. Thus, in this region, the structure functions are composed mainly of valence quarks.

(e) *Experimental parton distributions*

The experiments on deep inelastic electron- and (anti)neutrino-nucleon scattering via charged currents contain enough information to determine the quark distribution functions of the valence quarks $u(x)$, $d(x)$, and the sea quarks $s(x)$, $c(x)$, $\bar{u}(x)$, $\bar{d}(x), \ldots$ from the measured values for the structure functions (Table 1.7). Fig. 1.13 shows the typical momentum distribution for u, d, and sea quarks [Bu 78]. The valence quark contributions dominate for $0.2<x<1$ but the sea quarks give a significant contribution for $x<0.2$. The gluons mentioned in Section 1.4.3 (iii) have a momentum distribution which is concentrated for small values of x just like that of the sea quarks.

(f) *Neutral current reactions*

The sum rules (Section 1.4.3 (ii)) and the analysis of the quark distributions do not incorporate the structure functions of deep inelastic (anti)neutrino scattering via neutral currents. The reason for this is that the phenomenological analysis of the structure of these currents is not yet

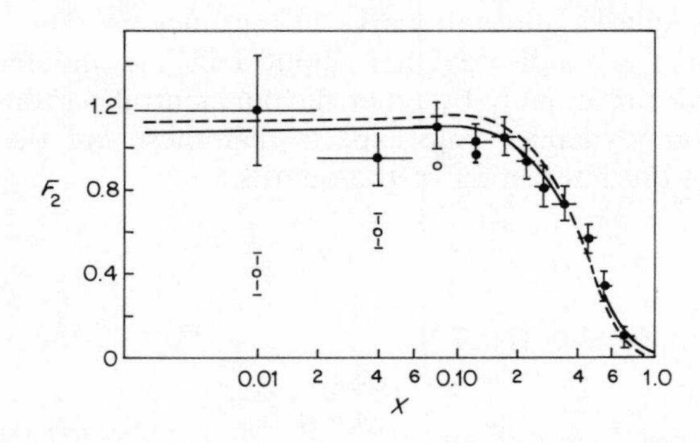

Figure 1.12 Comparison between $F_2^{\nu,\mathcal{N}}$ and $\frac{18}{5}F_2^{e,\mathcal{N}}$ [Pe 75]. ●, with closure correction; ○, without closure correction; ---, SLAC $3.6F_2^{e,\mathcal{N}}(X)$; ——, with smearing. $4<W^2<23$ GeV^2, $2.5<E<12$ GeV

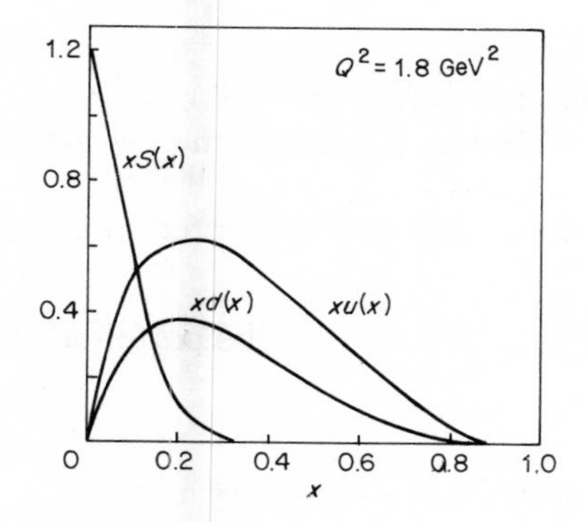

Figure 1.13 Momentum distribution of u, d, and sea quarks in the nucleon according to [Bu 78]. $S(x)$ is the distribution function of one type of light sea quark

completed. Instead one uses the parton model, which has been tested well in em and deep inelastic processes via charged currents in order to derive information on the structure of the neutral current. An analysis of the experimental data gives the following values for the couplings of the neutral current to the quarks [Dy 79]:

Experiment	Theory (Eqn. (1.3.25) with $\sin^2\theta_W = 0.23$
$\varepsilon_L^u = 0.32 \pm 0.03$	$\frac{1}{2} - \frac{2}{3}\sin^2\theta_W = 0.347$
$\varepsilon_R^u = -0.17 \pm 0.02$	$-\frac{2}{3}\sin^2\theta_W = -0.153$
$\varepsilon_L^d = -0.43 \pm 0.03$	$-\frac{1}{2} + \frac{1}{3}\sin^2\theta_W = -0.423$
$\varepsilon_R^d = -0.01 \pm 0.05$	$\frac{1}{3}\sin^2\theta_W = 0.077$

and to the electrons:

Experiment	Theory (Eqns. (1.3.7) and (1.3.8))
$\varepsilon_V = -0.03 \pm 0.04$	$-\frac{1}{4} + \sin^2\theta_W = -0.020$
$\varepsilon_A = -0.26 \pm 0.03$	$-\frac{1}{4} = -0.250$

1.4.4 Universality of the parton model

Apart from being applied to inclusive lepton scattering, the simple parton model can also be used for a phenomenological description of other deep inelastic processes, such as the e^+e^- annihilation into hadrons and the inclusive hadron production in electron–nucleon scattering. Just as in lepton–nucleon scattering, the cross-section is calculated from the following ingredients:

(a) the parton cross-sections (free, pointlike particles);
(b) the probabilities of finding partons in the hadrons, the parton distribution functions $f_i^h(x)$; and
(c) the probabilities that a parton is converted to a hadron h, the fragmentation functions $D_i^h(x)$.

Several examples are given to show how the parton model can be universally applied.

(a) *Electron–positron annihilation into hadrons*

The total cross-section for the e^+e^- annihilation into hadrons is calculated according to the above concepts from the pointlike $e^+e^- \rightarrow q\bar{q}$ cross-section similar to Eqn. (1.2.46) and from the probability that quarks are converted to hadronic states. However, according to the confinement hypothesis, this is equal to one, i.e. the cross-section is given by

$$\sigma(e^+e^- \rightarrow \text{hadrons}) = \frac{4\pi\alpha^2}{3s} \sum_i Q_i^2, \qquad s = (p_+ + p_-)^2 = E_{\text{CMS}}^2. \tag{1.4.27}$$

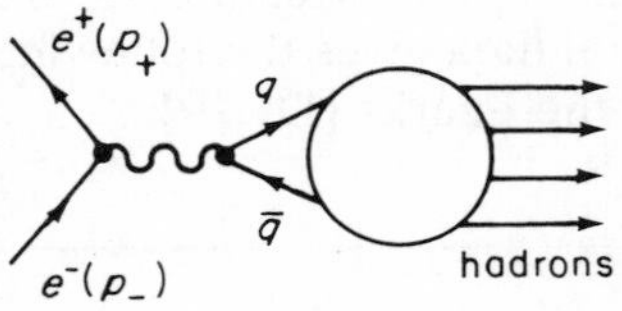

It shows scaling behaviour. The ratio

$$R = \frac{\sigma(e^+e^- \rightarrow \text{hadrons})}{\sigma(e^+e^- \xrightarrow[1\gamma]{} \mu^+\mu^-)} = \sum_i Q_i^2 \tag{1.4.28}$$

measures the hadron production in relation to μ pair production and results in the quark parton model as the sum of the charge squares Q_i^2 of the quarks with various flavour and colour. The different quark models give

$$\sum_i Q_i^2 = \begin{cases} \frac{2}{3} & \text{for } u, d, s \text{ quarks without colour} \\ 2 & \text{for } u, d, s \text{ quarks in three colours} \\ \frac{10}{3} & \text{for } u, d, s, c \text{ quarks in three colours} \end{cases} \tag{1.4.29}$$

The comparison with experiment shows that the factor 3, supplied by the colour degree of freedom, is needed (see Fig. 1.14) [Wo 80]. This can be regarded as one experimental confirmation of the hypothesis of coloured quarks.

The question as to why quarks which are created with high energies and momenta do not occur as free particles is tied up with the quark confinement problem (Section 2.6). In deep inelastic lepton scattering quarks are not seen as free particles but in the form of partons with quark quantum numbers. In e^+e^- annihilation quarks manifest themselves in the form of hadron jets, the angular distribution of which is given by (1.2.52). From this

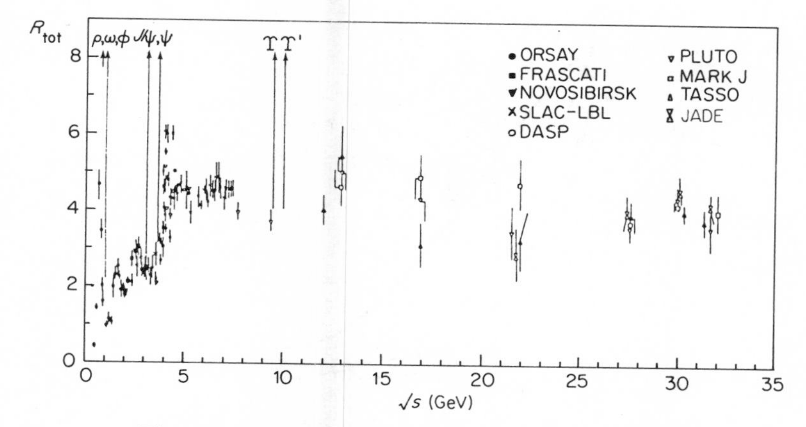

Figure 1.14 Measurements of R according to [Wo 80]

one may conclude that the primary process consists of the production of two high energy spin $\frac{1}{2}$ particles which then hadronize in subsequent stages [Ha 82N, Cr 82N].

(b) *Inclusive hadron prediction in e^+e^- annihilation*

If a specific hadron (e.g. $\pi^{\pm}$, $K^{\pm}$, $K^0, \ldots$) is detected in inclusive e^+e^- annihilation, then in the parton model, the cross-section is once again calculated from the possible quark pair production cross-section and the probability that a quark i will convert to a hadron h, i.e. fragmentation function $D_i^h(z)$ [Be 71, Gr 73a, Fi 77]. Experimentally, the energy distribution $\mathrm{d}N^h(z)/\mathrm{d}z$ is often given as ($z = 2p_h \cdot q/q^2 = E_h/E_{\text{beam}}$):

$$\frac{\mathrm{d}N^h(z)}{\mathrm{d}z} = \frac{\frac{\mathrm{d}\sigma}{\mathrm{d}z}(e^+e^- \to h + X)}{\sigma(e^+e^- \to h + X)}. \tag{1.4.30}$$

In the parton model, $\mathrm{d}N^h/\mathrm{d}z$ shows scaling behaviour and is calculated from the quark charges and fragmentation functions according to:

$$\frac{\mathrm{d}N^h(z)}{\mathrm{d}z} = \frac{\sum_i Q_i^2 D_i^h(z)}{\sum_i Q_i^2}. \tag{1.4.31}$$

From isospin and charge conjugation invariance one derives relations between fragmentation functions:

$$\begin{aligned} D_u^{\pi^+} &= D_d^{\pi^-} = D_{\bar{u}}^{\pi^-} = D_{\bar{d}}^{\pi^+} =: D^+, \\ D_u^{K^+} &= D_d^{K^0} = D_{\bar{u}}^{K^-} = D_{\bar{d}}^{K^0} =: K^+, \\ D_s^{\overline{K^0}} &= D_s^{K^-} = D_s^{K^0} = D_{\bar{s}}^{\overline{K^0}} =: K^+ + \delta. \end{aligned} \tag{1.4.32}$$

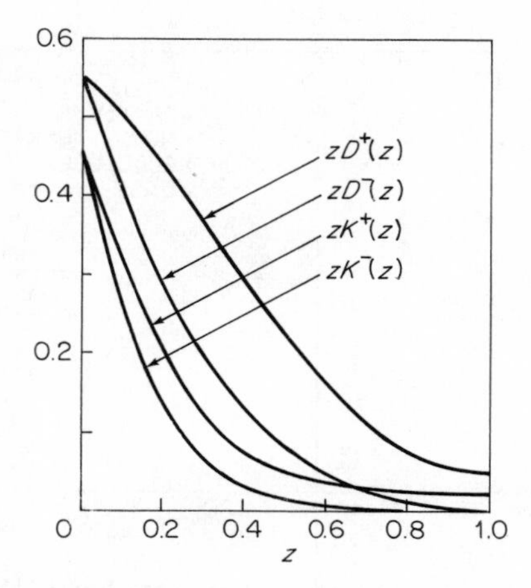

Figure 1.15 Quark fragmentation functions according to [Se 77]

Moreover, assuming that all sea quarks fragment in a similar way

$$\begin{aligned} D_s^{\pi^+} = D_{\bar{s}}^{\pi^+} = D_d^{\pi^+} = D_{\bar{u}}^{\pi^+} &=: D^-, \\ D_{\bar{u}}^{K^+} = D_d^{K^+} = D_{\bar{d}}^{K^+} = D_s^{K^+} &=: K^-, \end{aligned} \tag{1.4.33}$$

the fragmentation functions D^+, D^-, K^+, K^-, δ can be determined from the experimental data. From the results of a more recent analysis, shown in Fig. 1.15 [Se 77, Ha 77] one concludes that the valence quarks dominate the fragmentation, but that at low z values a significant amount of sea quarks is observed.

(c) *Inclusive electroproduction of hadrons*

This final example covers the inclusive hadron production in deep inelastic electron–nucleon scattering: $e + \mathcal{N} \to e + h + X$. In the parton model, this is determined by the parton distribution and fragmentation functions:

$$\frac{dN^h(z, x)}{dz} = \frac{1}{d\sigma/dx} \frac{d^2\sigma(e\mathcal{N} \to e + h + X)}{dx\, dz} = \frac{\sum_i Q_i^2 f_i^{\mathcal{N}}(x) D_i^h(z)}{\sum_i Q_i^2 f_i^{\mathcal{N}}(x)}. \tag{1.4.34}$$

A test of this result in inclusive electroproduction of neutral and charged pions is shown in Fig. 1.16 [Be 77].

The parton model has also been used for other deep inelastic reactions, e.g. the Drell–Yan process [Dr 70], i.e. the inclusive $\mu^+\mu^-$ creation in

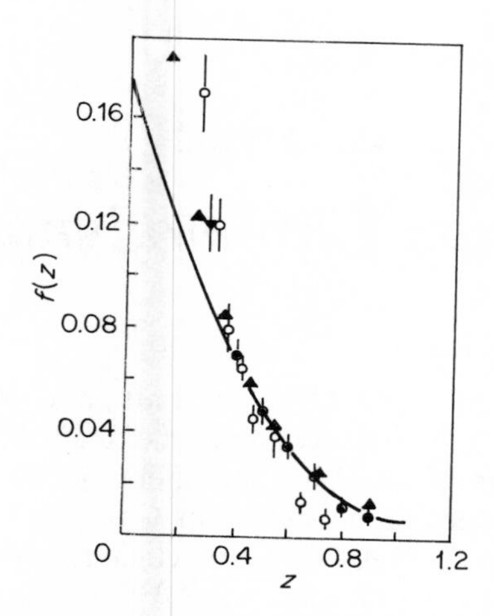

Figure 1.16 Test of the parton model with the fragmentation functions $D^+(z)$, $D^-(z)$ of Fig. 1.15; experimental data from [Be 77]

nucleon–nucleon scattering or hadron–hadron scattering with high transverse momentum [Be 71, El 74, Si 76] (see also [Ba 82N, Ak 82N]).

At small distances hadrons are made up of pointlike, to a good approximation, free constituents, which carry quark quantum numbers. Over and above this, there are hints that quarks are coloured and that flavour-neutral partons, gluons, are present in the hadrons.

CHAPTER 2

Quantum Chromodynamics

Quantum chromodynamics (QCD) is an attempt to provide a field-theoretical formulation of hadron dynamics, i.e. a strong interaction theory. This attempt is based on results from the phenomenological discussions on elementary particle physics, as sketched in Chapter 1. Quantum electrodynamics is the prototype for QCD. The spectacular precision which characterizes the description of dynamic effects in QED builds up confidence in the formal principles of quantized gauge field theories. The following two sections attempt to formulate QCD along these lines.

During the course of this description, QCD will be investigated in greater depth in Sections 2.3 and 2.4 in order to enable discussion of the two essential structural problems 'asymptotic freedom' (Section 2.5) and 'quark confinement' (Section 2.6). The asymptotic freedom aspect of QCD makes sense of the quark-parton model. The aim of quark confinement is to try to understand why no free quarks have been observed.

In contrast to the simple phenomenological ideas given in Chapter 1, QCD means that experimental questions can be extended and made more specific. Only the clear confirmation of the typical results of QCD will establish it as a physical theory. This 'new' phenomenology is discussed in Section 2.7 [QCD, PQ].

2.1 Quantum electrodynamics and local gauge invariance

Formally, the basic QCD equations are generalizations of the Dirac–Maxwell equations. So, further consideration will now be given to QED. A generally applicable principle which determines the structure of the electromagnetic interaction, is found in local gauge invariance. A precision experiment is analysed to demonstrate the validity of QED.

2.1.1 Basic concepts of quantum electrodynamics [QED, FT]

The field equations which describe the interaction between electrons e, muons μ, and photons γ are the Dirac equations

$$[\mathrm{i}\gamma^{\mu}(\partial_{\mu}-\mathrm{i}eA_{\mu}(x))-M_l]\psi^{(l)}(x)=0, \quad l=e,\mu, \tag{2.1.1}$$

and the Maxwell equations

$$\partial^{\mu}F_{\mu\nu}(x)=-e(\bar{\psi}^{(e)}(x)\gamma_{\nu}\psi^{(e)}(x)+\bar{\psi}^{(\mu)}(x)\gamma_{\nu}\psi^{(\mu)}(x)). \tag{2.1.2}$$

These field equations are the Euler–Lagrange equations of the Hamiltonian principle for the action:

$$S = \int d^4x \left(-\tfrac{1}{4} F_{\mu\nu} F^{\mu\nu} + \sum_{l=e,\mu} \bar{\psi}^{(l)} [i(\not\partial - ie\not{A}) - M_l] \psi^{(l)} \right) \tag{2.1.3}$$

with

$$F_{\mu\nu}(x) = \partial_\mu A_\nu(x) - \partial_\nu A_\mu(x). \tag{2.1.4}$$

In Section 1.2.2 we already referred to these basic QED equations and to the physical content of the quantum theory of these field equations. The most important points are repeated below:

(a) The quanta of the free basic fields $\psi^{(l)}$, A_μ describe the observed particles electrons, positrons, muons and photons (see Eqns. (1.2.25) and (1.2.21)).
(b) The non-linear terms in the field equations indicate an interaction between these particles which can be described by Feynman graphs representing a perturbation theory with respect to $\alpha = e^2/4\pi$.
(c) In the most simple case, a Feynman graph describes a reaction by a simple exchange of particles (tree graph). The graph in Fig. 1.6 for the reaction $e^+e^- \to \mu^+\mu^-$ was discussed including the calculation of the cross-section. Further examples are

Compton scattering $\gamma + e^- \to \gamma + e^-$:

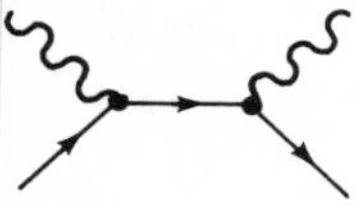

+1 graph with permuted lines,

Pair creation $\gamma + e^- \to e^- + e^+e^-$:

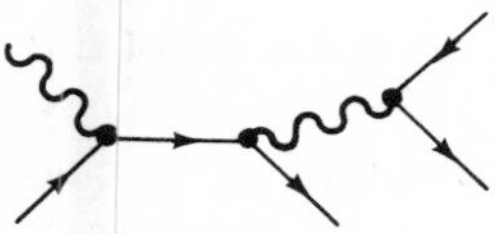

+graphs with permuted lines.

Tree graphs describe elementary particle reactions within the frame of a simple, relativistic collision theory. However, a complete field theory contains more subtle effects due to the retroaction of the fields. These are generally called 'internal radiative corrections'. In Feynman's perturbation theory [Fe 49] they are represented by loop diagrams. The following are given by way of example:

(1)

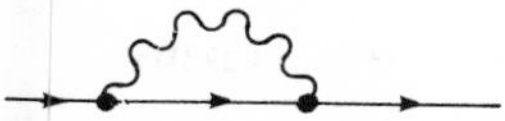

This graph describes the modification of the electron–positron field propagation by the self-interaction between electrons and the em field. This leads amongst other things to a change in the electron mass which must be compensated by renormalization.

(2) The same applies for the modification of the photon propagation by the vacuum polarization:

(3) The interaction between photons and electrons is modified by these and other graphs:

The effect of internal radiative corrections is that the electron (muon) receives, besides the Dirac coupling $e\gamma_\mu$, a Pauli coupling and a form factor [Sc 49], e.g.:

$$= +\mathrm{i}(2\pi)^4 e\left\{\gamma_\mu\left[1+\frac{\alpha}{3\pi}\frac{k^2}{M^2}\left(\log\frac{M}{\lambda}-\frac{3}{8}\right)\right]+\frac{\alpha}{2\pi\mathrm{i}}\frac{1}{2M}\sigma_{\mu\nu}k^\nu\right\} \times\delta^{(4)}(p_1+p_2-k) \quad (2.1.5)$$

This evaluation of the graph applies for $k^2 \ll M^2$, λ is an infrared cut-off parameter ([Bj 66], vol. 1, p. 184 onwards).

The experimental evidence of internal radiative corrections is the decisive test for the quantum field theoretic nature of a dynamics. The next section shows that this has been achieved for QED with highest precision. The central problem of hadron dynamics is to calculate measurable internal radiative corrections and to determine them by experiment. For this reason, further reference is made to the calculation of 1-loop diagrams in Section 2.4. In Section 2.7 the experimental confirmation of internal radiative corrections to the collision theory of the parton model is discussed.

An even more difficult question is whether all internal radiative corrections of QCD are perturbative. This is discussed in connection with the problem of quark confinement in Section 2.6.

2.1.2 A QED test: the (g-2) experiment

At least one of the precision experiments carried out to determine the validity of QED [Ki 78] is now considered in more detail: the measurement of the magnetic moment of the muon [Ba 79].

(a) First of all, a *short description of the experiment.* The fact that the magnetic moment $\boldsymbol{\mu}$ of muon

$$\boldsymbol{\mu} = \frac{e\hbar}{2mc} g\mathbf{s} \tag{2.1.6}$$

deviates from its Dirac value $g = 2$ means that when a muon circulates in a homogeneous magnetic field B, the direction of its spin rotates with the frequency

$$\omega_a = \frac{e}{mc} aB, \quad a = \tfrac{1}{2}(g\text{-}2) \tag{2.1.7}$$

against the direction of the momentum. At CERN, this frequency has been measured in a muon storage ring of the 'magic energy' $E = 3.094$ GeV. The muons obtained from the decay processes $\pi^+ \to \mu^+ + \nu_\mu$ and $\pi^- \to \mu^- + \bar{\nu}_\mu$ are polarized. The precession of the muon spin with frequency ω_a is measured as a time modulation of the frequency with which the muon decay electrons are detected in the storage ring plane (see Eqn. (1.3.12)). The experimental results are given in Fig. 2.1.

(b) For a_{μ^+} and a_{μ^-}, the experiment gives the values [Ba 79]

$$\begin{aligned} a_{\mu^+} &= 1165911(11) \times 10^{-9}, \\ a_{\mu^-} &= 1165937(12) \times 10^{-9}, \\ a_\mu = \tfrac{1}{2}(a_{\mu^+} + a_{\mu^-}) &= 1165924(8.5) \times 10^{-9}. \end{aligned} \tag{2.1.8}$$

Reference should be made to the original publication for a careful discussion of all systematic errors.

The theoretical calculation up to the fifth order in α as well as hadronic

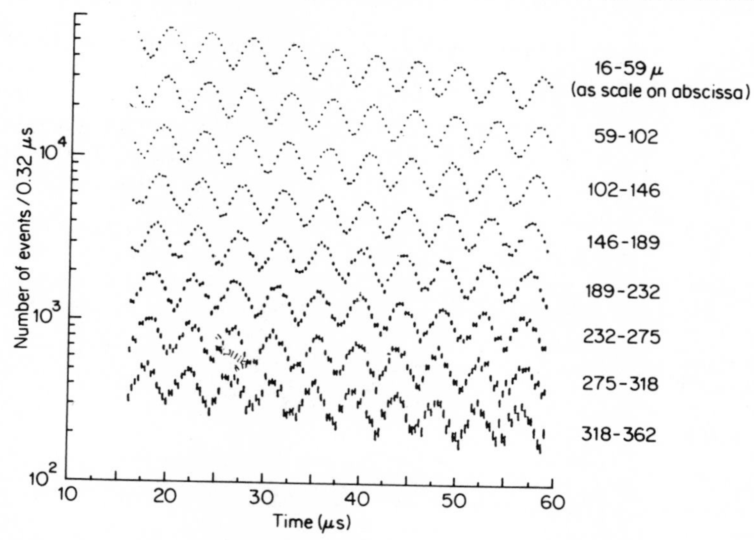

Figure 2.1 Time distribution of a total of 13 million electrons from the decay process $\mu \to \nu_\mu + e + \nu_e$ [Ba 77, Co 75a]

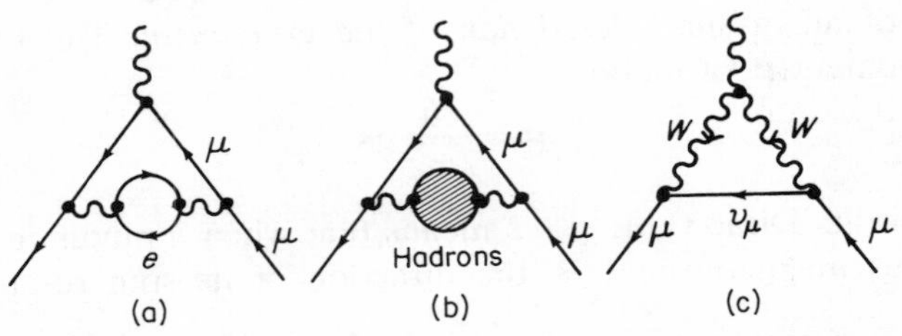

Figure 2.2 Contributions of electromagnetic (a), hadronic (b), and weak (c) interactions to the muon magnetic moment

corrections yield ([Ba 79] and reports [Ca 77] and [La 72]):

$$a_{\mu^+}^{\text{th}} = a_{\mu^-}^{\text{th}} = 1165921(8.3)\times 10^{-9}. \tag{2.1.9}$$

This means that the theoretical and the experimental result for the g-factor are in agreement within 9 significant digits.

(c) The importance of this result can be assessed by listing the different contributions to the theoretical result. According to the principle that ultimately everything interacts with everything else, one has to take into account the electromagnetic [Br 71, Ca 77, La 72], the hadronic [Ba 75, Ca 76], and the weak [Ca 77, Ja 72, Ba 72] interaction. Fig. 2.2 shows examples of Feynman graphs describing internal radiative corrections of these interactions. The numerical contributions of the different graphs are listed in Table 2.1.

Table 2.1 List of the individual contributions to a_μ. $\alpha^{-1} = 137.035987(29)$ [Co 73] is used

QED	Coefficient of $\left(\frac{\alpha}{\pi}\right)^n$		Contribution to a_μ ($\times 10^9$)	
$n=1$	0.5		1161409.84	±0.25
2	0.76578223		4131.77	±0.002
3	24.452±0.056		306.45	±0.70
4	135±63		3.9	±1.8
5	420±30		0.028	±0.002
		(+)		
QED total			1165852.0	±1.9
Hadronic				
4th order			70.2	±8.0
6th order			−3.5	±1.4
		(+)		
Hadronic total			66.7	±8.1
Weak				
Glashow–Salam–Weinberg theory			2.1	±0.2
Gravitation [Be 75]			4	$\times 10^{-32}$
Total result			$a_\mu^{\text{th}} = 1165921$	±8.3

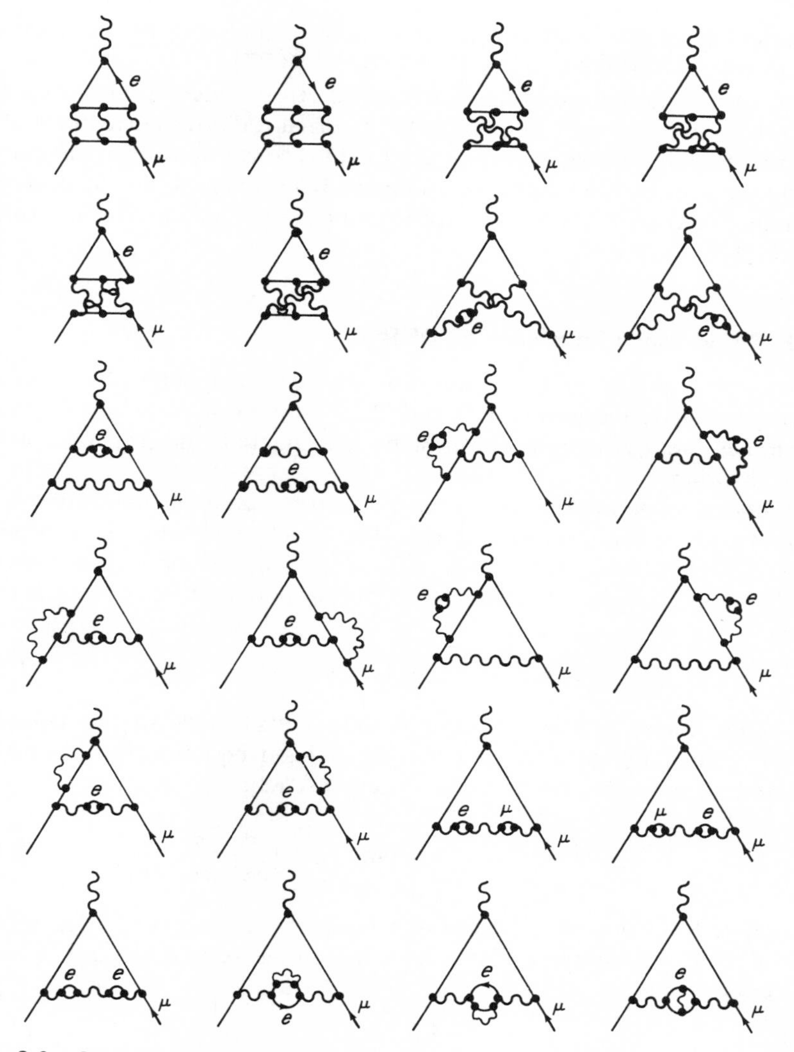

Figure 2.3 Contributions of order $e(\alpha/\pi)^3$ to the anomalous magnetic moment of the muon [La 72]

Pure QED supplies the greatest part. The contribution of the lowest order $n = 1$ comes from the Pauli term in (2.1.5) with the coefficient $\alpha/2\pi$. The graphs in Fig. 2.3 with seven vertices show the complexity of the higher-order internal radiative corrections described by QED. Their contribution is of the order $(\alpha/\pi)^3$.

According to Table 2.1, the coefficients of the powers of α/π in the QED perturbation series increase. The increase in the number of higher-order graphs makes this a divergent, asymptotic series [Hu 52, Th 53]. Methods for summing up divergent perturbation theory expansions are of interest in current research in field theory [Li 77, Br 77, It 77].

One surprising feature is how little effect strong interaction has on pure QED; it contributes only 60 ppm to the internal radiative corrections of the g-factor. Perhaps a field theory of the QED type can also form an example for the separation of phenomena connected with different flavour degrees of freedom. We come back to this problem under the heading 'Zweig's rule' in Section 2.7.4.

2.1.3 Local gauge invariance of QED

Knowing the quality of the description of electromagnetic phenomena by the simple basic equations (2.1.1) and (2.1.2), the principle which essentially determines these equations can now be investigated: the principle of local gauge invariance.

(a) *Local gauge invariance in non-relativistic quantum mechanics* is first encountered in the following form. The electromagnetic field strengths $F_{\mu\nu}(x)$ (electric field strength $\mathbf{E}=(F_{10}, F_{20}, F_{30})$, magnetic field strength $\mathbf{B}=(F_{23}, F_{31}, F_{12})$) do not uniquely define the potential $A_\mu(x)$ by the equation

$$F_{\mu\nu}(x)=\partial_\mu A_\nu(x)-\partial_\nu A_\mu(x). \tag{2.1.10}$$

The gauge transformation $A'_\mu(x)=A_\mu(x)+\partial_\mu\theta(x)$ leads to the same field strength tensor $F_{\mu\nu}(x)$. Consider the Schrödinger equation for a particle in an external, pure electric potential $V(x)=eA^0(x)$:

$$\frac{1}{\mathrm{i}}\frac{\partial}{\partial t}\psi=\left(-\frac{1}{2m}\boldsymbol{\partial}^2+V(x)\right)\psi, \quad \boldsymbol{\partial}=\left(\frac{\partial}{\partial x_1},\frac{\partial}{\partial x_2},\frac{\partial}{\partial x_3}\right). \tag{2.1.11}$$

The gauge transform of A_μ leads to an unphysical ambiguity. This problem is overcome by coupling the vector potential $\mathbf{A}=(A^1, A^2, A^3)$ by the substitution rule

$$\boldsymbol{\partial}\to\boldsymbol{\partial}-\mathrm{i}e\mathbf{A}(x). \tag{2.1.12}$$

Thus, the Schrödinger equation for a particle in the em field is given by

$$\frac{1}{\mathrm{i}}\frac{\partial}{\partial t}\psi=\left(-\frac{1}{2m}(\boldsymbol{\partial}-\mathrm{i}e\mathbf{A}(x))^2+eA^0(x)\right)\psi. \tag{2.1.13}$$

Then, the gauge transformation of the potential

$$A'^0(x)=A^0(x)+\partial^0\theta(x), \quad \mathbf{A}'(x)=\mathbf{A}(x)+\boldsymbol{\partial}\theta(x) \tag{2.1.14}$$

can be compensated in this equation by a gauge transformation of the Schrödinger wavefunction

$$\psi'(x)=\mathrm{e}^{\mathrm{i}e\theta(x)}\psi(x). \tag{2.1.15}$$

This means that a solution $\psi(x)$ for the potential $A_\mu(x)$ is transformed by (2.1.15) into a solution $\psi'(x)$ for the potential $A'_\mu(x)$:

$$\begin{aligned}\left(\frac{1}{\mathrm{i}}\frac{\partial}{\partial t}-eA'^0(x)+\frac{1}{2m}(\boldsymbol{\partial}-\mathrm{i}e\mathbf{A}'(x))^2\right)\psi'(x)\\ =\mathrm{e}^{\mathrm{i}\theta(x)}\left(\frac{1}{\mathrm{i}}\frac{\partial}{\partial t}+e\frac{\partial}{\partial t}\theta(x)-eA'^0(x)+\frac{1}{2m}(\boldsymbol{\partial}+\mathrm{i}e\,\boldsymbol{\partial}\theta(x)-\mathrm{i}e\mathbf{A}'(x))^2\right)\psi(x)\\ =\mathrm{e}^{\mathrm{i}\theta(x)}\left(\frac{1}{\mathrm{i}}\frac{\partial}{\partial t}-eA^0(x)+\frac{1}{2m}(\boldsymbol{\partial}-\mathrm{i}e\mathbf{A}(x))^2\right)\psi(x)=0.\end{aligned} \tag{2.1.16}$$

Schrödinger functions which are connected by such gauge transformations describe the same physical situation, just as various 4-potentials $A_\mu(x)$ connected by gauge transformations lead to the same em field strengths. The substitution rule relates the electric coupling uniquely with the magnetic one so that the Lorentz force acts on a moving charged particle in addition to the electric force.

(b) Moving on now to QED, given by the field equations (2.1.1)–(2.1.2)

$$[\mathrm{i}\gamma^\mu(\partial_\mu-\mathrm{i}eA_\mu(x))-M]\psi(x)=0,\quad \partial^\mu F_{\mu\nu}(x)=-e\bar{\psi}(x)\gamma_\nu\psi(x)$$

or by the Lagrangian density

$$\mathscr{L}=-\tfrac{1}{4}F_{\mu\nu}F^{\mu\nu}+\bar{\psi}[\mathrm{i}\gamma^\mu(\partial_\mu-\mathrm{i}eA_\mu)-M]\psi, \tag{2.1.17}$$

it is obvious that the interaction between photons and electrons is also given by the *substitution rule* in relativistic form:

$$\partial_\mu\to\partial_\mu-\mathrm{i}eA_\mu(x). \tag{2.1.18}$$

A calculation as above shows that these field equations and the Lagrangian density $\mathscr{L}$ are invariant under local gauge transformations:

$$A_\mu(x)\to A_\mu(x)+\partial_\mu\theta(x),\quad \psi(x)\to\mathrm{e}^{\mathrm{i}e\theta(x)}\psi(x),\quad \bar{\psi}(x)\to\bar{\psi}(x)\mathrm{e}^{-\mathrm{i}e\theta(x)}. \tag{2.1.19}$$

As in the calculation (2.1.16), the essential formula is:

$$(\partial_\mu-\mathrm{i}eA'_\mu(x)\psi'(x)=\mathrm{e}^{\mathrm{i}e\theta(x)}(\partial_\mu-\mathrm{i}eA_\mu(x))\psi(x). \tag{2.1.20}$$

(c) The local gauge invariance of QED can be considered from another point of view which will be useful for generalizations made later on: Lagrangian densities which contain a complex field in the form $\psi^\dagger\Gamma\psi$ are invariant under global gauge transformations:

$$\psi(x)\to\mathrm{e}^{\mathrm{i}\theta}\psi(x),\quad \psi^\dagger(x)=\psi^\dagger\mathrm{e}^{-\mathrm{i}\theta},\quad \theta\text{ space-time independent} \tag{2.1.21}$$

(see Section 1.2.3). This applies in particular for Lagrangian densities describing free fields $\mathscr{L}_0=\bar{\psi}(\mathrm{i}\gamma^\mu\,\partial_\mu-M)\psi$, $\mathscr{L}_0=(\partial_\mu\phi^\dagger)(\partial^\mu\phi)-M^2\phi^\dagger\phi$, etc. The coupling to the gauge field $A_\mu(x)$ according to the substitution rule makes the Lagrangian density invariant under local phase transformations

(2.1.19): this is called *gauge invariance of the second kind.* The gauge field itself becomes a dynamical quantity by adding the term $\mathscr{L}_{\text{gauge}} = -\frac{1}{4}F_{\mu\nu}F^{\mu\nu}$ to the Lagrangian density.

Thus, here is a principle which uniquely introduces an interaction into a free theory with charge symmetry: the minimal gauge-invariant coupling to a dynamical gauge field! This principle is highly successful in QED. Reference was made at the end of Section 1.2.3 to the fact that local gauge invariance in field theory seems more physical than global symmetry.

The structure of the principle of local gauge invariance is reminiscent of the theory of general relativity [Ei 22, Ad 65, Se 75] where the metric field $g_{\mu\nu}(x)$ describes the local geometry and acquires a dynamical meaning as does the gauge field $A_\mu(x)$. A geometrical description of gauge invariance of the second kind will be considered in the next section and generalized to more complicated charge symmetry groups. In this general form, local gauge invariance discovered in QED also determines the strong and electroweak interaction.

2.2 Formulation of quantum chromodynamics

In this section, quantum chromodynamics is approached in three stages. First of all, the understanding of a local symmetry is enlarged by the geometrical interpretation of the associated gauge fields. As in electrodynamics, the dynamical interpretation of the gauge fields leads in a second step to a field theory for the interaction between all particles which carry the charges of the internal symmetry. These Yang–Mills field theories form a general framework for the description of both the hadronic and the electroweak interaction. Stage three brings together all the aspects which are used to formulate hadron dynamics as a quantized Yang–Mills theory of local colour symmetry. [GT]

2.2.1 The geometry of local gauge symmetry

The importance of symmetry groups was repeatedly encountered during our earlier phenomenological considerations of elementary particle physics. In connection with a field theory a symmetry has important consequences. Section 1.2.3 showed the connection between 'global' gauge symmetry and conserved currents. It emerged that the principle of local action of field theory actually requires that the symmetry transformations of fields should be local rather than global. When applied to phase transformations of the electron field, this led to the understanding that the form of the em interaction in QED is determined almost uniquely by the principle of local gauge invariance (Section 2.1.2(c)).

(a) For this reason, the *local form of internal symmetry* will now be considered for general groups G, e.g. for the colour and flavour group. The necessary discussion for this is not given in full mathematical generality. Essentially the group theoretical concepts of Section 1.2.1 are used.

Nevertheless the considerations of this Section 2.2.1 go way beyond the general mathematical scope set for the book as a whole. Therefore, the most important concepts of the formulation of gauge theories [Ab 73, Ta 76, Ya 54] are listed in advance:

(1) the definition of the covariant derivative (Eqn. (2.2.8));
(2) the connection between field strength tensors $F^a_{\mu\nu}(x)$ and gauge fields $A^a_\mu(x)$ (Eqn. (2.2.12));
(3) the definition of gauge transformations (Eqn. (2.2.5)); and
(4) the transformation properties of the charged field (Eqn. (2.2.2)), field strength (Eqn. (2.2.14)), and covariant derivative (Eqn. (2.2.9)).

The starting point of our considerations are the global transformations of charged particle fields $\psi(x)$, e.g. quark fields as considered in Eqns. (1.2.61) and (1.2.63):

$$\psi_c(x) \to (\exp(-i\theta^a T_a))_{cc'}\psi_{c'}(x), \quad c, c' = 1, \ldots, n. \tag{2.2.1}$$

If these transformations leave the field equations or Lagrange function invariant, then the group G of transformations (2.2.1) generated by the charges T_a is a physical symmetry group. In the following, the components of the charged field ψ_c are regarded as the coordinates of a vector with respect to a basis $\{e_c\}$ in an n-dimensional vector space $\mathscr{H}$. The transformation (2.2.1) is interpreted as a coordinate transformation which reflects a change of the basis in $\mathscr{H}$. The symmetry is frequently described in the (passive) form: all coordinate systems which are transformed by the group G into each other are physically equivalent. For the sake of simplicity, only symmetries of the groups $SU(n)$ are considered in the following. These are physically realized by fields which transform according to the fundamental representation (quark fields). Thus all unitary, normalized coordinate systems are both permitted and equivalent in the charge space $\mathscr{H}$.

For the local form of a symmetry, the global transformations (2.2.1) are replaced by local transformations (gauge transformations of the second kind)

$$\psi_c(x) \to (\exp(-i\theta^a(x)T_a))_{cc'}\psi_{c'}(x) \equiv (g(x)\psi)_c(x). \tag{2.2.2}$$

T_a represent the infinitesimal generators of the symmetry group $SU(n)$ in the fundamental representation, e.g. $\frac{1}{2}\tau_a, \frac{1}{2}\lambda_a, \ldots$ for $SU(2)$, $SU(3)$, etc. The space dependence of the transformation $g(x)$ is a space dependence of the group parameters $\theta^a(x)$. In the following geometrical considerations, Eqn. (2.2.2) is again used to describe coordinate transformations, this time in a local charge space ${}^x\mathscr{H}$ at every point x. Using this approach, charged fields $\psi_c(x)$ are mappings of the space–time points x on vectors $(\psi_c(x)) \in {}^x\mathscr{H}$. The dynamics described by the field equations connects the fields $\psi(x)$ at different space–time points. If the conclusions drawn from symmetry transformations, e.g. the charge structure described by T_a, are to be physically significant, then there must be a connection between the local charge spaces ${}^x\mathscr{H}$ at different points.

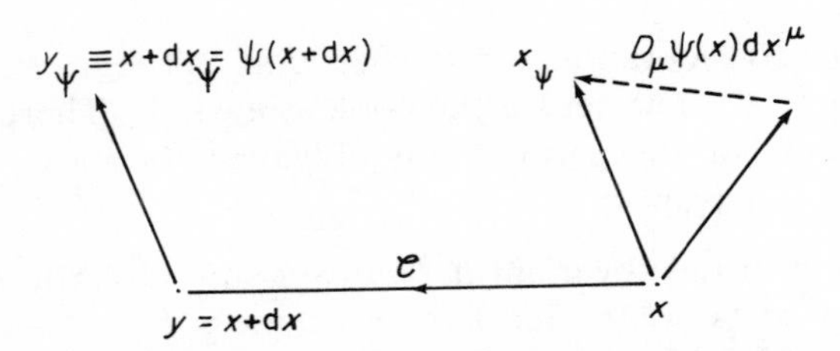

Figure 2.4 The geometrical interpretation of the covariant derivative and of an (infinitesimal) parallel displacement

(b) In the spirit of differential geometry [Ko 63, Te 68] this symmetry connection between infinitely near points $x=(x^\mu)$ and $y=(x^\mu+dx^\mu)$ is described with help of a *gauge field* $A^a_\mu(x)$. Geometrically, this connection is realized by the definition of the parallel displacement of vectors. Thereby, a vector ${}^y\psi$ is associated with a vector ${}^x\psi \in {}^x\mathscr{H}$ by parallel displacement from x to y along a curve $\mathscr{C}$ (Fig. 2.4).

Local coordinate systems are used in ${}^x\mathscr{H}$ and ${}^y\mathscr{H}$ to describe the parallel displacement explicitly. If ${}^x\psi_c$ are the coordinates of a vector from ${}^x\mathscr{H}$ with respect to an admissible, i.e. orthonormal basis, then the coordinates of the infinitesimally parallel displaced vector ${}^{x+dx}\psi_c$ are

$$ {}^{x+dx}\psi_c = {}^x\psi_c + \delta\, {}^x\psi_c = (\delta_{cc'} - i\mathscr{A}_{\mu,cc'}(x)\, dx^\mu)\, {}^x\psi_{c'}. \tag{2.2.3}$$

If one requires that the properties of vectors with respect to unitary symmetry, i.e. linearity, orthogonality, and normalization, are to be preserved during the parallel displacement, then $\mathscr{A}_{\mu,cc'}(x)\, dx^\mu$ is an infinitesimal unitary transformation. This can be represented by a linear combination of generators of the gauge group $U(n)$:*

$$\mathscr{A}_{\mu,cc'}(x) = \bar{g} A^a_\mu(x)(T_a)_{cc'}. \tag{2.2.4}$$

$\mathscr{A}_\mu$ as a linear combination of generators is mathematically a Lie algebra element of the symmetry group with commutation relations which are given by those of the charge operators T_a. Physically, $\mathscr{A}_\mu(x)$ is called a gauge field. $A^a_\mu(x)$ is the non-Abelian generalization of the em potentials. $\bar{g}$ is a proportionality constant, fixed by physical considerations and often differing conventions (see Section 2.2.2, page 74). Apart from later remarks on the theory of general relativity, the x^μ are regarded as the Cartesian coordinates of pseudo-Euclidean space–time. The symmetry connection $\mathscr{A}_{\mu,cc'}(x)$ is given with respect to a local coordinate system in ${}^x\mathscr{H}$ and ${}^y\mathscr{H}$ by Eqn. (2.2.3). If the relationship between the parallelism of vectors ψ at different points is to be independent of the local coordinate system, then $\mathscr{A}_{\mu,cc'}(x)$ must be transformed in the following way under general local transformations:

$$\mathscr{A}_{\mu,cc'}(x) \to \mathscr{A}'_{\mu,cc'}(x) = g_{cd}(x)\mathscr{A}_{\mu,dd'}(x)\overset{-1}{g}_{d'c'}(x) - i g_{cd}(x)\, \partial_\mu \overset{-1}{g}_{dc'}(x) \tag{2.2.5}$$

* $U(n)$ is frequently regarded as infinitesimally equivalent to $SU(n)\times U(1)$, which leads to independent $SU(n)$ and $U(1)$ gauge theories (see Eqn. (2.2.37)).

or, more concisely:

$$\mathscr{A}_\mu(x) \to \mathscr{A}'_\mu(x) = g(x)(\mathscr{A}_\mu - \mathrm{i}\,\partial_\mu)g^{-1}(x). \tag{2.2.5'}$$

This follows from the required commutativity of parallel displacement and local coordinate transformation

$$g(x+dx)(1-\mathrm{i}\mathscr{A}_\mu(x)\,dx^\mu)\,{}^x\psi = (1-\mathrm{i}\mathscr{A}'_\mu(x)\,dx^\mu)g(x)\,{}^x\psi,$$

if one considers the terms linear in dx^μ: $[-\partial_\mu g(x) + \mathrm{i}g(x)\mathscr{A}_\mu(x)]\,dx^\mu = \mathrm{i}\mathscr{A}'_\mu(x)\,dx^\mu g(x)$ and uses the following formula:

$$\partial_\mu(g(x)g^{-1}(x)) = 0, \quad (\partial_\mu g(x))g^{-1}(x) = -g(x)\,\partial_\mu g^{-1}(x). \tag{2.2.6}$$

The simultaneous transformation of the charged field $\psi(x)$ according to (2.2.2) and the gauge field $\mathscr{A}_\mu(x)$ according to (2.2.5) is called a gauge transformation. Because of the great importance of gauge transformations, their infinitesimal form is also given:

$$\begin{aligned} \delta\,{}^x\psi &= -\mathrm{i}T_a\,{}^x\psi\,\delta\theta^a(x), \\ \delta\,{}^x\mathscr{A}_\mu(x) &= -\mathrm{i}[T_a, \mathscr{A}_\mu(x)]\,\delta\theta^a(x) + T_a\,\partial_\mu\,\delta\theta^a(x). \end{aligned} \tag{2.2.7}$$

(c) The most important physical aspect of a symmetry connection $\mathscr{A}_{\mu,cc'}(x)$ described by a gauge field is that it can be used to define the *covariant derivative* of a charged field. A charged field describes the association of a vector from ${}^x\mathscr{H}$ to any space–time point $x; x \to {}^x\psi_c(x) \in {}^x\mathscr{H}$. According to our geometrical understanding of the components of the charged field these depend on the local basis of ${}^x\mathscr{H}$. Field vectors at different points can only be compared with each other if their components are referred to the same local coordinate system by means of a parallel displacement. Applied to the variation of the field vectors between neighbouring points $y^\mu = x^\mu + dx^\mu$ and x^μ (see Fig. 2.4) and with help of Eqn. (2.2.3) it follows:

$$\begin{aligned} {}^x\psi_c(x+dx) - {}^x\psi_c(x) &\approx (1+\mathrm{i}\mathscr{A}_\mu(x)\,dx^\mu)\,{}^y\psi_c(x+dx) - {}^x\psi_c(x) \\ &\approx [(\partial_\mu + \mathrm{i}\mathscr{A}_\mu)\psi(x)]_c\,dx^\mu =: (D_\mu\psi)_c(x)\,dx^\mu. \end{aligned}$$

This shows the intuitive meaning of the covariant derivative of a charged field:

$$(D_\mu\psi(x))_c \equiv \partial_\mu\psi_c(x) + \mathrm{i}\mathscr{A}_{\mu,cc'}\psi_{c'}(x). \tag{2.2.8}$$

Because of this geometrical relationship, the covariant derivative $D_\mu\psi$ has simple transformation properties with respect to general gauge transformations:

$$(\partial_\mu + \mathrm{i}\mathscr{A}_\mu)\psi(x) \to (\partial_\mu + \mathrm{i}\mathscr{A}'_\mu(x))\psi'(x) = (D_\mu\psi)'(x) = g(x)(D_\mu\psi)(x). \tag{2.2.9}$$

This is contrast to the ordinary gradient $\partial_\mu\psi$. Eqn. (2.2.9) follows from Eqns.

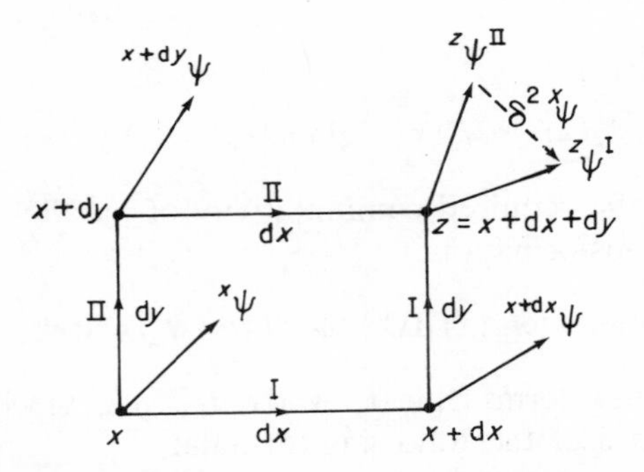

Figure 2.5 Geometrical interpretation of the field strength tensor

(2.2.2) and (2.2.5):

$$\begin{aligned}(D_\mu\psi)'(x) &= (\partial_\mu + \mathrm{i}g(x)\mathscr{A}_\mu(x)g^{-1}(x) + g(x)\,\partial_\mu g^{-1}(x))g(x)\psi(x)\\ &= g(x)(\partial_\mu + \mathrm{i}\mathscr{A}_\mu)\psi(x) + (\partial_\mu g(x) + g(x)(\partial_\mu g^{-1}(x))g(x))\psi(x)\\ &= g(x)(D_\mu\psi)(x),\end{aligned}$$

since the final term disappears due to Eqn. (2.2.6).

(d) Locally, the path dependence of a parallel displacement can be described by the *field strength tensor* $\mathscr{F}_{\mu\nu}(x)$. The displacement from x to y along the path I in Fig. 2.5 differs from the displacement along the path II by the following vector of the second order:

$$[(1-\mathrm{i}\mathscr{A}_\nu(x+\mathrm{d}x)\,\mathrm{d}y^\nu)(1-\mathrm{i}\mathscr{A}_\mu(x)\,\mathrm{d}x^\mu)]\,{}^x\psi \qquad \text{(I)}$$

$$-[(1-\mathrm{i}\mathscr{A}_\mu(x+\mathrm{d}y)\,\mathrm{d}x^\mu)(1-\mathrm{i}\mathscr{A}_\nu(x)\,\mathrm{d}y^\nu)]\,{}^x\psi \qquad \text{(II)}$$

$$\approx -\mathrm{i}\mathscr{F}_{\mu\nu}(x)\,\mathrm{d}x^\mu\,\mathrm{d}y^\nu\cdot{}^x\psi \equiv \delta^2\,{}^x\psi$$

with

$$\mathscr{F}_{\mu\nu}(x) = \partial_\mu\mathscr{A}_\nu(x) - \partial_\nu\mathscr{A}_\mu(x) + \mathrm{i}[\mathscr{A}_\mu(x), \mathscr{A}_\nu(x)]. \qquad (2.2.10)$$

Like $\mathscr{A}_\mu(x)$, $\mathscr{F}_{\mu\nu}(x)$ is Lie algebra valued. In components A^a_μ, $F^a_{\mu\nu}$ with

$$\mathscr{F}_{\mu\nu}(x) = \bar{g}F^a_{\mu\nu}(x)T_a \qquad (2.2.11)$$

Eqn. (2.2.9) can be written with help of the structure constants (1.2.1):

$$F^a_{\mu\nu}(x) = \partial_\mu A^a_\nu(x) - \partial_\nu A^a_\mu(x) - \bar{g}f^{abc}A^b_\mu(x)A^c_\nu(x). \qquad (2.2.12)$$

For the non-Abelian case, the field strength differs from the electromagnetic field strength by a non-linear term in the gauge field. Later on, this has extremely significant consequences.

The non-commutativity of the covariant derivative:

$$D_\mu D_\nu - D_\nu D_\mu = [D_\mu, D_\nu] = \mathrm{i}\mathscr{F}_{\mu\nu}(x) \qquad (2.2.13)$$

corresponds to the path dependence of the parallel displacement. The combination of this equation with Eqn. (2.2.9) shows in an easy way the behaviour of the field strength with respect to gauge transformations:

$$\mathcal{F}_{\mu\nu}(x) \to g(x)\mathcal{F}_{\mu\nu}(x)g^{-1}(x) \tag{2.2.14}$$

or

$$\delta\mathcal{F}_{\mu\nu}(x) = -\mathrm{i}[T_a, \mathcal{F}_{\mu\nu}(x)]\,\delta\theta^a(x) \tag{2.2.14'}$$

for infinitesimal gauge transformations. These equations indicate that the components of the field strength tensor are transformed according to the adjoint representation of the symmetry group under a coordinate transformation of the local charge space. Thus, if the same considerations which lead to the definition of the covariant derivative of charged fields (Eqn. (2.2.7)) are applied to the field strength tensor, then the following definition follows for the covariant derivative:

$$(D_\rho\mathcal{F}_{\mu\nu})(x) := \partial_\rho\mathcal{F}_{\mu\nu}(x) + \mathrm{i}[\mathcal{A}_\rho(x), \mathcal{F}_{\mu\nu}(x)] \tag{2.2.15}$$

or, in components:

$$(D_\rho\mathcal{F}_{\mu\nu})^a(x) = \partial_\rho F^a_{\mu\nu}(x) - \bar{g}f^{abc}A^b_\rho(x)F^c_{\mu\nu}(x). \tag{2.2.16}$$

The Bianchi identity for non-Abelian field strengths with respect to the covariant derivative

$$D_\rho\mathcal{F}_{\mu\nu}(x) + D_\mu\mathcal{F}_{\nu\rho}(x) + D_\nu\mathcal{F}_{\rho\mu}(x) = 0 \tag{2.2.17}$$

follows directly from the definition (2.2.10) of the field strength using the Jacobi identity for commutators. This equation generalizes the homogeneous Maxwell equations to non-Abelian gauge theories.

(e) So far, the symmetry connection has been considered in infinitesimal neighbouring domains. This part deals with the *parallel displacement* $U(\mathscr{C})$ of charge vectors between points x and y of finite distance. If ${}^x\psi_c$ are the components of a charge vector in ${}^x\mathcal{H}$ with respect to a local coordinate system, then the components ${}^y\psi_c$ of the vector displaced to y along the path $\mathscr{C}$ (Fig. 2.4) are

$${}^y\psi_c = U_{cc'}(\mathscr{C})\,{}^x\psi_{c'} \tag{2.2.18}$$

in the local coordinate system of ${}^y\mathcal{H}$.

If symmetry is conserved similar to Eqns. (2.2.3) and (2.2.4), then $U_{cc'}(\mathscr{C})$ is an element of the symmetry group G in the fundamental representation. When the local coordinates are changed in ${}^x\mathcal{H}$ or ${}^y\mathcal{H}$, the parallel displacement must be transformed according to

$$U(\mathscr{C}) \to g(y)U(\mathscr{C})g^{-1}(x) \tag{2.2.19}$$

in order to keep parallelism independent of coordinates.

The parallel displacement along a curve can be generated infinitesimally, i.e. the parallel displacement $U(\mathscr{C})$ can be calculated from the gauge fields $\mathcal{A}_\mu(x)$. For this purpose, the path $\mathscr{C}$ is described by a parameter s:

$$\mathscr{C} = \{t^\mu(s) \mid 0 \leqslant s \leqslant 1,\ t^\mu(0) = x^\mu,\ t^\mu(1) = y^\mu\}.$$

Then it is divided up into a large number N of segments along which infinitesimal parallel displacement can be done according to Eqn. (2.2.3). In the limit $N \to \infty$:

$$U(\mathscr{C}) = \lim_{N\to\infty} \prod_{m=0}^{N-1} \left[1 - \mathrm{i}\mathscr{A}_\mu\left(t\left(\frac{m}{N}\right)\right)\left(t^\mu\left(\frac{m+1}{N}\right) - t^\mu\left(\frac{m}{N}\right)\right)\right]$$
$$= \lim_{N\to\infty} \left[1 - \mathrm{i}\mathscr{A}_\mu\left(t\left(\frac{N-1}{N}\right)\right)\left(y^\mu - t^\mu\left(\frac{N-1}{N}\right)\right)\right]$$
$$\ldots \left[1 - \mathrm{i}\mathscr{A}_\mu(x)\left(t^\mu\left(\frac{1}{N}\right) - x^\mu\right)\right]$$
$$=: P \exp\left(-\mathrm{i}\int_{x^{\mathscr{C}}}^{y} \mathrm{d}t^\mu \mathscr{A}_\mu(t)\right). \tag{2.2.20}$$

With a general non-Abelian potential, the $\mathscr{A}_\mu(t)\,\Delta t^\mu$ of different path segments do not commute. Therefore, the product in Eqn. (2.2.20) can only be formally summed to an exponential function by using the path ordering operator P [Fe 51]. From the first line of Eqn. (2.2.20) one concludes that $U(\mathscr{C})$ is a solution of the differential equation

$$\mathrm{i}\frac{\mathrm{d}}{\mathrm{d}s}\bar{U}(s) = \mathscr{A}_\mu(t(s))\frac{\mathrm{d}t^\mu}{\mathrm{d}s}\bar{U}(s) \quad \text{with } \bar{U}(0) = 1 \tag{2.2.21}$$

namely:

$$U(\mathscr{C}) = \bar{U}(1). \tag{2.2.22}$$

The parallel displacements $U(\mathscr{C})$ can be used for introducing coordinates in the local charge spaces ${}^y\mathscr{H}$ for which $\mathscr{A}_\mu(x)$ has a more simple form. As an example we show that there exists a choice of coordinates for which

$$\mathscr{A}_0(x) \equiv 0 \tag{2.2.23}$$

(temporal gauge) (see e.g. [Co 77]). This applies when coordinate systems are used which emerge through parallel displacement along lines parallel to the time axis. The parallel displacements along these straight lines from $y = (t, \mathbf{x})$ to $x = (0, \mathbf{x})$:

$$U(y) = P \exp\left(-\mathrm{i}\int_t^0 \mathrm{d}t' \mathscr{A}_0(t', \mathbf{x})\right) \tag{2.2.24}$$

relate the components of $\psi(t, \mathbf{x})$ and $\mathscr{A}_\mu(t, \mathbf{x})$ for all t to parallel local coordinate systems. Therefore, the gauge transformation

$$\begin{aligned} {}^y\psi \to {}^y\psi' &= U(y)\,{}^y\psi(x) \\ \mathscr{A}_\mu(y) \to \mathscr{A}'_\mu(y) &= U(y)(\mathscr{A}_\mu - \mathrm{i}\,\partial_\mu)U^{-1}(y) \end{aligned} \tag{2.2.25}$$

describes the transition to the temporal gauge. This geometrical reasoning is confirmed by the following calculation:

$$\mathscr{A}_0' = U(y)(\mathscr{A}_0 - \mathrm{i}\,\partial_0)U^{-1}(y) = U(y)\left(\mathscr{A}_0 + \mathrm{i}U^{-1}\frac{\mathrm{d}U}{\mathrm{d}t}\right)U^{-1}(y)$$
$$= U(y)(\mathscr{A}_0 - \mathscr{A}_0)U^{-1}(y) = 0;$$

this is true, because in analogy to Eqn. (2.2.21) the relation $\mathrm{d}U/\mathrm{d}t = \mathrm{i}U\mathscr{A}_0(t, \mathbf{x})$ holds, according to Eqn. (2.2.24).

(f) The similarities between the geometry of local gauge symmetry and the description of space and time as a Riemannian manifold within the framework of the *theory of general relativity* [We 18, Ei 22, Ad 65, Se 75] are obvious. The following concepts correspond with each other:

(1) the local charge space ${}^x\mathscr{H}$ with unitary structure and the tangent space with local metric $g_{mn}(x)$ and Lorentz coordinate systems;
(2) the gauge transformations and the general basis transformations in tangent space;
(3) the gauge fields $\mathscr{A}_{\mu,cc'}(x)$ which define the symmetry connection and the Christoffel symbols

$$\Gamma^r_{ki}(x) = -\tfrac{1}{2}g^{rl}\left(\frac{\partial g_{kl}}{\partial x^i} + \frac{\partial g_{li}}{\partial x^k} - \frac{\partial g_{ik}}{\partial x^l}\right),$$

which describe the parallel displacement of tangent vectors;
(4) the covariant derivatives;
(5) the field strength $\mathscr{F}_{\mu\nu,cc'}(x)$ and the curvature tensor $R^i_{k;mn}$, etc.

Mathematically, the general structure of fibre bundles is the deep reason for these analogies (for further details see [Ko 63, Dr 77]). Physically, the principles of general relativity and internal symmetry can be combined by taking the Riemannian space rather than the flat Minkowski space as the space–time base space of the charged fields. This extended structure is again a fibre bundle, in which the space–time manifold not only has a tangent space at each point but also a charge space. As far as local colour symmetry is concerned, it might be tempting to regard this extended geometric structure as a 'coloured space–time'.

The most significant physical aspect of general relativity is that the geometric structure of space and time is determined by Einstein's dynamical equation. The corresponding step for local gauge geometry is described in the following section.

2.2.2 Yang–Mills field theories

The extension of local gauge transformations to non-Abelian symmetry groups makes it possible to transfer the principle of minimal gauge-invariant coupling to free fields with non-Abelian charge algebra. These field theories are called Yang–Mills field theories [Ya 54, Ut 56].

(a) For the description of this principle, consider Dirac fields $\psi(x) = (\psi_c(x))$, which form a multiplet with respect to the charges T_a of a symmetry group G. The Dirac fields propagate in empty space according to the free Dirac equation

$$i\gamma^\mu \partial_\mu \psi(x) - M\psi(x) = 0. \tag{2.2.26}$$

If the G symmetry of the charged fields $\psi(x)$ is compatible with the physical principle of local gauge covariance which is mediated by a gauge field $\mathscr{A}_\mu(x)$ according to the geometrical concepts outlined in the previous section, then the ordinary derivative ∂_μ in the Dirac equation must be replaced by the covariant derivative D_μ so that the field equation (2.2.26) according to Eqn. (2.2.9) remains invariant under local symmetry transformations. This gives the Dirac equation

$$i\gamma^\mu D_\mu \psi(x) \equiv \gamma^\mu (i\,\partial_\mu - \mathscr{A}_\mu(x))\psi(x) = M\psi(x), \tag{2.2.27}$$

in which the interaction with the gauge field $\mathscr{A}_\mu$ is introduced in analogy to QED by the substitution rule $\partial_\mu \to \partial_\mu + i\mathscr{A}_\mu(x)$. In a complete physical theory, $\mathscr{A}_\mu(x)$ cannot be an 'external' gauge field, but must itself be determined by the particle field $\psi(x)$. One obtains such a field equation for $\mathscr{A}_\mu(x)$ from a Lagrangian density (see Section 1.2.2(d)) which is made up of an extended Dirac part and a gauge field part $\mathscr{L}_{\text{gauge}}$:

$$\mathscr{L} = -\tfrac{1}{4}F^a_{\mu\nu}(x)F^{a,\mu\nu}(x) + \bar{\psi}(x)(i\gamma^\mu D_\mu - M)\psi(x). \tag{2.2.28}$$

The Dirac equation (2.2.27) is obtained by variation with respect to $\bar{\psi}(x)$. The variation with respect to the gauge field $A^a_\mu(x)$: $\mathscr{A}_\mu(x) = \bar{g}A^a_\mu T_a$ gives the generalized Maxwell equation

$$D^\mu F_{\mu\nu}(x) = \bar{g}\bar{\psi}(x)T\gamma_\nu\psi(x). \tag{2.2.29}$$

In these equations all indices which are associated with multiplets of the group G have been suppressed in order to emphasize the formal similarity to QED. The invariance under local gauge transformations of the Lagrangian density (2.2.28) and the field equations (2.2.27) and (2.2.29) follows directly from Eqns. (2.2.2), (2.2.9), and (2.2.14). The choice of the gauge field part $\mathscr{L}_{\text{gauge}}$ of the Lagrangian density

$$\mathscr{L}_{\text{gauge}} = -\tfrac{1}{4}F^a_{\mu\nu}(x)F^{a,\mu\nu}(x) = -\frac{1}{2\bar{g}^2}\operatorname{Tr}\mathscr{F}_{\mu\nu}(x)\mathscr{F}^{\mu\nu}(x) \tag{2.2.30}$$

(see Eqns. (2.2.11) and (1.2.5)) is motivated apart from gauge invariance, mainly by simplicity and similarity to QED.

The restriction to Dirac fields is not essential. By the same considerations an interaction is introduced for a multiplet of free scalar fields $\phi_c(x)$, which is mediated by the same dynamical gauge field $\mathscr{A}_\mu(x)$. For this, the free Lagrangian density

$$\mathscr{L}_0 = (\partial_\mu\phi)^\dagger(\partial^\mu\phi) - M^2\phi^\dagger\phi \tag{2.2.31}$$

is extended in gauge invariant form and is supplemented by the gauge field part. From the Lagrangian density

$$\mathscr{L} = -\tfrac{1}{4}F^a_{\mu\nu}F^{a,\mu\nu} + (D_\mu\phi)^\dagger(D^\mu\phi) - M^2\phi^\dagger\phi \tag{2.2.32}$$

the field equations

$$\begin{aligned} &D_\mu D^\mu\phi + M^2\phi = 0 \\ &D^\mu F^a_{\mu\nu} = \mathrm{i}\bar{g}(\phi^\dagger T^a D_\nu\phi - (D_\nu\phi)^\dagger T^a\phi) \end{aligned} \tag{2.2.33}$$

follow by variation with respect to $\phi^\dagger(x)$ or $\phi(x)$ and $A^a_\mu(x)$.

(b) A *universal interaction* between all particle fields which bear charges of the symmetry group G is set up in the way described. Further comparison for the description of the universal gravitational interaction within the theory of general relativity shows that:

(1) $\mathscr{L}_{\text{gauge}}$ corresponds to the Lagrangian function for the gravitational field $\mathscr{L}_G = R\sqrt{(\det g_{mn})}$; and
(2) the generalized Maxwell equations are the analogue of the Einstein equations

$$R_{mn} - \tfrac{1}{2}g_{mn}R = -\frac{8\pi\kappa}{c^2}T_{mn}.$$

(Here, $R_{mn} = R^k_{\ k;mn}$ is the contracted curvature tensor, $R = R^n_{\ n}$ the Riemann scalar and T_{mn} the energy–momentum tensor; κ is the gravitational constant.)

(c) A more detailed analysis of the generalized Maxwell equation (2.2.29) and its comparison with electrodynamics is useful. For this purpose it is written in components of the gauge field:

$$\begin{aligned} \partial^\mu G^a_{\mu\nu} &= \bar{g}f^{abc}(2A^b_\mu\,\partial^\mu A^c_\nu - A^b_\nu\,\partial^\mu A^c_\mu - A^{b,\mu}\,\partial_\nu A^c_\mu) \\ &\quad - \bar{g}^2 f^{abc}f^{cde}A^{\mu b}A^d_\mu A^e_\nu + \bar{g}\bar{\psi}T^a\gamma_\nu\psi \\ &\equiv ({}^A J^a_\nu(x) + {}^\psi J^a_\nu(x)) = J^a_\nu(x) \end{aligned} \tag{2.2.34}$$

with the 'linear' field strength tensor

$$G^a_{\mu\nu}(x) = \partial_\mu A^a_\nu(x) - \partial_\nu A^a_\mu(x). \tag{2.2.35}$$

It follows $\partial^\nu\,\partial^\mu G^a_{\mu\nu}(x) = 0$ and therefore the conservation of the currents $J^a_\nu(x)$:

$$\partial^\nu J^a_\nu(x) = 0. \tag{2.2.36}$$

Two aspects are important here: in non-Abelian gauge theories, not only the particle fields but also the gauge fields are charged. Accordingly, there is a self-coupling of the gauge fields $A^a_\mu(x)$ expressed by the non-linear terms in the Maxwell equations (2.2.34), i.e. by the charged gauge field current ${}^A J^a_\nu(x)$. The conserved currents $J^a_\nu(x)$ (Eqn. (2.2.34)), which form the sources of the gauge field, are proportional to the Noether currents which are derived from the symmetry of the Lagrangian function (see Section 1.2.3). According to Eqns. (1.2.57), (1.2.64), etc. this is a direct consequence of the substitution rule $\partial_\mu \to \partial_\mu + \mathrm{i}\mathscr{A}_\mu$ and the fact that the Yang–Mills equation (2.2.34) is derived by a variation of $A^a_\mu(x)$. The proportionality constant $\bar{g}$ which was introduced in Eqn. (2.2.4), represents a coupling constant for the quanta of the particle field in quantized field theory (see Section 1.2.2(c)). $\bar{g}$ determines the strength of the universal interaction of all particles which bear the charge of the local symmetry group G. This is known from QED, where $e = \bar{g} \geqslant 0$, ($T^a = -1$, $f^{abc} = 0$) is the absolute value of the elementary charge. The unfortunate, tradition-bound convention that the electric current generated by moving electrons flows against the particle current has led to different conventions for the introduction of the coupling constant $\bar{g}$. The connection between the classical convention of electrodynamics and the general formulation of Yang–Mills theories can be given by using $T = -1$ as the generator of gauge transformations of the electron field. However, for QCD we put $g = -\bar{g}$ in order to follow the prevalent convention. For a non-simple symmetry group G which is represented infinitesimally by a direct product of simple factors, e.g. $G = G_2 \times G_1$ with generators T_a and S_b for G_2 and G_1 respectively, several coupling constants can appear physically. In general, the Lie-algebra-valued potential of G can be written as

$$\mathscr{A}_\mu(x) = g_2 A^a_\mu(x) T_a + g_1 B^b_\mu(x) S_b \tag{2.2.37}$$

with different coupling constants g_2, g_1. Because $[T_a, S_b] = 0$, the gauge potentials $A^a_\mu(x)$ and $B^b_\mu(x)$ are independent of each other. In the Glashow–Salam–Weinberg theory of the electroweak interaction, $G = SU(2) \times U(1)$ and therefore this case is of importance.

2.2.3 Foundation of quantum chromodynamics

For the foundation of quantum chromodynamics [Fr 73, We 73, Gr 73b], the attempt of a field theoretical formulation of hadron dynamics, several aspects enter into play:

(i) the result of the phenomenological discussion;
(ii) the successful description of dynamical effects in QED; and
(iii) geometric aesthetics of Yang–Mills gauge theories.

Assuming that (ii) and (iii) provide sufficient motivation for trying to formulate hadron dynamics as a Yang–Mills theory, the phenomenological discussion suggests to start with coloured quark fields $\psi(x)$ as particle fields. As colour charges play an important part in the strong interaction of quarks (Section 1.1.2) they should be the source of the gauge fields which mediate the interaction. This means that the colour symmetry group $SU(3)_C$ becomes a gauge group of the second kind. According to the dimension of $SU(3)_C$: 8, there are eight gauge fields $A^a_\mu(x)$, $a = 1, \ldots, 8$, the so-called gluon fields, responsible for the interaction between the quarks. There is experimental evidence for these types of flavour-neutral constituents (see Eqn. (1.4.25)). The corresponding Lagrangian density takes the form:

$$\mathscr{L} = -\tfrac{1}{4} F^a_{\mu\nu}(x) F^{a,\mu\nu}(x) + \bar{\psi}(x)(\mathrm{i}\gamma^\mu D_\mu - M)\psi(x) \tag{2.2.38}$$

with

$\psi_{\alpha,f,c}(x)$ quark fields (see top of p. 27 for definition)

$F^a_{\mu\nu}(x)$ gluon fields; $F^a_{\mu\nu}(x) = \partial_\mu A^a_\nu(x) - \partial_\nu A^a_\mu(x) + gf^{abc}A^b_\mu(x)A^c_\nu(x)$

D_μ covariant derivative; $D_\mu = \partial_\mu - \mathrm{i}g\frac{\nu_a}{2}A^a_\mu(x)$; D_μ induces the minimal gauge invariant interaction of strength $g = -\bar{g}$ with $\bar{g}$ as in Eqn. (2.2.4) ff. (2.2.39)

The definitions of f^{abc}, λ_a etc. are given in Section 1.2.1; colour, flavour, and Dirac indices are summed up accordingly in $\mathscr{L}$. In order to describe the violation of flavour symmetry in QCD M can be regarded as a non-trivial mass matrix in flavour space, which induces the effective quark masses (1.1.12).

The Lagrangian density (2.2.38) allows the formulation of QCD as a quantum field theory. This can be described by the Feynman graphs of the perturbation series. A more accurate method of quantization is relatively complicated in view of the complex structure of local gauge transformations. Section 2.3 covers this more fully. As a follow-up, it can be shown that QCD has three general properties which can be considered as important arguments in its favour:

(1) QCD is renormalizable. This has far-reaching consequences for satisfying general physical principles [tH 72, Le 72].
(2) QCD is asymptotically free. This establishes the quark-parton model with internal radiative corrections which can be checked experimentally [Po 73, Gr 73].
(3) QCD is a promising starting point for the understanding of quark confinement [Ca 73, Wi 74].

These three general properties make QCD an impressive design of hadron dynamics.

Finally, using the Yang–Mills Lagrangian density the Feynman rules of QCD are given below without comment [Ma 78]. $\mathscr{L}$ is written out, separated into bilinear (↔ propagators), trilinear (↔ three vertices) and quadrilinear (↔ four vertices) terms, in order to find the elements of the graph calculus. In addition to this, 'gauge terms' (↔ ghost propagators and ghost vertices) are included, which will be explained later on,

$$\mathscr{L} = -\tfrac{1}{4}(\partial_\mu A^a_\nu - \partial_\nu A^a_\mu)^2 + \bar{\psi}(\mathrm{i}\gamma^\mu\,\partial_\mu - M)\psi - gf^{abc}A^a_\mu A^b_\nu \partial^\mu A^{c,\nu} \tag{2.2.40}$$

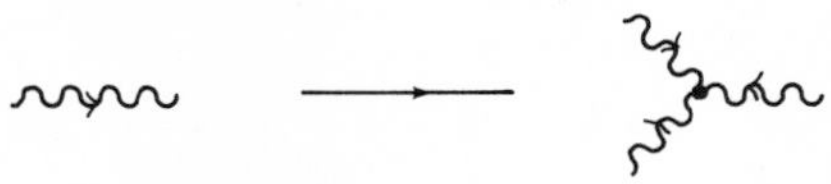

$$-\tfrac{1}{4}g^2 f^{bca}f^{ab'c'}A^b_\mu A^c_\nu A^{b',\mu}A^{c',\nu} + gA^a_\mu\bar{\psi}\gamma^\mu\frac{\lambda_a}{2}\psi + \text{gauge terms}$$

The following substitution rules apply:

Vertices

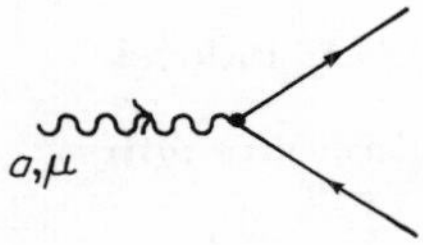

$$= \mathrm{i}g\gamma_\mu \frac{\lambda_a}{2}(2\pi)^4\,\delta(\textstyle\sum k)$$

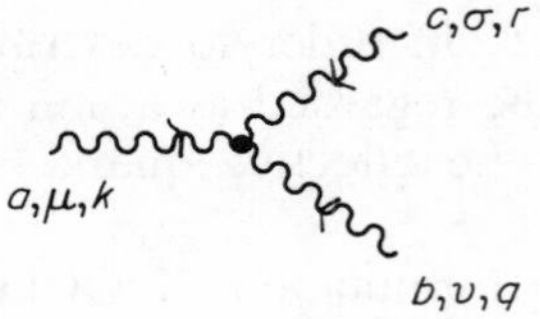

$$= gf^{abc}(g_{\mu\nu}(k-q)_\sigma + \text{cc.})(2\pi)^4\,\delta(\textstyle\sum k) \qquad (2.2.41)$$

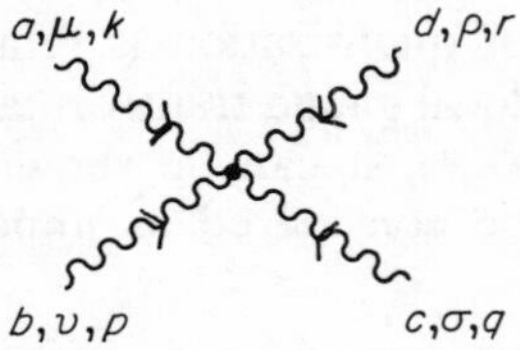

$$= -\mathrm{i}g^2(f^{abe}f^{cde}(g_{\mu\sigma}g_{\nu\rho} - g_{\mu\rho}g_{\nu\sigma}) + \text{symm.})(2\pi)^4\,\delta(\textstyle\sum k)$$

Internal lines

Quark propagator: $= \dfrac{\mathrm{i}}{(2\pi)^4}\dfrac{1}{\not{p} - M + \mathrm{i}\varepsilon}$

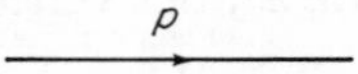

Gluon propagator: 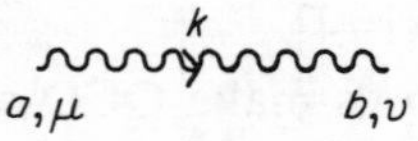 $=$ (2.2.42)

$$\mathrm{i}D(k)^{ab}_{\mu\nu} = \frac{-\mathrm{i}}{(2\pi)^4}\left[\left(g_{\mu\nu} - \frac{k_\mu k_\nu}{k^2}\right) + \xi\frac{k_\mu k_\nu}{k^2}\right]\frac{\delta_{ab}}{k^2 + \mathrm{i}\varepsilon}$$

Note: gluons have zero mass

Gauge terms

Ghost propagator: $= \dfrac{\mathrm{i}}{(2\pi)^4}\dfrac{\delta_{ab}}{k^2 + \mathrm{i}\varepsilon}$

(2.2.43)

Ghost gluon vertex: $= gf^{abc}P^\mu(2\pi)^4\,\delta(\textstyle\sum k)$

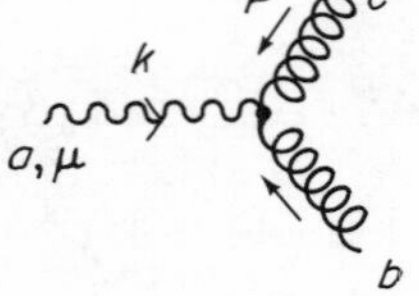

External lines

External quark and gluon lines play no significant rôle in QCD in view of the anticipated quark and gluon confinement. S-matrix elements for hadron scattering must be calculated from the Green functions as described in Section 2.3.1.

The use of the Feynman rules is not very different from that in QED. The self-interaction of gluons gives closed gluon graphs which have to be weighted with certain statistical factors. The non-Abelian nature of gauge transformations makes it necessary to treat gauge invariance in greater detail—see Sections 2.3 and 2.4. Gauge terms in the above formal Feynman rules are indicated by:

(1) the indefinite gauge parameter ξ in the gluon propagator, which cancels out in physical matrix elements;
(2) the propagation function of non-physical gauge degrees of freedom, described by the ghost propagator;
(3) the ghost gluon vertex. In the composition rules ghosts are treated as fermions (factor -1 for closed ghost lines, etc.).

The application of these rules is illustrated in Section 2.4.2.

2.3 The quantum theory of Yang–Mills fields

The result of the discussions so far is that the theory of both the strong and the electromagnetic interaction presumably takes the form of a quantized Yang–Mills theory. Further discussion in Section 3.1 will show that this also applies to the electroweak interaction. Thus, the structure of quantum field theory in general and of the Yang–Mills theory in particular must be more fully investigated [FT/ET].

So far, the quantum aspect of fields has been treated with help of the wave–particle dualism for linear field equations and with the recipe-like use of Feynman rules (Section 1.2.2). Although most applications are discussed on this basis, QCD does have problems for which this procedure is insufficient:

(1) Because quarks and gluons do not appear as free particles, it is necessary to give the general connection between fields and physical particles. This is done by applying general field theory which shows how the expectation values of field operators (Green functions in particular) describe particles and their interaction (see Section 2.3.1) [Ha 58, Zi 58, Ru 62].

(2) Quark confinement is a problem which can be grasped by non-perturbative techniques. Therefore it is important that there exists a closed, formal integral expression which determines the Green function of a field theory with help of the Lagrange function (Section 2.3.2). Perturbative evaluation of this Feynman path integral formula of quantum field theory results in the familiar Feynman rules for calculating S-matric elements. On

the other hand, a different approximation of the path integral formula of QCD will hopefully make the calculation of the quark confinement phenomenon possible (see Section 2.6).

(3) The closed representation of a Yang–Mills theory by the path integral formula is very complex due to local gauge invariance. However, this is the only possibility to describe the full structure of the theory, especially the consequences of gauge invariance. Thus, this formulation is the starting point for a deductive treatment of both QCD and the electroweak interaction.

2.3.1 Green functions and S-matrix elements

This section extends the field theoretical vocabulary which so far has been presumed in its very simplest form and which was summarized briefly in Section 1.2. The aim is to show how in a general field theory the S-matrix for relativistic particle reactions can be defined by quantum mechanical expectation values of field operators. At the same time, some general properties of the S-matrix are also discussed. The presentation is predominantly descriptive; in the literature there are systematic treatments of the scattering theory in the frame of general field theory [Kl 61, St 64, Jo 65].

(a) *The basic concepts of a quantum mechanical description* are states and observables. Within *field theory* the vacuum $|0\rangle$ and the states of one

$$\left|\begin{matrix} M & j \\ p & j_3 \end{matrix}\right\rangle,$$

or more free particles have been described in Section 1.2.2. The latter play an important role in interaction theories for the description of incoming and outgoing particle configurations. Quantum mechanical operators have already been mentioned in the form of quantized basic fields, e.g. as free fields according to Eqns. (1.2.22) and (1.2.23), or rather as important physical observables, e.g. currents derived from them as in Eqn. (1.2.68).

The basic physical principles can be formulated with help of these concepts in a rather general way.

(1) The principle of special relativity for closed systems requires a representation of the Poincaré group by unitary operators $U(\Lambda)$, $U(a)$ which transform the particle states according to Eqns. (1.2.18), (1.2.18′) and leave the vacuum invariant:

$$U(a)\,|0\rangle = |0\rangle, \quad U(\Lambda)\,|0\rangle = |0\rangle. \tag{2.3.1}$$

Accordingly, field operators must transform covariantly:

$$U(a)\psi_\alpha(x)U^{-1}(a) = \psi_\alpha(x+a), \tag{2.3.2}$$

$$U(\Lambda)\psi_\alpha(x)U^{-1}(\Lambda) = S^{-1}_{\alpha\alpha'}(\Lambda)\psi_{\alpha'}(\Lambda x). \tag{2.3.3}$$

The translation operators $U(a)$ and the Lorentz transformation operators

$U(\Lambda)$ are generated infinitesimally by the momentum operator P_μ and the generalized angular momentum operators $M_{\mu\nu}$, respectively.

(2) The energy and mass operator must be positive to ensure stability:

$$P^0 \geqslant 0, \quad P_\mu P^\mu \geqslant 0. \tag{2.3.4}$$

The physical vacuum $|0\rangle$ is the only state with lowest energy:

$$\mathrm{P}^0 \,|0\rangle = 0$$

(3) Causality is implemented by the requirement of locality. Observable fields at space-like distances must be quantum mechanically independent, i.e. they have to commute:

$$[\psi_\alpha(x), \psi_\beta(x')] = 0 \quad \text{for } (x-x')^2 < 0. \tag{2.3.5}$$

The anti-commutativity of Fermi fields at space-like distances implies that the (bilinear) observables formed from them commute.

(4) General charge conservation laws follow from the existence of charge operators Q_a, which generate infinitesimal symmetry transformations of the fields according to Eqns. (1.2.70), (1.2.71) and leave the vacuum invariant:

$$Q_a \,|0\rangle = 0. \tag{2.3.6}$$

The above represents the general framework of quantum field theory.

(b) The quantum mechanical expectation values with respect to the vacuum state of products of field operators, *vacuum expectation values* for short, describe the essential physical content of a quantum field theory. This is shown below. As far as possible the discussion will be restricted to scalar, real fields $A(x)$. The following notations are commonly used for vacuum expectation values

$$\langle 0| \, A(x_1) \ldots A(x_n) \,|0\rangle \equiv \langle A(x_1) \ldots A(x_n)\rangle =: W_n(x_1, \ldots, x_n). \tag{2.3.7}$$

The $W_n(x_1, \ldots, x_n)$ are also known as Wightman functions [Wi 56]. Using relativistic invariance, i.e. Eqns. (2.3.1)–(2.3.3), it follows that the vacuum expectation values are Lorentz-invariant functions of the coordinate differences:

$$\begin{aligned}\langle A(x_1+a) \ldots A(x_n+a)\rangle &= \langle 0| \, U(a)A(x_1)U^{-1}(a) \cdot U(a)A(x_2) \\ &\qquad \ldots A(x_n)U^{-1}(a) \,|0\rangle \\ &= \langle A(x_1) \ldots A(x_n)\rangle,\end{aligned} \tag{2.3.8}$$

$$\begin{aligned}\langle A(\Lambda x_1) \ldots A(\Lambda x_n)\rangle &= \langle 0| \, U(\Lambda)A(x_1) \ldots A(x_n)U^{-1}(\Lambda) \,|0\rangle \\ &= \langle A(x_1) \ldots A(x_n)\rangle.\end{aligned} \tag{2.3.8'}$$

This argumentation is an example of how properties of Wightman functions are derived from general principles. The analyticity properties which follow from Eqns. (2.3.4) and (2.3.5), are discussed in the quoted literature.

Free fields were defined completely by field equations and commutation relations in Section 1.2.2, so that their W_n functions can now be calculated. Beforehand, another type of vacuum expectation values, the Green functions, will be defined because they play the main rôle in our 'practical' considerations. To this end, the time-ordered product of field operators $TA(x)\ldots A(x_n)$ is defined by the convention that factors are arranged in increasing times from right to left; thus, the following applies for the time-ordered product of two factors

$$TA(x)A(y)=TA(y)A(x)=\begin{cases}A(x)A(y) & \text{for } x^0>y^0,\\ A(y)A(x) & \text{for } y^0>x^0.\end{cases} \tag{2.3.9}$$

This definition is relativistically invariant due to the commutativity of field operators at space points with equal times or spacelike distances (Eqn. (2.3.5)). The Green functions are the vacuum expectation values of T-products:

$$\tau(x_1,\ldots,x_n)=\langle TA(x_1)\ldots A(x_n)\rangle. \tag{2.3.10}$$

Just like the Wightman functions these are relativistically invariant functions of the coordinate differences, which are symmetrical in the coordinates because of the symmetry of the T-product. Because of the anti-commutativity of Fermi fields $\psi_\alpha(x)$ for spacelike distances, their time ordering is defined with a sign change

$$T\psi_\alpha(x)\psi_\beta(y)=\begin{cases}\psi_\alpha(x)\psi_\beta(y) & \text{for } x^0>y^0,\\ -\psi_\beta(y)\psi_\alpha(x) & \text{for } y^0>x^0.\end{cases} \tag{2.3.9$'$}$$

The Green functions are again the vacuum expectation values of T-products according to Eqn. (2.3.10). They are antisymmetrical in the arguments (x, α).

(c) The representation of the *free fields* by creation and annihilation operators makes for a simple calculation of the vacuum expectation values. Thus, it follows from (1.2.25) for scalar fields

$$A(x)=(2\pi)^{-3/2}\int\frac{d^3p}{2E_p}(a(p)e^{-ipx}+a^\dagger(p)e^{ipx}), \tag{2.3.11}$$

with help of Eqn. (1.2.20): $a(p)\,|0\rangle=0$, $\langle 0|\,a^\dagger(p)=0$ and Eqn. (1.2.19): $a(p)a^\dagger(p')=a^\dagger(p')a(p)+2E_p\,\delta(\mathbf{p}-\mathbf{p}')$ that

$$\begin{aligned}\langle A(x)A(y)\rangle&=\frac{1}{(2\pi)^3}\iint\frac{d^3p}{2E_p}\frac{d^3p'}{2E_{p'}}\langle a(p)a^\dagger(p')\rangle e^{-ipx}e^{ip'y}\\ &=\frac{1}{(2\pi)^3}\int\frac{d^3p}{2E_p}e^{-ip(x-y)}=\Delta_+(x-y,M).\end{aligned} \tag{2.3.12}$$

From this, using (1.2.38) the two-point-Green function is as follows:

$$\begin{aligned}\langle TA(x)A(y)\rangle&=\theta(x^0-y^0)\langle A(x)A(y)\rangle+\theta(y^0-x^0)\langle A(y)A(x)\rangle\\ &=i\,\Delta_F(x-y,M).\end{aligned} \tag{2.3.13}$$

The Feynman propagator is the 2-point Green function of the free field!

The 4-point Wightman function can be calculated for the scalar free field in the same way:

$$\langle A(x)A(x')A(x'')A(x''')\rangle = \Delta_+(x-x')\,\Delta_+(x''-x''') + \Delta_+(x-x'')\,\Delta_+(x'-x''') + \Delta_+(x-x''')\,\Delta_+(x'-x''). \tag{2.3.14}$$

Note that in the product of four field operators, each particle created by an $a^\dagger(p)$ must be annihilated by an $a(p')$. Thus, only the Wightman functions with an even number of arguments are different from zero. Time ordering similar to Eqn. (2.3.13) gives:

$$\langle TA(x)A(x')A(x'')A(x''')\rangle = i^2\,\Delta_F(x-x')\,\Delta_F(x''-x''') + i^2\,\Delta_F(x-x'')\,\Delta_F(x'-x''') + i^2\,\Delta_F(x-x''')\,\Delta_F(x'-x''). \tag{2.3.15}$$

for the 4-point Green function.

As $\Delta_F(x-y) = \Delta_F(y-x)$ applies, the 4-point Green function is in fact symmetrical in the four space–time points $x^{(i)}$. The general expression for the symmetrical n-point Green function of the free scalar field is as follows:

$$\begin{aligned}\langle TA(x_1)\ldots A(x_n)\rangle &= i^n \sum_{\substack{\text{partitions}\\ \text{for } n \text{ even}}} \Delta_F(x_{i_1}-x_{i_2})\,\Delta_F(x_{i_3}-x_{i_4})\ldots\Delta_F(x_{i_{n-1}}-x_{i_n})\\ &\equiv 0 \quad \text{for } n \text{ odd}.\end{aligned} \tag{2.3.16}$$

The general representation of free quantized fields by creation and annihilation operators according to (1.2.25) leads to the factorization of the Green functions of the fields with arbitrary spin into 2-point functions according to (2.3.16). Accordingly, the Green functions of the Dirac field are sums of products of $\langle T\psi_\alpha(x)\bar{\psi}_\beta(y)\rangle = iS_F(x-y)_{\alpha\beta}$. In view of the importance of spin-1 particles in gauge theories, the 2-point function of massive vector bosons is considered explicitly. In accordance with convention, $u_\mu(p, j_3) = \varepsilon_\mu(p, j_3)$ is adopted for the spin-1 polarization vector of (1.2.25):

$$A_\mu(x) = (2\pi)^{-3/2} \sum_{j_3=0,\pm 1} \int \frac{d^3p}{2E_p}\left(\varepsilon_\mu(p, j_3)a_{j_3}(p)e^{-ipx} + \varepsilon^*_\mu(p, j_3)a^\dagger_{j_3}(p)e^{ipx}\right). \tag{2.3.17}$$

Using the definition of one-particle states (1.2.21), the calculation (2.3.12) above can now be written as:

$$\begin{aligned}\langle A_\mu(x)A_\nu(y)\rangle &= (2\pi)^{-3}\iint \frac{d^3p}{2E_p}\frac{d^3p'}{2E_{p'}} \sum_{j_3j_3'} \left\langle \begin{matrix} M & 1 \\ p & j_3 \end{matrix} \middle| \begin{matrix} M & 1 \\ p' & j_3' \end{matrix} \right\rangle\\ &\quad \times \varepsilon_\mu(p, j_3)\varepsilon^*_\nu(p', j_3')e^{-ip(x-y)}\\ &= (2\pi)^{-3}\int \frac{d^3p}{2E_p}\left(-g_{\mu\nu} + \frac{p_\mu p_\nu}{M^2}\right)e^{-ip(x-y)}\\ &= -\left(g_{\mu\nu} + \frac{1}{M^2}\partial_\mu\,\partial_\nu\right)\Delta_+(x-y, M).\end{aligned} \tag{2.3.18}$$

Here the following orthogonality relation

$$\left\langle \begin{matrix} M & 1 \\ p & j_3 \end{matrix} \middle| \begin{matrix} M & 1 \\ p' & j_3' \end{matrix} \right\rangle = 2E_p\, \delta(\mathbf{p}-\mathbf{p}')\, \delta_{j_3 j_3'} \tag{2.3.19}$$

was used, which follows from the commutation relation (1.2.19). Irrespective of the various representations of the polarization vectors $\varepsilon_\mu(p, j_3)$, the relation

$$\sum_{j_3} \varepsilon_\mu(p, j_3)\varepsilon^*_\nu(p, j_3) = -g_{\mu\nu} + p_\mu p_\nu/M^2 \tag{2.3.19'}$$

holds, because this polarization sum must be Lorentz-covariant and in the centre of gravity system $p=(M, \mathbf{0})$ it has to describe a state with angular momentum 1:

$$\sum \varepsilon_\mu \varepsilon^*_\nu = \delta_{\mu\nu} \quad \text{for } \mu, \nu = 1, 2, 3, \qquad \sum \varepsilon_\mu \varepsilon^*_\nu = 0 \quad \text{for } \mu = 0 \text{ or } \nu = 0.$$

By time ordering in (2.3.18), the Green function

$$\langle TA_\mu(x)A_\nu(y)\rangle = \frac{i}{(2\pi)^4}\int d^4p e^{-ip(x-y)} \frac{-g_{\mu\nu} + p_\mu p_\nu/M^2}{p^2 - M^2 + i\varepsilon} = \int d^4p e^{-ip(x-y)} i\,\tilde{\Delta}_F(p)_{\mu\nu}$$

$$= -i\left(g_{\mu\nu} + \frac{1}{M^2}\partial_\mu\, \partial_\nu\right)\Delta_F(x-y). \tag{2.3.20}$$

is obtained, disregarding the term $M^{-2}\,\delta_{\mu 0}\,\delta_{\nu 0}\,\delta^{(4)}(x-y)$.

One comment on the vector particle propagator: in the Feynman rules of QED (see Eqn. (1.2.40)) the photon propagator is given by $i\,\tilde{\Delta}_F(p)_{\mu\nu} = -ig_{\mu\nu}/(2\pi)^4(p^2 + i\varepsilon)$. According to Eqn. (2.3.18), this corresponds to a Wightman function

$$\langle A_\mu(x)A_\nu(y)\rangle = -(2\pi)^{-3} g_{\mu\nu}\int d^3p\, \exp[-ip(x-y)]/2\,|\mathbf{p}|.$$

Now consider a state created from the vacuum by application of $A_0(x)$:

$$|\psi\rangle = \int d^4x f(x) A_0(x)\,|0\rangle.$$

Then if

$$\tilde{f}(\mathbf{p}) = (2\pi)^{-3/2}\int d^4x\, \exp[i(E_p x^0 - \mathbf{px})]f(x),$$

the norm of this state is

$$\|\psi\|^2 = -\int d^4x \int d^4y \langle A_0(x)A_0(y)\rangle f^*(x)f(y)(2\pi)^{-3}$$

$$= -\int \frac{d^3p}{2\,|\mathbf{p}|}\tilde{f}^*(\mathbf{p})\tilde{f}(\mathbf{p}) < 0, \tag{2.3.21}$$

i.e. a vector propagator of the form (1.2.40) implies the appearance of states with negative norm. This is not the case for a propagator of the form (2.3.20) because of Eqn. (2.3.19′). The formal occurrence of negative norm states is a frequent phenomenon in the relativistic covariant treatment of quantized gauge field theories. Care must be taken with gauge symmetry to prevent these states having any physical significance and violating the unitarity of the S-matrix (see Sections 2.4.3(c), (d)).

(d) The relation between the Feynman propagator and the 2-point Green function of free fields, (Eqn. (2.3.13)), suggests the *extension of the Feynman rules to cover the calculation of Green functions.* To this end, graphs with field points ●x [tH 73] are introduced which correspond to the arguments of the Green functions. The analytical expression

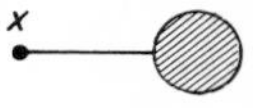

$$\overset{x}{\bullet}\!\!-\!\!\!-\!\!\!-\!\!\bigcirc : \quad \begin{array}{ll} e^{-ipx} & \text{for momentum towards } x \\ e^{ipx} & \text{for momentum from } \quad x \end{array} \tag{2.3.22}$$

is assigned to a field point. Lines which connect field points with other field points or with a vertex of the graph are internal lines. Otherwise, the Feynman rules as already described in connection with QED and QCD apply. Accordingly, the 2-point Green function (2.3.13), can be represented graphically as

$$\langle TA(x)A(y)\rangle = \overset{x}{\bullet}\!\!-\!\!\!-\!\!\!-\!\!\overset{y}{\bullet} = \frac{i}{(2\pi)^4}\int d^4p \frac{e^{-ip(x-y)}}{p^2-M^2+i\varepsilon} = i\,\Delta_F(x-y, M) \tag{2.3.23}$$

(compare Eqns. (1.2.38′), (1.2.40)). The 4-point function (2.3.15) is written as

$$\langle TA(x)A(x')A(x'')A(x''')\rangle = \begin{array}{c} x \;\; x' \\ \bullet\!-\!\bullet \\ \bullet\!-\!\bullet \\ x'' \;\; x''' \end{array} + \begin{array}{c} x \;\; x'' \\ \bullet\!-\!\bullet \\ \bullet\!-\!\bullet \\ x' \;\; x''' \end{array} + \begin{array}{c} x \;\; x''' \\ \bullet\!-\!\bullet \\ \bullet\!-\!\bullet \\ x' \;\; x'' \end{array} \tag{2.3.24}$$

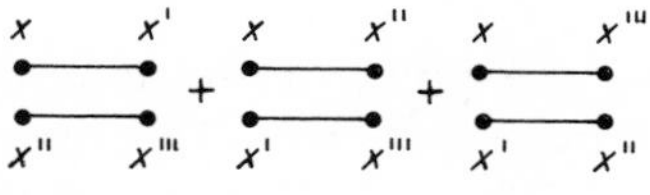

The general Green function of free fields (2.3.16) can be represented accordingly.

It is essential that the extension of the 'Feynman rules for the calculation of S-matrix elements' by field points allows the calculation of Green functions for theories with interaction. A field theory for charged, scalar particles in interaction with neutral scalars is given by way of example. The interaction part of the Lagrange function $\mathcal{L}_I$ corresponds to the vertex:

$$\mathcal{L}_I = -g\psi^*(x)\psi(x)A(x): \quad \begin{array}{l} p_1 \searrow \\ \quad\; \rangle\text{- - - -}\,k \\ p_2 \nearrow \end{array} = -i(2\pi)^4 g\,\delta^{(4)}(p_1+p_2-k)$$

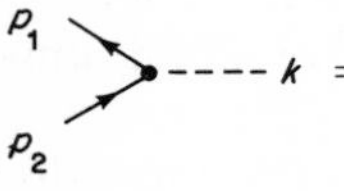

From this, the lowest order 4-point Green function can be calculated:
$\langle 0|\, T\psi^*(x_1)\psi(x_2)\psi^*(x_3)\psi(x_4)\,|0\rangle$

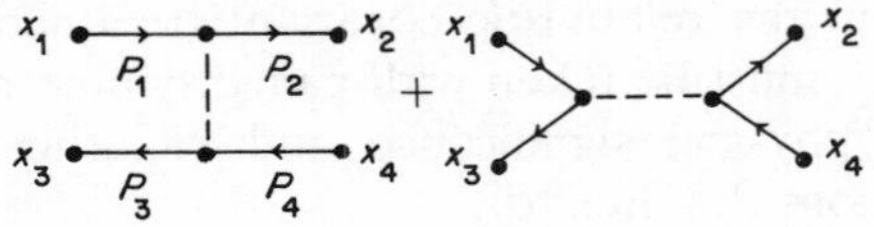

$$=\frac{1}{\mathrm{i}(2\pi)^{12}}\int\dots\int \mathrm{d}^4p_1\dots\mathrm{d}^4p_4\,\delta^{(4)}(p_1+p_3-p_2-p_4)$$

$$\times\exp[-\mathrm{i}(p_1x_1+p_3x_3-p_2x_2-p_4x_4)]$$

$$\times\frac{1}{(p_1^2-m^2)(p_3^2-m^2)}\left(\frac{g^2}{(p_1-p_2)^2-\mu^2}+\frac{g^2}{(p_1+p_3)^2-\mu^2}\right)\frac{1}{(p_2^2-m^2)(p_4^2-m^2)}. \tag{2.3.25}$$

The extended Feynman rules result in Green functions which have all the properties following from the general principles outlined in (a) above. Thus, they represent a meaningful extrapolation of an S-matrix defined by Feynman rules to Green functions. The ultimate justification for these rules is that they form a systematic approximation of a closed representation of field theory (see Section 2.3.2).

Conversely, how do Green functions determine S-matrix elements? The answer to this question can be obtained by comparing the Feynman rules for S-matrix elements and Green functions. It is illustrated by the example above: take the Fourier transform of the corresponding τ-function

$$\tilde{\tau}(p_1,\dots,p_n)=\int \mathrm{d}^4x_1\dots\mathrm{d}^4x_n\,\exp[\mathrm{i}(p_1x_1+\dots+p_nx_n)]\tau(x_1,\dots,x_n)$$
$$=(2\pi)^4\,\delta^{(4)}(p_1+\dots+p_n)G(p_1,\dots,p_n) \tag{2.3.26}$$

and remove the poles of the internal lines which end in field points by multiplying by $-\mathrm{i}(p_i^2-M_i^2)$. This operation is called amputation (see Eqns. (2.3.33), (2.3.39), etc.). If the amputated τ-functions are multiplied by the wavefunctions of the external lines, this gives the S-matrix elements for particles with momenta of the field points p_i, where $p_i^2=M_i^2$. The p_i are said to lie on the mass shell. Conversely, the amputated Green functions represent S-matrix elements which are continued to momenta p_i which do not lie on the mass shell. A more accurate presentation of an extended form of these facts, incorporating the description of bound states, is made in the following considerations.

(e) The representation of Green functions and S-matrix elements by graphs describes some structural properties—*vacuum and one-particle structure*—which are physically of great importance beyond perturbation theory.

The graphs for Green functions consisting of points (field points and

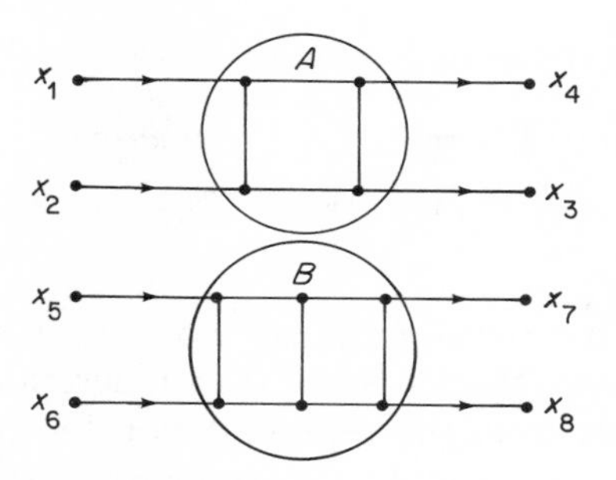

Figure 2.6 Contribution of a disconnected Feynman graph to the 8-point Green function of a field theory with A^3 interaction (see Eqn. (2.3.82))

vertices) and (internal) lines decompose in connected subgraphs. A graph is said to be connected when two points are always connected to each other via internal lines [De 74]. In a free theory, only the graph of the 2-point function (2.3.23) is connected due to lack of interaction vertices. Fig. 2.6 gives another example of a disconnected graph of an 8-point function.

The contributions of disconnected graphs to a Green function factorize according to the Feynman rules. Trivial examples are the Green functions of free fields (Eqns. (2.3.15), (2.3.16) and (2.3.24)). As a simple example consider the 4-point function of a theory with interaction. The graphs which contribute can be classified as follows:

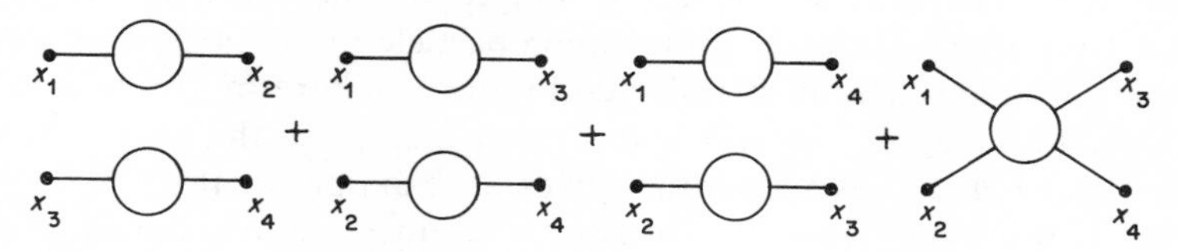

where •—○—• and ⟩○⟨ represent the sum of graphs in which two or four field points are connected, respectively. If $\varphi(x_1, x_2) = \tau(x_1, x_2)$ and $\varphi(x_1, \ldots, x_4)$ are the corresponding analytical expressions, then

$$\begin{aligned}\tau(x_1, x_2, x_3, x_4) = \varphi(x_1, x_2)\varphi(x_3, x_4) + \varphi(x_1, x_3)\varphi(x_2, x_4)\\ + \varphi(x_1, x_4)\varphi(x_2, x_3) + \varphi(x_1, x_2, x_3, x_4).\end{aligned} \quad (2.3.27)$$

$\varphi(x_1, \ldots, x_4)$ is called the connected 4-point Green function. In the general cases, corresponding considerations give the following representation of n-point Green functions by connected parts:

$$\begin{aligned}\tau(x_1, \ldots, x_n) = \varphi(x_1, \ldots, x_n) + \sum \varphi(x_{i_1}, \ldots, x_{i_\nu})\varphi(x_{i_{\nu+1}}, \ldots, x_{i_n})\\ + \sum \varphi(\ldots)\varphi(\ldots)\varphi(\ldots) + \ldots\end{aligned} \quad (2.3.28)$$

The sums indicated must be extended over all partitions of the indices $1, \ldots, n$. Eqn. (2.3.28) can be inverted iteratively to give the general definition of connected Green functions:

$$\varphi(x_1, \ldots, x_n) := \tau(x_1, \ldots, x_n) - 1! \sum \tau(\ldots)\tau(\ldots)$$
$$+ 2! \sum \tau(\ldots)\tau(\ldots)\tau(\ldots) - 3! \sum \tau(\ldots)\tau(\ldots)\tau(\ldots)\tau(\ldots) \pm \ldots \tag{2.3.29}$$

The factorization of the Green functions $\tau(x_1, \ldots, x_n)$ into connected parts, called vacuum structure [Wa 53, Sy 54, Zi 59] describes an important physical point. Quite apart from the representation by Feynman diagrams, the following equations are valid in the framework of general field theory [Jo 65]:

$$\lim_{|a^2| \to \infty} \langle TA(x_1) \ldots A(x_n) A(x_{n+1} + a) \ldots A(x_{n+m} + a) \rangle$$
$$= \langle TA(x_1) \ldots A(x_n) \rangle \langle TA(x_{n+1}) \ldots A(x_{n+m}) \rangle$$

or

$$\lim_{|a^2| \to \infty} \varphi(x_1, \ldots, x_n, x_{n+1} + a, \ldots, x_{n+m} + a) = 0.$$

This means that processes which take place far away from each other are quantum mechanically independent. In this respect, the graph shown in Fig. 2.6 contributes to that part of the 8-point Green function $\tau(x_1, \ldots, x_8)$ which involves the two independent scattering processes A and B. The part of the S-matrix derived from the φ's describes connected processes.

Two processes can be connected simply by the fact that an outgoing particle from one reaction is an incoming particle in the next. The possibility of these multiple reactions is reflected by the structure of graphs and Green functions, too. Fig. 2.7 in which the graph describes the two processes A and B from Fig. 2.6 as sequential, gives an example of this.

Several additional concepts of graph techniques are introduced to account for this situation. A graph is called 1-particle-reducible if it decomposes into two parts by leaving out one internal line. In graphs for Green functions, the field point $\bullet x_i$ is separated from the rest of the graph by one internal line. This type of reducability is treated by 'amputation' [Sy 60, Sy 67, Zi 59]. To this end, the largest subgraph of a connected graph G is considered which is connected to a field point and which can be separated from the other field points by cutting a single internal line (graph C in Fig. 2.8).

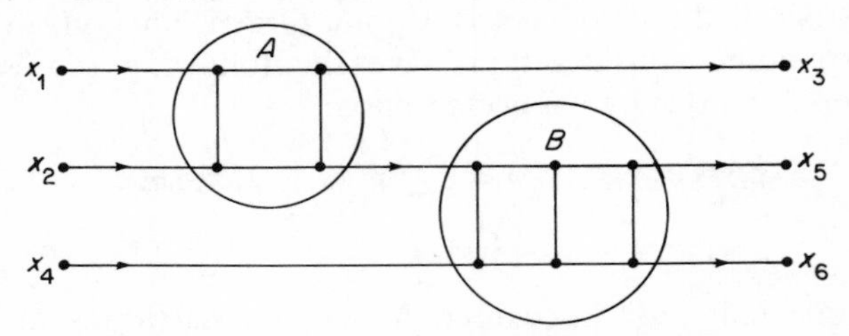

Figure 2.7 Example of a 1-particle-reducible Feynman graph

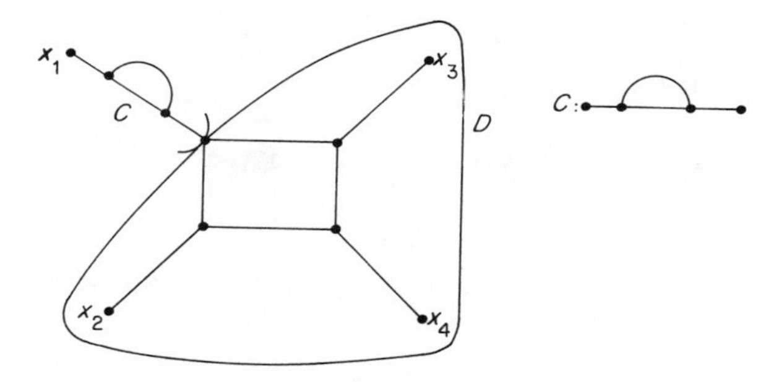

Figure 2.8 Graphical representation of the amputation procedure

This subgraph takes the form of a contribution to the 2-point Green function, i.e. to the particle propagator with internal radiative corrections (see Section 2.1.1(1)). If $\tau_C(x_1, x_2)$ is the analytical expression for this 2-point graph, then the expression for the whole graph G can be written according to the Feynman rules as:

$$\tau_G(x_1, \ldots, x_n) = \int d^4z \tau_C(x_1, z) \underline{\tau}(\underline{z}, x_2, \ldots, x_n). \tag{2.3.31}$$

$\underline{\tau}(\underline{z}, x_2, \ldots, x_n)$ is the analytical expression for the graph G amputated at point z, graph D in Fig. 2.8. This amputation procedure can be carried out at all field points.

If a connected, completely amputated graph is 1-particle-irreducible, it is called a proper vertex graph. The subgraphs A and B in Fig. 2.7 are of this type. 1-particle-reducible, amputated graphs consist of several proper vertex graphs A, B which are connected by propagator graphs C. If $\Gamma_A(x_1, \ldots, x_\nu, z)$, $\Gamma_B(z', x_{\nu+1}, \ldots, x_n)$, $\tau_C(z, z')$ are the analytical expressions for these graphs, then according to the Feynman rules, the analytical expression for the connected graph G is

$$\Gamma_G(x_1, \ldots, x_n) = \int d^4z \Gamma_A(x_1, \ldots, x_\nu, z) \tau_C(z, z') \Gamma_B(z', x_{\nu+1}, \ldots, x_n) \, d^4z'. \tag{2.3.32}$$

This structure of Feynman graphs can be transferred to Green functions [Zi 59, Sy 60]:

(1) A Green function is amputated in a field point (shown by underlining the space coordinate) by convolution with the inverse 2-point Green function:

$$\begin{aligned} \tau(\underline{x}_1, x_2, \ldots, x_n) &= \int d^4z \tau^{-1}(x_1, z) \tau(z, x_2, \ldots, x_n), \\ \int d^4z \tau(x_1, z) \tau^{-1}(z, x_2) &= \delta^{(4)}(x_1 - x_2), \end{aligned} \tag{2.3.33}$$

i.e. the Fourier transform of $\tau(z, \ldots, x_n)$ with respect to z is divided by the Fourier transform $\mathrm{i}\,\Delta'(p)$ of $\tau(x_1, x_2)$:

$$\tilde{\tau}(p_1, p_2) = (2\pi)^4\,\delta^{(4)}(p_1+p_2)\mathrm{i}\,\Delta'(p_1) = \iint \mathrm{d}^4x_1\,\mathrm{d}^4x_2\tau(x_1, x_2)\exp[\mathrm{i}(p_1x_1+p_2x_2)] \tag{2.3.26'}$$

(2) The sum of all contributions of proper vertex graphs gives the n-point vertex function symbolized by:

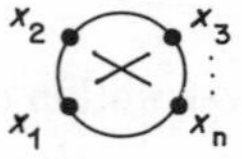

x_2 x_3 x_1 x_n

(3) The (connected) Green functions are made up of vertex functions and propagators which can be illustrated by means of 'structural graphs'. A recursive procedure can be used. The 3-point function can be obtained by reversing the amputation:

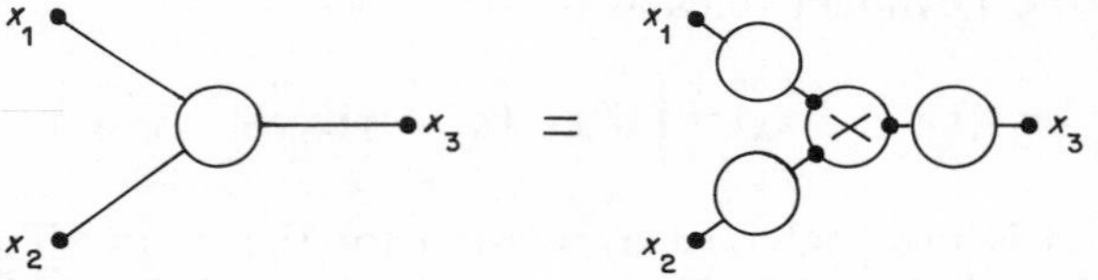

x_1 x_3 x_2 = x_1 x_3 x_2 (2.3.34)

$$\varphi(x_1, x_2, x_3) = \iiint \mathrm{d}z_1\,\mathrm{d}z_2\,\mathrm{d}z_3\tau(x_1, z_1)\tau(x_2, z_2)\tau(x_3, z_3)\Gamma(z_1, z_2, z_3).$$

With help of the combinatorics of Feynman graphs the composition of Green functions with n points from those with $(n-1)$ points can be obtained in the following way: the additional field point x_n is connected via a propagator line to either an m-vertex $(n-1 \geqslant m \geqslant 3)$ or to a propagator line (via a 3-vertex) of a structural graph of the $(n-1)$-vertex function. This has to be carried out in all possible ways on the different $(n-1)$ structural graphs. The following pattern emerges for the connected 4-point function:

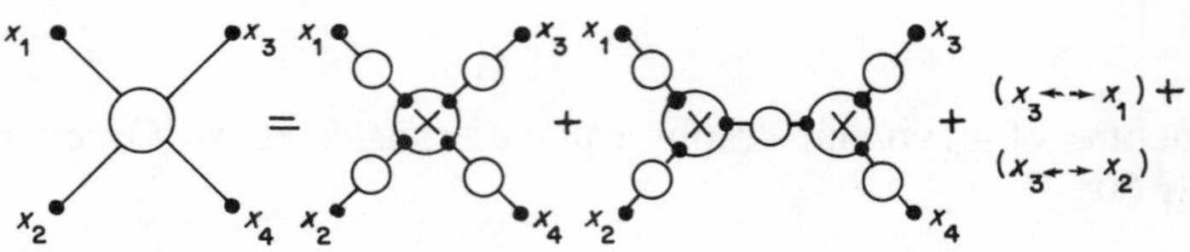

x_1 x_3 x_2 x_4 = x_1 x_3 x_2 x_4 + x_1 x_3 x_2 x_4 + $(x_3 \leftrightarrow x_1)$ + $(x_3 \leftrightarrow x_2)$

This recursive construction can be inverted to a recursive definition of the vertex functions $\Gamma(x_1, \ldots, x_n)$ by the connected Green functions $\varphi(x_1, \ldots, x_n)$ (see Section 2.3.2(c)).

The representation of Green functions from general propagators and vertex functions describes a general collision process as made up of individual processes connected over a long range by propagating particles. The individual processes are described by the vertex functions and this is the reason for their great physical importance. Important physical parameters like effective coupling constants can be expressed by values of the vertex functions as described later on. Because of energy–momentum conservation, the particle propagation between individual processes can only be virtual. In any case, the mass pole of the Fourier transform of the 2-point function determines the long-range component of the particle propagation. However, in general, radiative corrections cause a shift of the pole in comparison with that of the free propagator (1.2.38), which means that the mass must be renormalized.

(f) Poles in the momentum coordinates of the Green functions point to the existence of free particles and the corresponding 1-particle structure of their reactions. This was demonstrated with help of a simple example of a field theory in which the field quanta describe the free particles directly.

However, even *particles which are bound states of field quanta* and their reactions can be described by Green functions. In order to formulate this, the field-theoretical wavefunctions for one-particle states composed of field quanta, the so-called Bethe–Salpeter (BS) amplitudes, must be introduced. For a $\psi^*\psi$ bound state they are defined by

$$(2\pi)^{-3/2}\chi_k(p_1, p_2)\,\delta^{(4)}(p_1+p_2-k)$$
$$=\frac{1}{(2\pi)^4}\int \mathrm{d}^4x_1\,\mathrm{d}^4x_2\langle 0|\,T\psi(x_1)\psi^*(x_2)\left|\begin{matrix}M & j\\ k & j_3\end{matrix}\right\rangle \exp[\mathrm{i}(p_1x_1+p_2x_2)]. \quad (2.3.36)$$

Here,

$$\left|\begin{matrix}M & j\\ k & j_3\end{matrix}\right\rangle$$

according to Eqn. (1.2.18) denotes a one-particle state, i.e. an eigenvector of a discrete eigenvalue of the mass operator:

$$P_\mu P^\mu\left|\begin{matrix}M & j\\ k & j_3\end{matrix}\right\rangle = M^2\left|\begin{matrix}M & j\\ k & j_3\end{matrix}\right\rangle,\quad M^2=k^2. \quad (2.3.37)$$

The BS amplitude χ has a certain relationship to the Schrödinger wavefunction of a bound state in non-relativistic quantum mechanics. Like the Schrödinger equation, a field theoretical BS equation determines both mass and wavefunction of the bound state. Therefore, the BS equation can be used to provide a relativistic generalization of the phenomenological quark model for mesons and baryons according to Section 1.1.3 [Bö 73]. As this plays no role in QCD up to now, any further details on the BS equation should be gleaned from the literature [Be 51, Ge 51, Sy 60].

The general relation between S-matrix and Green functions can now be given by two theorems of general field theory:

(1) The Fourier transformed τ-functions $\tilde{\tau}(p_1, \ldots, p_n)$ have poles when a partial sum of the momenta $k=p_{i_1}+\ldots+p_{i_\nu}$ lies on the mass shell of a

particle (e.g. $p_i^2 = m^2$, $(p_1 - p_2)^2 = \mu^2$ in Eqn. (2.3.25)). Thus, in a field theory the mass spectrum is found from the position of the poles in the Green functions.

(2) If the momenta of a $\tilde{\tau}$-function are divided up into partial sums k_l:

$$k_l = p_{i_1} + \ldots + p_{i_{\nu_l}}, \quad \sum_{l=1}^{r} k_l = 0,$$

and the k_l lie on mass shells, then the residue of the $\tilde{\tau}$-function factorizes into S-matrix elements and BS-amplitudes:

$$\begin{aligned}(-\mathrm{i})^r (k_1^2 - m_1^2) \ldots (k_r^2 - m_r^2)\tilde{\tau}(p_1, \ldots, p_n) \\ = (2\pi)^{3r/2} \chi_{k_1}(p_{i_1}) \ldots \chi_{k_s}(p_{i_s}) \langle k_1, \ldots, k_s | \, S \, | k_{s+1}, \ldots, k_r \rangle \\ \times \bar{\chi}_{k_{s+1}}(p_{i_{s+1}}) \ldots \bar{\chi}_{k_r}(p_{i_r}) \end{aligned} \quad (2.3.38)$$

for

$$k_l^2 = m_l^2, \quad (k^0)_l \gtrless 0 \quad \text{for } l = \begin{cases} 1, \ldots, s, \text{ outgoing} \\ s+1, \ldots, r, \text{ incoming} \end{cases} \text{configuration.}$$

This general 'LSZ formula' [Le 55, Zi 59] of field theoretical scattering theory defines generally the relation between Green functions and S-matrix which was demonstrated by the example of Eqn. (2.3.25). Notice that the BS amplitude for one field is:

$$(2\pi)^{-3/2} \chi_k(p) \, \delta^{(4)}(p-k) = \frac{1}{(2\pi)^4} \int \mathrm{d}^4x \langle 0 | \, \psi(x) \, | k \rangle \mathrm{e}^{\mathrm{i}px}. \quad (2.3.39)$$

From relativistic covariance it follows that $\chi(p)$ is a constant (normalized to one).

On the left-hand side of Eqn. (2.3.38), $\tilde{\tau}(p_1, \ldots, p_n)$ is amputated by the inverse free propagator $(p^2 - m^2) = (2\pi)^{-4} \Delta_F^{-1}(p)$. According to theorem (1), the full propagator $\Delta'(p)$ (2.3.26′) also has a pole, so $\Delta'^{-1}(p_1)\tilde{\tau}(p_1, \ldots, p_n) = (p_1^2 - m^2)\tilde{\tau}(p_1, \ldots, p_n)$, applies on the mass shell, and amputation according to the general definition (2.3.33), leads to the same result.

The fundamental significance of the general formula (2.3.28) for the quark model of hadrons is illustrated by the following example of $\pi + \pi \to \pi + \pi$ scattering: as pions are described in the quark model as $q\bar{q}$ bound states, the pion BS amplitudes in configuration space take the form:

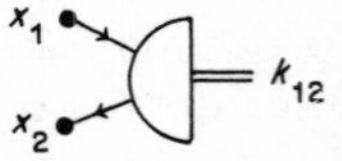

x_1, x_2, k_{12} $= \langle 0 | \, T\psi(x_1)\bar{\psi}(x_2) \, | k_{12} \rangle$.

The S-matrix element is contained in the pole structure of the 8-point function, e.g. according to (2.3.38) as a factor in the residue of the simultaneous poles $[(k_{12}^2 - m_\pi^2)(k_{34}^2 - m_\pi^2)(k_{56}^2 - m_\pi^2)(k_{78}^2 - m_\pi^2)]^{-1}$, $k_{12} = p_1 + p_2, \ldots$, etc. This can be represented symbolically by the following graph:

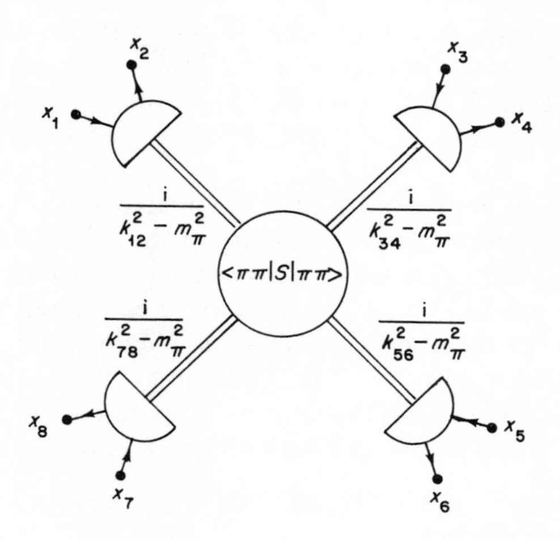

(g) An important property of the S-matrix is its *unitarity*:

$$S \cdot S^\dagger = 1 \tag{2.3.40}$$

which states that a normalized initial state is transformed into a normalized final state during a collision process. This conservation of probability is essential for the consistency of the quantum-mechanical description. The conditions under which the S-matrix defined by Feynman rules is unitary are described below. This investigation is important in gauge theories [tH 79] because it shows how uncontrolled contributions from negative-norm states in the propagator (see Eqn. (2.3.21)) can violate unitarity.

First of all, the Feynman rules for S-matrix elements and Green functions are rewritten in configuration space. To this end, space–time coordinates are ascribed to all points, not just field points but vertices as well, e.g.:

$$\begin{aligned} &= G(x_1, x_2, x_3, x_4, x_5) \\ &= (-\mathrm{i}g)\, \Delta_\mathrm{F}(x_1 - x_5)\, \Delta_\mathrm{F}(x_2 - x_5)\, \Delta_\mathrm{F}(x_3 - x_5)\, \Delta_\mathrm{F}(x_4 - x_5) \end{aligned} \tag{2.3.41}$$

The following correspondences then apply between graph elements and analytical expressions (using a real, scalar field theory with $\mathscr{L}_\mathrm{I} = -(g/4!)A^4(x)$ as an example):

field point:	factor 1	
vector:	factor $-\mathrm{i}g$	(2.3.42)
line from x to y:	factor $\mathrm{i}\,\Delta_\mathrm{F}(x-y)$.	

The composition rules correspond to those in Section 1.2.2f(4), with the exception that one has to integrate not over the momenta of the internal lines but over the space coordinates of the vertices. Using the Fourier representation of the δ-function $(2\pi)^4\,\delta^{(4)}(p)=\int d^4x\,\exp(ipx)$ and of the propagator (1.2.38), it is easy to see that these rules for Green functions give a result which corresponds to that given in Section 1.2.2(f) and Eqn. (2.3.22).

In order to check the unitarity of the S-matrix with help of graph rules one has to enlarge and refine the structure of the graphs [tH 73] such that it is possible to represent complex conjugate S-matrix elements. For this, part of the vertices is marked by circles. The space coordinates of these vertices are overlined in the corresponding analytical expression $F(x_1, \ldots, x_n)$ before vertex integration (see Fig. 2.9). Only graphs consisting of vertices and internal lines, i.e. those contributing to amputated Green functions, are considered. The graph rules are then extended as follows

vertex: factor $-\mathrm{i}g$ (not marked), $\mathrm{i}g$ (overlined)
internal line:

$$\begin{aligned}
\Delta_{ij} &\equiv \mathrm{i}\,\Delta_F(x_i - x_j) && \text{for } x_i,\ x_j,\\
\Delta_{ij}^{+} &&& \text{for } \overline{x_i},\ x_j,\\
\Delta_{ij}^{-} &&& \text{for } x_i,\ \overline{x_j},\\
\Delta_{ij}^{*} &&& \text{for } \overline{x_i},\ \overline{x_j}.
\end{aligned} \tag{2.3.43}$$

The invariant functions $\Delta_F(x_i - x_j)$ and $\Delta_{\pm}(x_i - x_j) \equiv \Delta_{ij}^{\pm}$ were defined in Eqn. (1.2.38). With help of the relations

$$\begin{aligned}
\Delta_{ij}^{\pm} &= (\Delta_{ij}^{\mp})^* = \Delta_{ji}^{\mp},\\
\Delta_{ij} &= \theta(x_i^0 - x_j^0)\,\Delta_{ij}^{+} + \theta(x_j^0 - x_i^0)\,\Delta_{ij}^{-}\\
\Delta_{ij}^{*} &= \theta(x_i^0 - x_j^0)\,\Delta_{ij}^{-} + \theta(x_j^0 - x_i^0)\,\Delta_{ij}^{+}
\end{aligned} \tag{2.3.44}$$

the following theorems hold:

(1) If x_1 is the x coordinate with the greatest time component $x_1^0 > x_i^0$ for all i, then

$$F(x_1, \ldots) + F(\overline{x_1}, \ldots) = 0. \tag{2.3.45}$$

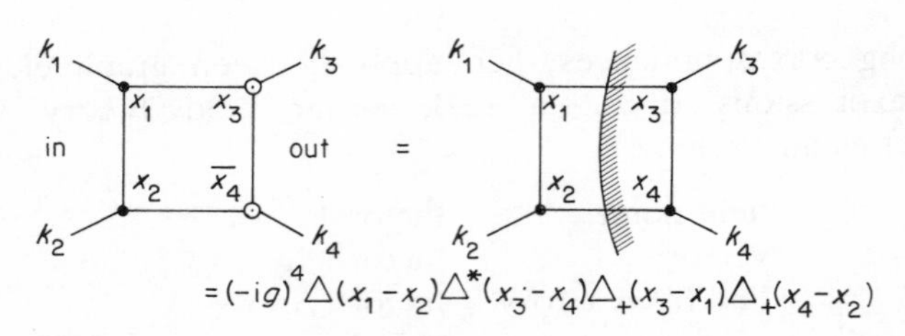

Figure 2.9 Example illustrating Cutkosky's cutting rule (see Eqn. (2.3.47))

Here, all other coordinates are overlined or not in the same way for both terms.

Proof. Because of $\theta(x_1^0 - x_j^0) = 1$ for all j, it follows from (2.3.44):

$$\Delta_{1j} = \Delta_{1j}^{+} \quad \text{i.e.}$$

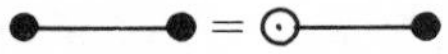

$$\Delta_{\overline{1}j}^{-} = \Delta_{1j}^{*} \quad \text{i.e.}$$

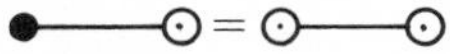

Therefore, the line contributions are equal for both terms. If the vertex x_1 is overlined, this causes a sign change.

(2) The following applies for all values of the coordinates x_i:

$$\sum_{\substack{\text{all possible} \\ \text{overlinings}}} F(x_1, \ldots, x_n) = 0 \tag{2.3.46}$$

Proof. There is always an x with the greatest time component. Thus, the diagrams can be arranged such that they cancel pairwise according to (2.3.45).

The Fourier transforms of the $F(x_1, \overline{x_2}, \ldots)$ are now examined. If part of the corresponding momenta is fixed as that of incoming ($p^0 < 0$) or outgoing ($p^0 > 0$) particles and the rest is integrated over, then contributions to S-matrix elements are obtained. Many sum terms in Eqn. (2.3.46) disappear because of the $\theta(\pm k^0)$ functions in $\tilde{\Delta}_{\pm}(k)$, (Eqn. (1.2.38)), and the momentum conservation in the vertices. For example, the positive energy always flows into the encircled points. If such a point is connected only to non-marked points or incoming lines, a contradiction follows from energy conservation. The graph makes no contribution. The extension of this argument leads to the conclusion that the only contributing diagrams are those in which the encircled points form a connected region with at least one outgoing line. Correspondingly, the non-marked points must form a connected region with at least one incoming line. Both regions can be characterized by a cut which makes the marking of points superfluous (see Fig. 2.9). If the contribution of the fully marked $\mathring{F}(k_1, \ldots, k_n)$ and the fully non-marked graph is removed, then Eqn. (2.3.46) can be written as:

$$\tilde{F}(k_1, \ldots, k_n) + \mathring{\tilde{F}}(k_1, \ldots, k_n) = -\sum_{\text{cuts}} \tilde{F}_{\text{cuts}}(k_1, \ldots, k_n). \tag{2.3.47}$$

This is Cutkosky's cutting rule [Cu 60, Ve 63].

Under certain conditions for the coupling constant g and the propagator, Eqn. (2.3.47) implies the unitarity of the S-matrix. In order to explain this, the first step is to write for the T-matrix

$$S = 1 + \mathrm{i}T, \quad \langle k_1, \ldots, k_n | \, T \, | k_1', \ldots, k_m' \rangle = \mathcal{M} \, \delta(\textstyle\sum k - k') \tag{2.3.48}$$

(see Section 1.2.2(g)), the unitarity condition (2.3.40) explicitly:

$$\frac{1}{i}[\langle k_1,\ldots,k_n|\,T\,|k'_1,\ldots,k'_m\rangle-\langle k_1,\ldots,k_n|\,T\,|k'_1,\ldots,k'_m\rangle^*]$$

$$=\sum_l\int\ldots\int\frac{d^3q_1}{2E_{q_1}}\cdots\frac{d^3q_l}{2E_{q_l}}\langle k_1,\ldots,k_n|\,T\,|q_1,\ldots,q_l\rangle \times\langle q_1,\ldots,q_l|\,T^\dagger\,|k'_1,\ldots,k'_m\rangle. \tag{2.3.49}$$

Eqn. (2.3.47) has the same structure for individual graphs as the unitarity equation under the following conditions:

(1) the coupling constant g must be real; then, the graph with marked vertices describes the complex conjugate matrix element: $\langle k_1,\ldots,k_n|\,iT\,|q_1,\ldots,q_l\rangle^*=\langle q_1,\ldots,q_l|\,(iT)^\dagger\,|k_1,\ldots,k_n\rangle$;

(2) the Δ_+-functions are determined according to Eqn. (2.3.18) by integration over the physical one-particle intermediate states. The integration over the physical intermediate states on the right-hand side of the unitary equation (2.3.49) then agrees with the integration over the internal lines of the graphs in Eqn. (2.3.47), which connect the marked and the non-marked vertices across the cut (see Eqn. (2.3.43)).

If the physical restrictions on the momenta are imposed, the summation of all graphs then leads from the cutting rule to the unitarity equation [Ve 63]. The cutting rule generalizes the unitary relation to non-physical energy–momentum values.

2.3.2 The functional integral representation of quantum field theory

(a) The set of Green functions describes the physical content of a field theory. In the same way as in combinatorics, where number sequences are combined to generating functions [Er 55], it is useful in field theory to represent the set of symmetrical Green functions $\{\tau_n(x_1,\ldots,x_n)\}$ as coefficients of a Volterra series of a functional $T\{j\}$:

$$T\{j\}=\sum_{n=0}^{\infty}\frac{i^n}{n!}\int\ldots\int d^4x_1\ldots d^4x_n\tau_n(x_1,\ldots,x_n)j(x_1)\ldots j(x_n). \tag{2.3.50}$$

For example, if the expressions (2.3.16) are substituted for $\tau_n(x_1,\ldots,x_n)$, then the closed expression

$$T\{j\}=\exp\Bigl(-\frac{i}{2}\int d^4x\int d^4y\,j(x)\,\Delta_F(x-y)j(y)\Bigr) \tag{2.3.51}$$

is obtained for the *generating functional of the Green functions* of the free scalar field. Several rules for the manipulation of functionals are summarized below before this technique is applied to field theory.

(b) In order to *calculate with functionals* [Be 66, Fr 72] concepts from the analysis of functions of several variables such as power series expansion, partial differentiation, and integration must be extended to functionals, i.e. to functions of an infinite number of variables. A systematic treatment of these methods is way beyond the scope of this book. However, we hope that our examples will demonstrate the formal parallels with analysis and thus give a first pragmatic orientation.

The power series expansion of a function $F(y_1, \ldots, y_k)$ of k variables y_i

$$F(\mathbf{y}) = F(y_1, \ldots, y_k) = \sum_{n=0}^{\infty} \sum_{i_1=1}^{k} \cdots \sum_{i_n=1}^{k} \frac{1}{n!} T_n(i_1, \ldots, i_n) y_{i_1} \ldots y_{i_n}, \tag{2.3.52}$$

$T_n(i_1, \ldots, i_n)$ symmetrical, yields the representation of a functional by means of its Volterra series

$$F\{y\} = \sum_{n=0}^{\infty} \int dx_1 \ldots \int dx_n \frac{1}{n!} T_n(x_1, \ldots, x_n) y(x_1) \ldots y(x_n), \tag{2.3.53}$$

$T_n(x_1, \ldots, x_n)$ symmetrical, after the formal transition to an infinite (continuous) number of variables: $i = 1, \ldots, k \to -\infty < x < \infty$; $y_i \to y(x)$; $\sum_i \to \int dx$. The partial differential operator $\partial/\partial y_i$ applied to polynomials or power series, is defined as a linear operator, for which the following equations apply:

$$\begin{gathered} \frac{\partial}{\partial y_k} 1 = 0, \quad \frac{\partial}{\partial y_k} y_i = \delta_{ik}, \\ \frac{\partial}{\partial y_k} (F(\mathbf{y})G(\mathbf{y})) = \left(\frac{\partial}{\partial y_k} F(\mathbf{y})\right) G(\mathbf{y}) + F(\mathbf{y}) \left(\frac{\partial}{\partial y_k} G(\mathbf{y})\right) \quad \text{(product rule)} \end{gathered} \tag{2.3.54}$$

Accordingly, the functional differentiation $\delta/\delta y(x)$ is a linear operation with product rule:

$$\begin{gathered} \frac{\delta}{\delta y(x)} 1 = 0, \quad \frac{\delta}{\delta y(x)} y(x') = \delta(x - x'), \\ \frac{\delta}{\delta y(x)} (F\{y\}G\{y\}) = \left(\frac{\delta}{\delta y(x)} F\{y\}\right) G\{y\} + F\{y\} \left(\frac{\delta}{\delta y(x)} G\{y\}\right). \end{gathered} \tag{2.3.55}$$

All formal differentiation rules are then valid. Thus, for example, the coefficients of the Volterra series (2.3.53) can be obtained as derivatives at the origin:

$$T_n(x_1, \ldots, x_n) = \frac{\delta}{\delta y(x_1)} \cdots \frac{\delta}{\delta y(x_n)} F\{y\}\big|_{y \equiv 0}. \tag{2.3.56}$$

The methods for functional integration as used in the following text are based on the Gaussian integral

$$\pi^{-1/2} \int_{-\infty}^{+\infty} dy \exp(-ay^2) = 1/\sqrt{a}.$$

If

$$(y, Ay) = \sum_{i,l=1}^{k} y_i A_{il} y_l$$

is a positive definite, quadratic form in k variables, then

$$\int_{-\infty}^{+\infty} \cdots \int_{-\infty}^{+\infty} \frac{d^k y}{\pi^{k/2}} e^{-(y,Ay)} = (\det A)^{-1/2}. \tag{2.3.57}$$

This is proved by diagonalization with help of an orthogonal transformation. This formula can be generalized for the function space $\{y(x)\}$:

$$\begin{aligned} \int \mathscr{D}[y] e^{-(y,Ay)} &= (\det A)^{-1/2}, \\ (y, Ay) &= \iint dx\, dx' y(x) A(x, x') y(x'). \end{aligned} \tag{2.3.58}$$

The determinant $\det A$ is defined as the (infinite) product of the eigenvalues of the integral operator $A(x, x')$. The formula

$$\det A = \exp[\mathrm{Tr}(\log A)] \tag{2.3.59}$$

which is valid for finite k allows often an approximate calculation of $\det A$.

Just like in the case of usual integrals, expressions of the type $\int \mathscr{D}[y] F\{y\} \exp(-(y, Ay))$, $F\{y\}$ = power series, can be evaluated by algebraic manipulations of the Gaussian integral. This is extensively used in Section (c). Thereby, the formal (infinite-dimensional) differential $\mathscr{D}[y]$ is regarded as translation invariant:

$$\int \mathscr{D}[y] F\{y - y'\} \exp[-(y - y', A(y - y'))] = \int \mathscr{D}[y] F\{y\} e^{-(y,Ay)}. \tag{2.3.60}$$

In general, the substitution rule is used with an infinite-dimensional Jacobi determinant:

$$\int \mathscr{D}[y] \det \frac{\delta G\{y\}}{\delta y(x)} F\{G\{y\}\} e^{-(G\{y\},AG\{y\})} = \int \mathscr{D}[y] F\{y\} e^{-(y,Ay)}. \tag{2.3.61}$$

For a linear transformation, the Jacobi determinant $\det \delta G\{y\}/\delta y(x)$ is the determinant of a linear integral operator as already described above:

$$\begin{aligned} z = G\{y\}: z(x) &= \int dx' G(x, x') y(x'), \\ \det \frac{\delta G\{y\}}{\delta y(x')} &= \det(G(x, x')). \end{aligned} \tag{2.3.62}$$

A systematic foundation for this calculus which defines the functional integrals as limits of finite-dimensional integrals or which is even based on a complete measure theory, is not available at the moment [Sk 74, Fr 53, Wi 58]. Many difficulties of field theory are connected with these open mathematical questions.

(c) As an example for the use of functionals in field theory the *generating functionals* [Sc 51, Fr 72] of *Green functions:*

$$T\{j\} = \sum_{n=0}^{\infty} \frac{i^n}{n!} \int \dots \int \mathrm{d}x_1 \dots \mathrm{d}x_n \tau_n(x_1, \dots, x_n) j(x_1) \dots j(x_n), \quad (2.3.63)$$

connected Green functions

$$\Phi\{j\} = \sum_{n=0}^{\infty} \frac{i^n}{n!} \int \dots \int \mathrm{d}x_1 \dots \mathrm{d}x_n \varphi_n(x_1, \dots, x_n) j(x_1) \dots j(x_n), \quad (2.3.64)$$

vertex functions

$$\Gamma\{a\} = \sum_{n=0}^{\infty} \frac{1}{n!} \int \dots \int \mathrm{d}x_1 \dots \mathrm{d}x_n \Gamma_n(x_1, \dots, x_n) a(x_1) \dots a(x_n) \quad (2.3.65)$$

and their interrelation are discussed. Comparison of coefficients in $j(x)$ shows that the equation (2.3.27) can be written in functional form:

$$T\{j\} = \mathrm{e}^{\Phi\{j\}} \quad (2.3.66)$$

According to Eqns. (2.3.55), (2.3.56) it follows that:

$$\begin{aligned}
\tau_4(x_1, x_2, x_3, x_4) &= \frac{\delta^4}{\delta j(x_1) \dots \delta j(x_4)} \mathrm{e}^{\Phi\{j\}}\Big|_{j(x)\equiv 0} \\
&= \Bigg(\frac{\delta^4 \Phi}{\delta j(x_1) \dots \delta j(x_4)} + \frac{\delta \Phi}{\delta j(x_1)} \frac{\delta^3 \Phi}{\delta j(x_2)\, \delta j(x_3)\, \delta j(x_4)} + \text{symm.} \\
&\quad + \frac{\delta^2 \Phi}{\delta j(x_1)\, \delta j(x_2)} \frac{\delta^2 \Phi}{\delta j(x_3)\, \delta j(x_4)} + \text{symm.} + \frac{\delta \Phi}{\delta j(x_1)} \frac{\delta \Phi}{\delta j(x_2)} \frac{\delta \Phi}{\delta j(x_3)} \frac{\delta \Phi}{\delta j(x_4)}\Bigg) \mathrm{e}^{\Phi}\Big|_{j=0} \\
&= \varphi_4(x_1, x_2, x_3, x_4) + \varphi_1(x_1)\varphi_3(x_2, x_3, x_4) + \text{symm.} \\
&\quad + \varphi_2(x_1, x_2)\varphi_2(x_3, x_4) + \text{symm.} \\
&\quad + \varphi_1(x_1) \dots \varphi_1(x_4).
\end{aligned} \quad (2.3.67)$$

For $\varphi_1(x) = 0$ this agrees with Eqn. (2.3.27). The inversion of Eqn. (2.3.66) gives

$$\Phi\{j\} = \log T\{j\}. \quad (2.3.68)$$

This is the functional form of Eqn. (2.3.29). For the free field it follows that

$$\Phi\{j\} = -\frac{i}{2} \iint \mathrm{d}^4x\, \mathrm{d}^4y\, j(x)\, \Delta_\mathrm{F}(x-y) j(y)$$

according to Eqn. (2.3.51).

The relationship between the functional of vertex functions $\Gamma\{a\}$ and $\Phi\{j\}$ takes the form of a Legendre transformation:

$$\Gamma\{a\} = -\mathrm{i}\int \mathrm{d}^4x j(x) a(x) + \Phi\{j\} \tag{2.3.69}$$

with

$$a(x) = \frac{\delta\Phi\{j\}}{\mathrm{i}\,\delta j(x)}, \tag{2.3.70}$$

as will be shown below. (There, the inessential simplifying assumption $\langle A(x)\rangle \equiv 0$ is made, i.e. $a(x) \equiv 0$ follows from $j(x) \equiv 0$.)

The inversion of Eqn. (2.3.70) is obtained by differentiating Eqn. (2.3.69):

$$\frac{\delta\Gamma}{\delta a(x)} = -\mathrm{i}j(x) - \mathrm{i}\int \mathrm{d}^4y \frac{\delta j(y)}{\delta a(x)} a(y) + \int \mathrm{d}^4y \frac{\delta j(y)}{\delta a(x)} \frac{\delta\Phi\{j\}}{\delta j(y)} = -\mathrm{i}j(x),$$

hence:

$$j(x) = \mathrm{i}\frac{\delta\Gamma\{a\}}{\delta a(x)}. \tag{2.3.71}$$

Differentiation of Eqn. (2.3.70) and application of the chain rule results in

$$\begin{aligned} \delta^{(4)}(x-y) &= \frac{\delta a(x)}{\delta a(y)} = \frac{\delta}{\delta a(y)} \frac{\delta\Phi\{j\}}{\mathrm{i}\,\delta j(x)} \\ &= \int \mathrm{d}^4z \frac{\mathrm{i}\,\delta j(z)}{\delta a(y)} \frac{\delta^2\Phi\{j\}}{\mathrm{i}\,\delta j(x) \mathrm{i} j(y)} \\ &= -\int \mathrm{d}^4z \frac{\delta^2\Gamma\{a\}}{\delta a(y)\,\delta a(z)} \frac{\delta^2\Phi\{j\}}{\mathrm{i}\,\delta j(z) \mathrm{i}\,\delta j(x)}. \end{aligned} \tag{2.3.72}$$

If the integration with respect to z in Eqn. (2.3.72) is formally regarded as 'matrix multiplication' of the forms $\delta^2\Gamma/\delta a(x)\,\delta a(y)$ and $\delta^2\Phi/\mathrm{i}\,\delta j(x)\mathrm{i}\,\delta j(y)$, then one is the negative inverse of the other;

$$\frac{\delta^2\Phi\{j\}}{\mathrm{i}\,\delta j(x)\mathrm{i}\,\delta j(y)} = X(x, y), \quad \frac{\delta^2\Gamma\{a\}}{\delta a(x)\,\delta a(y)} = -X^{-1}(x, y). \tag{2.3.73}$$

As $X(x, y)|_{j=0} = \langle TA(x)A(y)\rangle$, $X(x, y)$ is generally called a propagator in the presence of sources (see explanation after Eqn. (2.3.81)).

According to Eqns. (2.3.72) and (2.3.73),

$$\frac{\delta^2\Phi\{j\}}{\mathrm{i}\,\delta j(x_1)\mathrm{i}\,\delta j(x_2)} = -\iint \mathrm{d}^4z_1\,\mathrm{d}^4z_2 X(x_1, z_1)X(x_2, z_2)\frac{\delta^2\Gamma\{a\}}{\delta a(z_1)\,\delta a(z_2)}$$

is an identity. Differentiation with respect to $\mathrm{i}j(x_3)$ gives

$$\begin{aligned} \frac{\delta^3\Phi\{j\}}{\mathrm{i}\,\delta j(x_1)\mathrm{i}\,\delta j(x_2)\mathrm{i}\,\delta j(x_3)} &= \int\ldots\int \mathrm{d}^4z_1\,\mathrm{d}^4z_2\,\mathrm{d}^4z_3 X(x_1, z_1)X(x_2, z_2)X(x_3, z_3) \\ &\quad\times\frac{\delta^3\Gamma\{a\}}{\delta a(z_1)\,\delta a(z_2)\,\delta a(z_3)}. \end{aligned} \tag{2.3.74}$$

Here, Eqn. (2.3.73) has been used several times as well as the chain rule combined with Eqn. (2.3.70):

$$\frac{\delta}{\delta j(x)}=\int \mathrm{d}^4 z \frac{\delta a(z)}{\delta j(x)} \frac{\delta}{\delta a(z)}=\mathrm{i}\int \mathrm{d}^4 z X(x, z)\frac{\delta}{\delta a(z)}. \tag{2.3.75}$$

Eqn. (2.3.74) agrees with Eqn. (2.3.34) for $j(x)=a(x)\equiv 0$. Further differentiation of Eqn. (2.3.74) with respect to $j(x)$ gives with help of (2.3.75) the inductive definition of vertices as discussed after Eqn. (2.3.34). This shows that $\Gamma\{a\}$ is its generating functional.

(d) By far the most important application of functional calculus in field theory is the *representation of Green functions by a path integral* [Fe 48, Fe 65, Sc 51]. Consider the free field first and define:

$$Z\{j\}=\int \mathcal{D}[A]\exp\left(\mathrm{i}\int \mathrm{d}^4x[\tfrac{1}{2}((\partial^\mu A(x))\,\partial_\mu A(x)-M^2A^2(x))+j(x)A(x)]\right). \tag{2.3.76}$$

Then, formally, the relation

$$T\{j\}=Z\{j\}/Z\{0\}. \tag{2.3.77}$$

holds.

In order to establish this, $Z\{j\}$, which is represented by a Gaussian functional integral, is formally manipulated. For this, the following quadratic completion is performed in the exponent of the functional integral:

$$\begin{aligned}
&\int \mathrm{d}^4x[\tfrac{1}{2}A(x)(-\partial^\mu\,\partial_\mu-M^2)A(x)+j(x)A(x)]\\
&=\int \mathrm{d}^4x\frac{1}{2}\left(A(x)+\int \mathrm{d}^4x'\,\Delta_\mathrm{F}(x-x')j(x')\right)(-\partial^\mu\,\partial_\mu-M^2)\\
&\quad\times\left(A(x)+\int \mathrm{d}^2x'\,\Delta_\mathrm{F}(x-x')j(x')\right)\\
&\quad-\frac{1}{2}\iint \mathrm{d}^4x\,\mathrm{d}^4x' j(x)\,\Delta_\mathrm{F}(x-x')j(x'),
\end{aligned} \tag{2.3.78}$$

since

$$(-\partial^\mu\,\partial_\mu-M^2)\,\Delta_\mathrm{F}(x-x',M)=\delta^{(4)}(x-x'). \tag{2.3.79}$$

The linear substitution $A'(x)=A(x)+\int \mathrm{d}^4x'\,\Delta_\mathrm{F}(x-x')j(x')$ in the functional integral then gives:

$$\begin{aligned}
Z\{j\}&=\exp\left(-\frac{\mathrm{i}}{2}\int \mathrm{d}^4x\,\mathrm{d}^4x' j(x)\,\Delta_\mathrm{F}(x-x')j(x')\right)Z\{0\},\\
Z\{0\}&=\int \mathcal{D}[A']\exp\left(\frac{\mathrm{i}}{2}\int \mathrm{d}^4x((\partial^\mu A'(x))\,\partial_\mu A'(x)-M^2A'^2(x))\right).
\end{aligned} \tag{2.3.80}$$

Thus, $Z\{j\}/Z\{0\}$ represents the generating functional $T\{j\}$ for free fields according to (2.3.51). The choice of the Feynman propagator in Eqn. (2.3.79) follows from a damping factor $\exp[-\varepsilon \int d^4x (A(x))^2]$ which is needed in principle in Eqn. (2.3.76) but which has been omitted here for the sake of simplicity.

The Lagrangian density of the free scalar field $A(x)$ is $\mathcal{L}_0 = \frac{1}{2}((\partial^\mu A(x))\,\partial_\mu A(x) - M^2 A^2(x))$. Thus, the Z-functional can also be written as

$$Z\{j\} = \int \mathcal{D}[A] \exp\left(i \int d^4x (\mathcal{L}_0 + j(x)A(x))\right)$$

$$= \int \mathcal{D}[A] e^{iS\{A\}} \exp\left(i \int d^4x\, j(x)A(x)\right) \tag{2.3.81}$$

$$= \int \mathcal{D}[A] \exp(iS_j\{A, j\}), \tag{2.3.81'}$$

i.e. as a functional Fourier transformation of the exponential of the action $S\{A\}$ of the free field (Eqn. (2.3.81)) or as a functional average over the exponential of the action with external source $\mathcal{L} = \mathcal{L}_0 + j(x)A(x)$ (Eqn. (2.3.81')). The field equation derived from $\mathcal{L}$ is $(-\partial^\mu \partial_\mu - M^2)A(x) = j(x)$. This is why the argument of the functionals $T\{j\}$, $\Phi\{j\}$ is called a source.

This representation of the generating functional of the Green functions can be extended to interacting fields:

$$Z\{j\} = \int \mathcal{D}[A] \exp\left(i \int d^4x (\mathcal{L}(A(x)) + j(x)A(x))\right),$$

$$T\{j\} = Z\{j\}/Z\{0\},$$

The Lagrangian density $\mathcal{L}$ is given by

$$\mathcal{L} = \mathcal{L}_0 + g\mathcal{L}_I, \tag{2.3.82}$$

e.g.

$$\mathcal{L}_I = -\frac{1}{3!} A^3(x), \quad \mathcal{L}_I = -\frac{1}{4!} A^4(x), \quad \text{etc.}$$

These formulae form the basic equations of a quantum field theory with arbitrary Lagrangian density. The formulation of a quantum field theory with help of this Feynman path integral formula is an equivalent alternative to the quantization with help of commutation relations and Hamiltonian function in the framework of the canonical formalism. It forms the starting point for the non-perturbative calculation of Green functions (see Sections 2.6.3–4). In the next section, QCD will be defined in this compact and complete way.

(e) *The derivation of Feynman graph rules* from Eqns. (2.3.82) is described first of all, in order to establish a connection between the treatment of theories with interaction as discussed in the preceding chapters and the path integral representation of quantum field theory. To this end, the field variable in the interaction term $\mathcal{L}_I$ is replaced by a differentiation with respect to the source j. The result of Gaussian integration over free fields,

Eqn. (2.3.80), is then used and the exponential of $\mathscr{L}_I$ is expanded in powers of the coupling constant:

$$Z\{j\} = \int \mathscr{D}[A] \exp\left(ig\int d^4x \mathscr{L}_I(A(x))\right) \exp\left(i\int d^4x(\mathscr{L}_0(A(x)) + j(x)A(x))\right)$$

$$= \exp\left[ig\int d^4x \mathscr{L}_I\left(\frac{1}{i}\frac{\delta}{\delta j(x)}\right)\right] \int \mathscr{D}[A] \exp\left(i\int d^4x(\mathscr{L}_0(A(x)) + j(x)A(x))\right)$$

$$= \left[1 + ig\int d^4x \mathscr{L}_I\left(\frac{1}{i}\frac{\delta}{\delta j(x)}\right) - \frac{g^2}{2}\int d^4x_1 \mathscr{L}_I\left(\frac{1}{i}\frac{\delta}{\delta j(x_1)}\right) \times \int d^4x_2 \mathscr{L}_I\left(\frac{1}{i}\frac{\delta}{\delta j(x_2)}\right) + \ldots\right]$$

$$\times \exp\left(-\frac{i}{2}\int d^4x\, d^4x' j(x)\, \Delta_F(x - x') j(x')\right) Z_0\{0\}. \tag{2.3.83}$$

Here, $Z_0\{0\}$ is the Z functional of the free field for vanishing sources. The evaluation of the functional differentiations leads to the perturbation expansion of the Green functions according to the Feynman rules.

A simple example is used by way of illustration. For this purpose, the $\mathscr{L}_I = -(1/4!)A^4(x)$ interaction of the scalar field is considered and the 4-point Green function $\tau^{(1)}(x_1, \ldots, x_4)$ is calculated up to the first order in g. Because of

$$\mathscr{L}_I\left(\frac{1}{i}\frac{\delta}{\delta j(x)}\right) = -\frac{1}{4!}\frac{\delta^4}{(\delta j(x))^4},$$

the corresponding expansion of Eqn. (2.3.83) up to the contributing order in g and $j(x)$ results in

$$\left.\frac{\delta^4 Z\{j\}}{\delta j(x_1) \ldots \delta j(x_4)}\right|_{j=0} \cdot Z_0\{0\}^{-1}$$

$$= -\frac{\delta^4}{\delta j(x_1) \ldots \delta j(x_4)} \frac{1}{2!}\left(\frac{-i}{2}\right)^2 \left(\int d^4z\, d^4z' j(z)\, \Delta_F(z - z') j(z')\right)^2$$

$$- \frac{\delta^4}{\delta j(x_1) \ldots \delta j(x_4)} \left(\frac{-ig}{4!}\right) \int d^4x \frac{\delta^4}{(\delta j(x))^4} \frac{1}{4!}\left(\frac{-i}{2}\right)^4 \times \left(\int d^4z\, d^4z' j(z)\, \Delta_F(z - z') j(z')\right)^4$$

$$\text{(a)} \quad = -\sum_{\text{partitions}} \Delta_F(x_i - x_j)\, \Delta_F(x_k - x_l)$$

$$\text{(b)} \quad -ig\int d^4x\, \Delta_F(x - x_1)\, \Delta_F(x - x_2)\, \Delta_F(x - x_3)\, \Delta_F(x - x_4)$$

$$\text{(c)} \quad -ig\sum_{\text{partitions}} \Delta_F(x_i - x_j) \frac{1}{2}\int d^4x\, \Delta_F(x - x)\, \Delta_F(x - x_k)\, \Delta_F(x - x_l) \tag{2.3.84}$$

$$\text{(d)} \quad -ig\sum_{\text{partitions}} \Delta_F(x_i - x_j)\, \Delta_F(x_k - x_l) \frac{1}{2^3}\int d^4x\, \Delta_F(x - x)\, \Delta_F(x - x),$$

and

$$Z\{0\}/Z_0\{0\}=1+\mathrm{i}g\frac{1}{2^3}\int \mathrm{d}^4x\, \Delta_{\mathrm{F}}(x-x)\, \Delta_{\mathrm{F}}(x-x). \tag{2.3.85}$$

From this it follows that:

$$\tau^{(1)}(x_1,\ldots,x_4)=(\delta^4 Z\{j\}/\delta j(x_1)\ldots\delta j(x_4))/Z\{0\}$$

(a) $$=-\sum_{\text{partitions}} \Delta_{\mathrm{F}}(x_i-x_j)\,\Delta_{\mathrm{F}}(x_k-x_l)$$

(b) $$-\mathrm{i}g\int \mathrm{d}^4x\,\Delta_{\mathrm{F}}(x-x_1)\,\Delta_{\mathrm{F}}(x-x_2)\,\Delta_{\mathrm{F}}(x-x_3)\,\Delta_{\mathrm{F}}(x-x_4)$$

(c) $$-\mathrm{i}g\sum_{\text{partitions}} \Delta_{\mathrm{F}}(x_i-x_j)\frac{1}{2}\int \mathrm{d}^4x\,\Delta_{\mathrm{F}}(x-x)\,\Delta_{\mathrm{F}}(x-x_k)\,\Delta_{\mathrm{F}}(x-x_l). \tag{2.3.86}$$

The term (d) in Eqn. (2.3.84) is cancelled by the product of $(\mathrm{i}g/2^3)\int \mathrm{d}^4x(\Delta(0))^2$ in the expansion of $1/Z\{0\}$ in powers of g with the term (a). The disconnected vacuum loops make no contribution! According to the Feynman graph rules in configuration space (2.3.42), Eqn. (2.3.86) corresponds to the graphs:

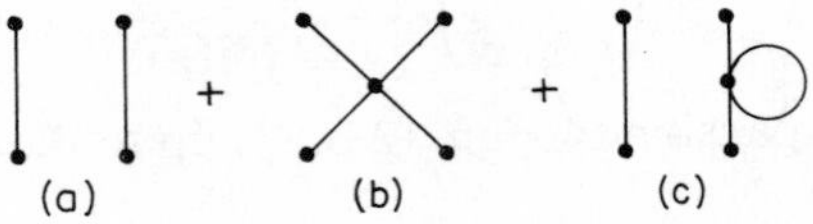

(a) + (b) + (c)

Connected and disconnected contributions are obtained. Only the graphs (b) contribute to the connected Green function $\varphi^{(1)}(x_1,\ldots,x_4)$. Loops occur in the disconnected graphs (c), and these cause divergent expressions $\Delta_{\mathrm{F}}(0)$. Section 2.4 deals with these divergences.

We conclude the derivation of the Feynman rules from the path integral representation with a short remark [Na 66]. For this, the Planck constant $\hbar$ which has been used so far as a unit, is written explicitly in Eqn. (2.3.83):

$$Z\{j\}=\int \mathscr{D}[A]\exp\left(\frac{\mathrm{i}}{\hbar}\int \mathrm{d}^4x(\mathscr{L}(A(x))+\hbar j(x)A(x))\right)$$

$$=\exp\left[\mathrm{i}\frac{g}{\hbar}\int \mathrm{d}^4x\mathscr{L}_{\mathrm{I}}\left(\frac{1}{\mathrm{i}}\frac{\delta}{\delta j(x)}\right)\right]\exp\left(-\frac{\mathrm{i}}{2}\hbar\iint \mathrm{d}^4x\,\mathrm{d}^4x'j(x)\,\Delta_{\mathrm{F}}(x-x')j(x')\right). \tag{2.3.87}$$

Thus, each vertex is linked with a factor $\hbar^{-1}$ and each propagator with a factor $\hbar$ in the perturbation series. The number of loops for any connected graph consisting of N points and I lines is

$$l=I-N+1=I-V+1-F \tag{2.3.88}$$

[De 74]. $N-1$ lines are required to connect the N points simply, by a tree graph; any further line will then generate a loop. The N points of a graph for Green functions are made up of V vertices and F field points. As each source function $j(x)$ is linked with a field point, the above counting of powers of $\hbar$ gives the following result for the perturbative expansion of the generating functional of connected Green functions:

$$\Phi\{j\} = \hbar \log(Z\{j\}/Z\{0\}) = \sum_l \Phi_l\{\hbar j(x)\}\hbar^l, \tag{2.3.89}$$

where Φ_l is the contribution of the graphs with l loops. This result points to a close relation between the order of quantum corrections and the number of loops in Feynman diagrams. Especially, one would expect that the tree graph approximation without loops has a close connection with the solutions of the classical field equations. Actually, for the most simple, semiclassical approximation (method of stationary phase [De 61], WKB method):

$$\Phi\{j\} = \hbar \log Z\{j\} \approx \mathrm{i} \int \mathrm{d}^4x(\mathcal{L}(a(x)) + \hbar j(x)a(x)) \tag{2.3.90}$$

and $a(x)$ is the solution of the classical equation of motion:

$$\frac{\delta}{\delta a(x)} \int \mathrm{d}^4y \mathcal{L}(a(y)) = -\hbar j(x). \tag{2.3.91}$$

If both these equations are compared with the introduction of the functional of vertex functions (Eqns. (2.3.69), (2.3.71)), one concludes that in this approximation, the vertices are generated by

$$\Gamma\{a\} = +\mathrm{i} \int \mathrm{d}^4x \mathcal{L}(a(x)), \tag{2.3.92}$$

i.e. by the action. These are the vertices of the tree graph approximation. This will be used in Section 2.4.3(a), p. 133.

(f) Section 1.2.3 referred to the connection between symmetry of the Lagrange function and charge conservation. These considerations are now extended to the path integral representation of quantum field theory. *Ward identities* for Green functions are thus obtained, the generalizations of which are of great importance for the treatment of gauge field theories [Wa 50, Ta 57].

As an example, consider real, scalar fields $A_i(x)$ with an orthogonal symmetry group $SO(N)$ of the Lagrangian function $\mathcal{L}(A, \partial_\mu A)$ (gauge invariance of the first kind):

$$(gA)_i(x) = g_{ik}(\theta)A_k(x) \quad \text{or} \quad \delta A_i(x) = (T_a)_i^k A_k(x)\, \delta\theta^a$$

with (2.3.93)

$$g = \exp(\theta^a T_a), \quad (T_a)_i^k = -(T_a)_k^i, \qquad i, k = 1, \ldots, N,$$

$$\mathcal{L}(gA, \partial_\mu gA) = \mathcal{L}(A, \partial_\mu A) \tag{2.3.94}$$

or

$$\partial \mathscr{L} = 0 \tag{2.3.95}$$

(see Eqns. (1.2.56), (1.2.61), (1.2.62)). From the invariance of the functional measure $\mathscr{D}[gA] = \mathscr{D}[A]$ (see Eqn. (2.3.62)) with respect to orthogonal transformations it follows from the substitution rule (2.3.61) that:

$$\begin{aligned} Z\{J_k\} &= \int \mathscr{D}[A] \exp\left(\mathrm{i} \int \mathrm{d}^4x (\mathscr{L}(A, \partial_\mu A) + J_k(x) A_k(x))\right) \\ &= \int \mathscr{D}[gA] \exp\left(\mathrm{i} \int \mathrm{d}^4x (\mathscr{L}(gA, \partial_\mu gA) + J_k(x) g_{ki} A_i(x))\right) \\ &= \int \mathscr{D}[A] \exp\left(\mathrm{i} \int \mathrm{d}^4x (\mathscr{L}(A, \partial_\mu A) + J_k(x) g_{ki} A_i(x))\right). \end{aligned} \tag{2.3.96}$$

If the first and third line in the power series expansion of the generating functionals, (2.3.63), are compared the result is

$$\langle A_{i_1}(x_1) \ldots A_{i_n}(x_n) \rangle = g_{i_1 k_1} \ldots g_{i_n k_n} \langle A_{k_1}(x_1) \ldots A_{k_n}(x_n) \rangle. \tag{2.3.97}$$

This same symmetry of the Green functions also follows from the existence of a symmetry operator $U(g) = \exp(-\mathrm{i}Q_a\theta^a)$, where Q_a are changes of $SO(N)$, according to the prescription of (2.3.8) if $U(g)$ transforms the fields covariantly (Eqn. (1.2.71)) and leaves the vacuum invariant (Eqn. (2.3.6)).

The local form of the charge conservation law, i.e. the vanishing of the divergence of the current $j^a_\mu(x)$ follows from the invariance of the Lagrange function and the equations of motion:

$$j^a_\mu(x) = \frac{\partial \mathscr{L}}{\partial(\partial_\mu A_i(x))} (T^a)_{ik} A_k(x), \quad \partial^\mu j^a_\mu(x) = 0 \tag{2.3.98}$$

(see Eqns. (1.2.57), (1.2.64)). $(T^a)_{ik}$ is a basis of the space of infinitesimal generators of $SO(N)$. The analogous consideration applied in the path integral formulation of quantum field theory gives the Ward identities. For this, the field equations are derived from the translation invariance of the measure $\mathscr{D}[A + \delta A] = \mathscr{D}[A]$. The expression

$$\begin{aligned} Z\{J_k\} = \int \mathscr{D}[A] \exp\Big(\mathrm{i} \int \mathrm{d}^4x (\mathscr{L}(A + \delta A, \partial_\mu (A + \delta A)) + J_k(x) \\ \times (A_k(x) + \delta A_k(x)))\Big) \end{aligned} \tag{2.3.99}$$

is expanded up to terms linear in $\delta A(x)$ and compared with Eqn. (2.3.96). It follows that this term vanishes and therefore also the expectation value of the Euler–Lagrange equations with external source:

$$\int \mathscr{D}[A] \left(\frac{\partial \mathscr{L}}{\partial A_i(y)} - \frac{\partial}{\partial y^\mu} \frac{\partial \mathscr{L}}{\partial(\partial_\mu A_i(y))} + J_i(y) \right) \exp\left(\mathrm{i} \int \mathrm{d}^4x (\mathscr{L} + J_i A_i)\right) = 0. \tag{2.3.100}$$

When this 'equation of motion' is used together with the invariance equation

$$\frac{\partial \mathscr{L}}{\partial A_i}(T^a)_{ik}A_k + \frac{\partial \mathscr{L}}{\partial(\partial_\mu A_i)}(T^a)_{ik}\,\partial_\mu A_k = 0, \tag{2.3.101}$$

which follows from Eqn. (2.3.94) by differentiation with respect to θ then a direct calculation according to Eqn. (1.2.57) gives the result:

$$\int \mathscr{D}[A](\partial^\mu j^a_\mu(y, A) + J_i(y)(T^a)_{ik}A_k(y)) \exp\left(i\int d^4x(\mathscr{L} + J_iA_i)\right) = 0 \tag{2.3.102}$$

or:

$$\left[\frac{\partial}{\partial y_\mu} j^a_\mu\left(y, \frac{\delta}{i\,\delta J}\right) + J_i(y)(T^a)_{ik}\frac{\delta}{i\,\delta J_k(y)}\right]T\{J\} = 0. \tag{2.3.103}$$

The symbol $j^a_\mu(y, A)$ is used to indicate that the current is a local functional of the fields $A_i(y)$ according to (2.3.98). Eqn. (2.3.103) represents the Ward identities in functional form. If this equation is expanded up to second order with respect to the sources, for example, then the result is as follows:

$$\begin{aligned}\partial^\mu_y\langle Tj^a_\mu(y)A(x_1)A(x_2)\rangle = \delta^{(4)}(y - x_1)&\langle T(T^aA(x_1))A(x_2)\rangle \\ &+ \delta^{(4)}(y - x_2)\langle TA(x_1)(T^aA(x_2))\rangle.\end{aligned} \tag{2.3.104}$$

Integration with respect to space–time in Eqn. (2.3.104) gives Eqn. (2.3.97) for infinitesimal symmetry transformations. A different derivation is based on the equation of current conservation $\partial_\mu j^\mu_a(x) = 0$ and the local form of Eqn. (1.2.71):

$$[j^a_0(y), A_i(x)]|_{x^0=y^0} = (T^a)_{ik}A_k(x)\,\delta(\mathbf{x} - \mathbf{y}).$$

The calculation is done according to the scheme:

$$\begin{aligned}\partial^\mu_y Tj^a_\mu(y)A_i(x) &= \partial^\mu_y(\theta(y^0 - x^0)j^a_\mu(y)A_i(x) + \theta(x^0 - y^0)A_i(x)j^a_\mu(y)) \\ &= \delta(y^0 - x^0)[j^a_0(y), A_i(x)] = (T^a)_{ik}A_k(x)\,\delta^{(4)}(x - y).\end{aligned} \tag{2.3.105}$$

(g) The formulation of quantum field theory with help of the Feynman path integral formula brings field theory close to *classical statistical mechanics* [Sy 66, Wi 75]. This will be touched upon briefly below, but it will be brought up again in more detail in connection with the confinement problem which in many different ways involves the use of methods of statistical mechanics.

Eqn. (2.3.80) has already established that the Gaussian functional integral (2.3.76) or (2.3.81)) requires a damping factor before it is mathematically well defined. From a mathematical standpoint, one particularly satisfactory solution of this problem is obtained by considering the fields $A(x)$ at imaginary times: $x_0 = -ix_4$, x_4 real. This makes the action ((2.3.81), (2.3.82)) imaginary and the exponent negative definite:

$$iS\{A\} = i\int dx_0\, d^3x \left\{\frac{1}{2}\left[\left(\frac{\partial}{\partial x_0}A\right)\frac{\partial}{\partial x_0}A - (\boldsymbol{\partial}A)\cdot\boldsymbol{\partial}A - M^2A^2\right] - \mathcal{L}_I(A)\right\}$$

$$\to -\int dx_4\, d^3x \left\{\frac{1}{2}\left[\left(\frac{\partial}{\partial x_4}A\right)^2 + (\boldsymbol{\partial}A)^2 + M^2A^2\right] + \mathcal{L}_I(A)\right\} = S_E\{A\} \leqslant 0. \tag{2.3.106}$$

Fields in space–time points with imaginary time are called Euclidean fields; accordingly $S_E\{A\}$ is the Euclidean action. The generating functional of Green functions of Euclidean field theory [Sy 66, Ru 61, Sc 51] can now be obtained by analytic continuation of Eqn. (2.3.82):

$$\begin{aligned} Z\{J\} &= \int \mathcal{D}[A] \exp\left(S_E\{A\} + \int d^4x J(x) A(x)\right), \\ T\{J\} &= Z\{J\}/Z\{0\}. \end{aligned} \tag{2.3.107}$$

As $S_E\{A\} < 0$, $d\mu\{A\} \equiv \mathcal{D}[A]\exp(S_E\{A\})$ is approximately a Gaussian measure. From a physical point of view, the analytic continuation of Green functions to imaginary times may seem somewhat artificial. However, the exact conditions under which Euclidean Green functions can be continued analytically back to physical times are known [Os 75]. Above all, the pole structure which is important for the physical interpretation of Green functions (Section 2.3.1(f)) remains essentially preserved in the Euclidean region (see Eqn. (2.6.8)).

The formulae of Euclidean field theory for the calculation of Green functions correspond to the basic equations of statistical mechanics. In order to show this, put $S_E\{A\} = -\beta\mathcal{H}_E\{A\}$; in many cases, $\beta \approx 1/g^2$, where g is the field theoretical coupling constant. If the functional differentiation $\delta/\delta J(x)\ldots$ for the calculation of the Green functions is carried out under the integral, it follows that:

$$\langle A(x_1)\ldots A(x_n)\rangle = \frac{\int \mathcal{D}[A] A(x_1)\ldots A(x_n)\exp(-\beta\mathcal{H}_E\{A\})}{\int \mathcal{D}[A]\exp(-\beta\mathcal{H}_E\{A\})}. \tag{2.3.107'}$$

Hence, Euclidean Green functions are the statistical expectation values of products of field operators with respect to the canonical distribution $\exp(-\beta\mathcal{H}_E\{A\})$ [Hu 64]. The negative action is interpreted as the classical Hamilton function of a four-dimensional field theory. The coupling constant $g \approx 1/\beta^{1/2}$ gets the meaning of a temperature and $Z\{0\} = Z(\beta) = \int \mathcal{D}[A]\exp(-\beta\mathcal{H}_E\{A\})$ that of a partition function. The notions 'free energy' $F \sim -\beta^{-1}\log Z(\beta)$, 'internal energy' $U \sim \delta F/\delta\beta$, etc. are used accordingly.

(h) Point (b) above gave a description of the simplest notions and operations in functional analysis in order to formulate the functional integral representation for the quantum theory of boson fields. The formal similarity with the corresponding operations in the analysis of functions of several variables was stressed thereby. In order to generalize the *Feynman path integral representation* to *anticommuting Fermi fields*, these mathematical methods must be extended to the analysis of functionals of Grassmann algebra valued functions [Be 66]. In this instance, only formal algebraic aspects form the guidelines for the corresponding definitions and relationships.

The starting point of this calculus is the introduction of $2n$ generators of a Grassmann algebra, i.e. anticommuting objects y_i:

$$y_i y_j + y_j y_i = 0, \quad y_i^* y_j^* + y_j^* y_i^* = 0, \quad y_i y_j^* + y_j^* y_i = 0, \qquad i, j = 1, \ldots, n, \ldots .$$
$$\text{(especially: } y_i^2 = y_j^{*2} = 0). \tag{2.3.108}$$

The elements of the Grassmann algebra are the formal power series in y_i, y_j^* with complex coefficients $T_{m,n}(l_1, \ldots, l_m \mid k_1, \ldots, k_n)$ which are antisymmetrical in k_i and l_j:

$$F(y^*, y) = \sum_{n,m} \sum_{k_i, l_j} y_{l_m}^* \ldots y_{l_1}^* T_{m,n}(l_1, \ldots, l_m \mid k_1, \ldots, k_n) y_{k_1} \ldots y_{k_n}. \tag{2.3.109}$$

Integration and differentiation of these quantities are defined as linear algebraic operations, that is differentiation by

$$\frac{\partial^{L,R}}{\partial y_k} y_i = \delta_{ik}, \quad \frac{\partial^{L,R}}{\partial y_k^*} y_i = 0, \quad \frac{\partial^{L,R}}{\partial y_k^*} y_i^* = \delta_{ik}, \quad \frac{\partial^{L,R}}{\partial y_k} 1 = 0,$$
$$\frac{\partial^L}{\partial y_k} y_k y_{i_1} \ldots y_{i_l} = y_{i_1} \ldots y_{i_l}, \quad \frac{\partial^R}{\partial y_k} y_{i_1} \ldots y_{i_l} y_k = y_{i_1} \ldots y_{i_l}, \quad k \neq i_\nu, \tag{2.3.110}$$

(left derivative ∂^L, right derivative ∂^R) and integration by

$$\int dy_i = 0, \quad \int dy_i y_i = 1. \tag{2.3.111}$$

For the differentials the corresponding anticommutation relations are given by

$$[dy_i, dy_k]_+ = [dy_i, y_k]_+ = [dy_i, dy_k^*]_+ = [dy_i^*, y_k]_+ = \ldots = 0.$$

These definitions can be used to derive the usual, analogous rules of calculation in which, however, $y_i^2 = 0$ and care is needed with the various signs which occur from the anticommutativity of the y_i. Two rules are quoted by way of example:

(1) the product rule of differentiation:

$$\frac{\partial^L}{\partial y_i}(F_1 \cdot F_2) = \left(\frac{\partial^L}{\partial y_i} F_1\right) F_2 \pm F_1 \left(\frac{\partial^L}{\partial y_i} F_2\right), \quad \text{if } F_1 \begin{cases} \text{is even} \\ \text{is odd} \end{cases}$$

(2) the substitution rule of integration:

$$z_i = \sum_k a_{ik} y_k, \quad A = (a_{ik}),$$
$$\int \ldots \int dz_n \ldots dz_1 F(z) = (\det A^{-1}) \int \ldots \int dy_n \ldots dy_1 F(z(y)). \tag{2.3.112}$$

This corresponds to a transformation law for the differentials $dz_i = \sum_k \tilde{a}_{ik}\, dy_k$ with the inverse matrix $A^{-1} = (\tilde{a}_{ik})$.

The conjugate quantities y_i^* play the same role as y_j in all these calculations. However, their introduction allows the definition of a conjugation for the elements (2.3.109) of the Grassmann algebra by

$$F^*(y^*, y) = \sum_{n,m} \sum_{k_i, l_j} y_{k_n}^* \ldots y_{k_1}^* T_{m,n}^*(l_1, \ldots, l_m \mid k_1, \ldots, k_n) y_{l_1} \ldots y_{l_m}. \tag{2.3.113}$$

A formula similar to the Gaussian integral plays an important role in the practical application of integration in field theory, namely:

$$\int \ldots \int dy_1 \ldots dy_n\, dy_n^* \ldots dy_1^* \exp\Big(\sum_{i,k} y_i^* a_{ik} y_k\Big) = \det A. \tag{2.3.114}$$

Formulae from the analysis of a finite number of variables (Eqns. (2.3.52), (2.3.54), (2.3.57)) for the definition of power series expansion, differentiation and integration were transferred to functionals (see Eqns. (2.3.53), (2.3.55), (2.3.58)). Similarly, the corresponding Grassmann algebra formulae are extended to Grassmann functionals.

Power series:

$$T\{y^*, y\} = \sum_{m,n} \int \ldots \int dx_m' \ldots dx_1'\, dx_1 \ldots dx_n$$
$$\times y^*(x_m') \ldots y^*(x_1') T_{m,n}(x_1', \ldots, x_m' \mid x_1, \ldots, x_n) y(x_1) \ldots y(x_n), \tag{2.3.115}$$

$$[y(x), y(x')]_+ = [y^*(x), y(x')]_+ = \ldots = 0,$$

$(T(x_1', \ldots, x_m' \mid x_1, \ldots, x_n)$ antisymmetrical in x_i' and $x_j)$.

Differentiation:

$$\frac{\delta^L}{\delta y(x)} y(x_1) \ldots y(x_n) = \delta(x - x_1) y(x_2) \ldots y(x_n) - \delta(x - x_2) y(x_1) y(x_3) \ldots y(x_n)$$
$$\pm \ldots + (-1)^{n+1} \delta(x - x_n) y(x_1) \ldots y(x_{n-1}). \tag{2.3.116}$$

'Gaussian integral':

$$\iint \mathscr{D}[y] \mathscr{D}[y^*] e^{y^* A y} = \det A,$$
$$y^* A y := \iint dx\, dx'\, y^*(x) A(x, x') y(x'). \tag{2.3.117}$$

These formulae can be used to give the generating functional of the Green functions of a free Dirac field in analogy to Eqns. (2.3.50), (2.3.51):

$$T\{\bar{\eta}, \eta\} = \sum_n \frac{(-1)^n}{(n!)^2} \int \ldots \int d^4x_n \ldots d^4x'_n \bar{\eta}(x_n) \ldots \bar{\eta}(x_1)$$
$$\times \langle 0 | T\psi(x_1) \ldots \psi(x_n)\bar{\psi}(x'_n) \ldots \bar{\psi}(x'_1) |0\rangle \eta(x'_1) \ldots \eta(x'_n)$$
$$= \exp\left(-i \int\int d^4x\, d^4x' \bar{\eta}(x) S_F(x-x')\eta(x')\right). \qquad (2.3.118)$$

The sources $\bar{\eta}(x)$, $\eta(x')$ are anticommuting, i.e. elements of an infinite-dimensional Grassmann algebra.

With help of the formula (2.3.117) the path integral representation for the free Dirac field is derived similar to Eqns. (2.3.76), (2.3.77):

$$T\{\bar{\eta}, \eta\} = Z\{\bar{\eta}, \eta\}/Z\{0, 0\},$$
$$Z\{\bar{\eta}, \eta\} = \int\int \mathscr{D}[\psi]\mathscr{D}[\bar{\psi}] \exp\left(i \int d^4x(\bar{\psi}(i\gamma^\mu \partial_\mu - M)\psi + \bar{\eta}\psi + \bar{\psi}\eta)\right)$$
$$= \int\int \mathscr{D}[\psi]\mathscr{D}[\bar{\psi}] \exp\left(i \int d^4x(\mathscr{L}_0 + \bar{\eta}\psi + \bar{\psi}\eta)\right). \qquad (2.3.119)$$

2.3.3 Path integral formulation of quantum chromodynamics

a) The application of the Feynman path integral formulation of quantum field theory to QCD, leads to the following ansatz for the generating functional of the Green functions:

$$T\{j; \bar{\eta}, \eta\} = Z\{j; \bar{\eta}, \eta\}/Z\{0; 0, 0\}$$
$$Z\{j; \bar{\eta}, \eta\} = \int \mathscr{D}[A^a_\mu]\mathscr{D}[\psi]\mathscr{D}[\bar{\psi}]$$
$$\times \exp\left(i \int d^4x(-\tfrac{1}{4}F^a_{\mu\nu}(x)F^{a,\mu\nu}(x) + \bar{\psi}(x)(i\gamma^\mu D_\mu - M)\psi(x)\right.$$
$$\left. + j^\mu_a(x)A^a_\mu(x) + \bar{\psi}(x)\eta(x) + \bar{\eta}(x)\psi(x))\right). \qquad (2.3.120)$$

However, this formula contains mathematical discrepancies which make even a perturbative evaluation of $Z\{j; \bar{\eta}, \eta\}$ impossible. This problem can already be seen in the case of the free photon field. An attempt to evaluate the corresponding functional integral:

$$Z\{j\} = \int \mathscr{D}[A_\mu] \exp\left(i \int d^4x(\tfrac{1}{2}A_\mu(x)(g^{\mu\nu}\,\partial_\sigma\,\partial^\sigma - \partial^\mu\,\partial^\nu)A_\nu(x) + j^\mu(x)A_\mu(x))\right) \qquad (2.3.121)$$

In analogy to Eqn. (2.3.78), using Gaussian integration

$$Z\{j\} \sim \exp\left(-\frac{i}{2}\int d^4x\, d^4y j^\mu(x)K^{-1}_{\mu\nu}(x-y)j^\nu(y)\right),$$

fails because the operator $K^{\mu\nu} = g^{\mu\nu}\,\partial_\sigma\,\partial^\sigma - \partial^\mu\,\partial^\nu$ has zero eigenvalues in the space of 4-potentials $[A_\mu(x)]$. For example for pure gauges $A_\nu(x) = \partial_\nu\theta(x)$: $K^{\mu\nu}A_\nu(x) = (g^{\mu\nu}\,\partial_\sigma\,\partial^\sigma - \partial^\mu\,\partial^\nu)\,\partial_\nu\theta(x) = 0$. Thus, the condition for Gaussian integration $(A_\mu, K^{\mu\nu}A_\nu) > 0$ is not satisfied and the result is divergent, as $K^{-1}_{\mu\nu}$ does not exist.

For the correct quantization of any gauge theory it is necessary to solve this problem, which is caused by the gauge invariance of the Lagrangian density $\mathscr{L}$. The following method, proposed by L. D. Faddeev and V. N. Popov [Fa 67], is used for this purpose.

If $\mathscr{A}^\theta_\mu(x)$ is the gauge field transformed with $g(x) = \exp(-i\theta^a(x)T_a)$ according to Eqn. (2.2.5) then the integration over the gauge degrees of freedom $\theta^a(x)$, from which $\mathscr{L}$ is independent, results in an infinite factor. This is gauge invariant and can be factored off in a unique way. When forming the ratio $Z\{j;\bar{\eta},\eta\}/Z\{0;0,0\}$ this factor cancels out. In order to accomplish this, Eqn. (2.3.120) is multiplied by a constant factor which is represented as a Gaussian integral with respect to $C^a(x)$:

$$\text{constant} = \int \mathscr{D}[C^a]\exp\left(-\tfrac{1}{2}i\int d^4x(C^a(x))^2\right). \tag{2.3.122}$$

If $C^a(x)$ is regarded as a functional of the gauge fields $\mathscr{A}_\mu(x)$ with the property that $C^a[\mathscr{A}_\mu(x)]$ fixes the gauge, i.e. that from $C^a[\mathscr{A}^\theta_\mu(x)] = C^a[\mathscr{A}^{\theta'}_\mu(x)]$ it follows that $\theta(x) = \theta'(x)$,* then the transformation of variables $C^a[\mathscr{A}^\theta_\mu(x)] \to \theta^a(x)$ can be carried out:

$$\text{constant} = \int_{\text{gauge group}} \mathscr{D}[\theta^a]\ \det\left(\frac{\delta C^a[\mathscr{A}^\theta_\mu(x)]}{\delta\theta^b(y)}\right)\exp\left(-\tfrac{1}{2}i\int d^4x(C^a[\mathscr{A}^\theta_\mu(x)])^2\right). \tag{2.3.123}$$

Our special gauge-fixing term chosen is:

$$C^a[\mathscr{A}_\mu(x)] = \frac{1}{\sqrt{\xi}}\,\partial^\mu A^a_\mu(x). \tag{2.3.124}$$

After multiplying Eqn. (2.3.120) with the constant (2.3.123), the integrations are interchanged and the following is obtained by using the gauge invariance of $\mathscr{L}$ (Eqn. (2.3.38)) and $\mathscr{D}[A^a_\mu]\mathscr{D}[\psi]\mathscr{D}[\bar{\psi}]$:

$$\begin{aligned} Z\{j;\bar{\eta},\eta\} \sim \int_{\text{gauge group}} \mathscr{D}[\theta^a] \int \mathscr{D}[A^a_\mu]\mathscr{D}[\psi]\mathscr{D}[\bar{\psi}] \\ \times\det\left(\frac{\delta C^a[\mathscr{A}_\mu(x)]}{\delta\theta^b(y)}\right)\exp\left(i\int d^4x\{\mathscr{L} - \tfrac{1}{2}(C^a[\mathscr{A}_\mu(x)])^2\right. \\ \left. + j^{a,\mu}(x)A^{(-\theta)}_{a,\mu}(x) + \bar{\psi}^{(-\theta)}(x)\eta(x) + \bar{\eta}(x)\psi^{(-\theta)}(x)\}\right). \end{aligned} \tag{2.3.125}$$

* There was doubt [Gr 78] as to whether the structure of all gauge potentials in a non-Abelian gauge theory [Mi 79, Si 78] in fact permits a general gauge fixing by $C^a(\mathscr{A}_\mu)$, and in particular as to whether such simple $C^a(\mathscr{A}_\mu)$ as in Eqn. (2.3.124) satisfy this condition for all $\mathscr{A}_\mu$. Despite this, the method described here is still regarded as suitable for the application in perturbation theory.

For physical matrix elements, i.e. matrix elements of gauge-invariant quantities, the gauge dependence of the sources in $j^{a,\mu}A^{(-\theta)}_{a,\mu}$ etc. is irrelevant. These quantities can be calculated by putting $\theta(x)\equiv 0$ in Eqn. (2.3.125). Then the integration $\int \mathscr{D}[\theta^a]$ over the gauge group gives only an insignificant factor and according to construction $Z\{j; \bar{\eta}, \eta\}$ is independent of $C^a[\mathscr{A}_\mu(x)]$, i.e. especially independent of the gauge parameter ξ.

For an infinitesimal gauge transformation it follows according to Eqn. (2.2.7):

$$\delta C^a[\mathscr{A}_\mu(x)] = \frac{1}{\sqrt{\xi}}\,\partial^\mu\,\delta A^a_\mu(x)$$

$$= \frac{1}{\sqrt{\xi}}\,\partial^\mu\left(-\frac{1}{g}\,\delta^{ab}\,\partial_\mu + f^{abc}A^c_\mu(x)\right)\delta\theta^b(x). \qquad (2.3.126)$$

From this, the functional determinant can be calculated (see Eqns. (2.3.61), (2.3.62)):

$$\det \frac{\delta C_a[\mathscr{A}_\mu(x)]}{\delta\theta^b(y)} \sim \det K_{ab}(x-y, A)$$

with

$$K_{ab}(x-y, A) = \partial^\mu_x(\delta_{ab}\,\partial^x_\mu - g f_{abc} A^c_\mu(x))\,\delta^{(4)}(x-y). \qquad (2.3.127)$$

This can be expressed by a functional integration over anticommuting ghost fields $u^a(x)$, $\bar{u}^a(x)$ according to Eqn. (2.3.117):

$$\det K_{ab}(x-y, A) \sim \int \mathscr{D}[u^a]\mathscr{D}[\bar{u}^a]\exp\left(\mathrm{i}\int \mathrm{d}^4x\,\mathrm{d}^4y(\bar{u}^a(x)K_{ab}(x-y, A)u^b(y))\right). \qquad (2.3.128)$$

The *path integral representation of the generating functional of the Green functions of QCD* is finally obtained after introducing sources $\bar{\omega}_a(x)$, $\omega_a(x)$ for the ghost fields and changing $Z\{j; \bar{\eta}, \eta\}$ by infinite constant factors which are irrelevant for the determination of Green functions:

$$T\{j; \bar{\eta}, \eta; \bar{\omega}, \omega\} = Z\{j; \bar{\eta}, \eta; \bar{\omega}, \omega\}/Z\{0; 0, 0; 0, 0\},$$

$$Z\{j; \bar{\eta}, \eta; \bar{\omega}, \omega\} = \int \mathscr{D}[A^a_\mu]\mathscr{D}[\psi]\mathscr{D}[\bar{\psi}]\mathscr{D}[u^a]\mathscr{D}[\bar{u}^a]$$

$$\times\exp\Big(\mathrm{i}\int \mathrm{d}^4x(\mathscr{L} + \mathscr{L}_{\text{fix}} + \mathscr{L}_{\text{ghost}} + j^\mu_a(x)A^a_\mu(x)$$

$$+\bar{\psi}(x)\eta(x) + \bar{\eta}(x)\psi(x) + \bar{u}^a(x)\omega_a(x) + \bar{\omega}_a(x)u^a(x))\Big) \qquad (2.3.129)$$

with

$$\mathscr{L} = -\tfrac{1}{4}F^a_{\mu\nu}(x)F^{\mu\nu}_a(x) + \bar{\psi}(x)(\mathrm{i}\gamma^\mu D_\mu - M)\psi(x), \qquad (2.3.129')$$

$$\mathscr{L}_{\text{fix}} = -\tfrac{1}{2}(C^a[\mathscr{A}_\mu(x)])^2 = -\frac{1}{2\xi}(\partial^\mu A^a_\mu(x))^2, \qquad (2.3.129'')$$

$$\mathscr{L}_{\text{ghost}} = \bar{u}^a(x)\,\partial^\mu(\partial_\mu\,\delta_{ab} - g f_{abc}A^c_\mu(x))u^b(x). \qquad (2.3.129''')$$

As stressed several times, this is the compact formulation of QCD. A perturbative evaluation according to Eqn. (2.3.83) leads to the Feynman rules of Section 2.2.3. The gauge fixing term $C^a[\mathscr{A}_\mu(x)] = \partial^\mu A^a_\mu(x)/\sqrt{\xi}$ determines the gauge parameter ξ in the gluon propagator. The Faddeev–Popov ghost fields $u^a(x)$, $\bar{u}^a(x)$ lead to the appearance of ghost propagators and ghost–gluon vertices. However, the ghost states have no physical significance. Later on, in Section 2.4.3 it will be shown that due to gauge invariance the contributions of unphysical ghost states to the unitarity equation (2.3.49) are cancelled according to the cutting rule (2.3.47) by contributions of non-gauge-invariant, unphysical gluon states.

Mathematically, the functional integral (2.3.129) is not well defined. Divergences which result in the perturbation expansion must be overcome by regularization in the frame of renormalization theory. Section 2.4 deals with this aspect in more detail. The question as to how far the imperfectly defined functional integral can be evaluated non-perturbatively after regularization and whether the physical results obtained will be meaningful, is one of the central points of research in QCD at the present time. The problem is confronted again in Section 2.6 in connection with quark confinement.

(b) The manifest gauge invariance of the formulation of QCD is lost by the introduction of the gauge-fixing term $C^a[\mathscr{A}_\mu(x)]$. However, a symmetry of the full Lagrangian function $\mathscr{L} + \mathscr{L}_{\text{fix}} + \mathscr{L}_{\text{ghost}}$ with all the consequences of gauge invariance for all physical results can be defined again by an extension of the gauge transformations to the ghost fields.

The extended gauge transformations are the Becchi–Rouet–Stora (BRS) transformations [Be 74]. They are defined infinitesimally by

$$\begin{aligned}
\delta_s A^a_\mu(x) &= -f^{abc} A^b_\mu(x) u^c(x)\, \delta\bar{\lambda} - \frac{1}{g}\, \partial_\mu u^a(x)\, \delta\bar{\lambda}, \\
\delta_s \psi(x) &= -\mathrm{i} T_a \psi(x) u^a(x)\, \delta\bar{\lambda}, \\
\delta_s \bar{\psi}(x) &= \mathrm{i} \bar{\psi}(x) T_a u^a(x)\, \delta\bar{\lambda}, \\
\delta_s u^a(x) &= -\tfrac{1}{2} f^{abc} u_b(x) u_c(x)\, \delta\bar{\lambda}, \\
\delta_s \bar{u}^a(x) &= \frac{1}{g\sqrt{\xi}}\, C^a[\mathscr{A}_\mu(x)]\, \delta\bar{\lambda} = \frac{1}{g\xi}\, \partial^\mu A^a_\mu(x)\, \delta\bar{\lambda}.
\end{aligned} \tag{2.3.130}$$

$u^a(x)$ is the ghost field introduced above; $\delta\bar{\lambda}$ is an infinitesimal constant which anticommutes with the ghost fields $u^a(x)$, $\bar{u}^a(x)$.

A comparison with Eqn. (2.2.7) shows that the BRS transformations (2.3.130) for quark and gluon fields are gauge transformations with $\delta\theta^a(x) = u^a(x)\, \delta\bar{\lambda}$.* It follows that the Lagrangian density $\mathscr{L}$ (Eqn. (2.3.129′)) is BRS invariant: $\delta_s \mathscr{L} = 0$. The transformation behaviour of ghost fields is chosen so

* If a ghost number characterizing the ghost fields is introduced in analogy to the fermion number, then $u^a(x)$, $\bar{u}^a(x)$ and $\delta\bar{\lambda}$ have ghost numbers $+1$, -1 and -1, respectively. $\delta\theta^a(x) = u^a(x)\, \delta\bar{\lambda}$ has ghost and fermion number zero.

that $\mathscr{L}_{\text{fix}}+\mathscr{L}_{\text{ghost}}$ is also BRS-invariant although both terms are transformed non-trivially:

$$\delta_s(\mathscr{L}_{\text{fix}}+\mathscr{L}_{\text{ghost}})=0. \tag{2.3.131}$$

This shows the importance of the Faddeev–Popov ghost fields which re-establish the gauge invariance of a theory destroyed by $\mathscr{L}_{\text{fix}}$.

In order to prove Eqn. (2.3.131), we notice that in view of the Jacobi identity of the structure constants f^{abc} and the anticommutativity of the ghost fields

$$\delta_s(\partial_\mu u_a(x)+gf_{abc}A^b_\mu(x)u^c(x))=0.$$

From this it follows that

$$\delta_s\left(\int d^4y K_{ab}(x-y,A)u^b(y)\right)=\delta_s(\partial^\mu(\partial_\mu u_a(x)+gf_{abc}A^b_\mu(x)u^c(x)))=0$$

and

$$\begin{aligned}\delta_s\mathscr{L}_{\text{ghost}}&=\delta_s\left(\int d^4x\, d^4y\bar{u}^a(x)K_{ab}(x-y,A)u^b(y)\right)\\&=(\delta_s\bar{u}^a(x))\,\partial^\mu(\partial_\mu u_a(x)+gf_{abc}A^b_\mu(x)u^c(x)).\end{aligned} \tag{2.3.132}$$

On the other hand,

$$\begin{aligned}\delta_s\mathscr{L}_{\text{fix}}=\delta_s(-\tfrac{1}{2}(C^a[\mathscr{A}_\mu(x)])^2)&=-\frac{1}{\xi}(\partial^\nu A^a_\nu(x))\,\partial^\mu\,\delta_s A^a_\mu(x)\\&=\frac{1}{g\xi}(\partial^\nu A^a_\nu(x))\,\partial^\mu(\partial_\mu u_a(x)+gf_{abc}A^b_\mu(x)u^c(x))\,\delta\bar{\lambda}.\end{aligned}$$

is obtained for the variation of $\mathscr{L}_{\text{fix}}$. From this Eqn. (2.3.131) follows with help of Eqns. (2.3.130) and (2.3.132) because of the anticommutativity of $\delta\bar{\lambda}$ with $u_a(x)$.

(c) In Section 2.3.2 a functional differential equation for the generating functional of the Green functions (Eqn. (2.3.103)) was derived from the gauge symmetry of the first kind of the Lagrangian density $\mathscr{L}$ of a scalar field $A(x)$ and the invariance of the measure $\mathscr{D}[A]$ in the functional integral (2.3.96). Carrying these considerations over to BRS symmetry gives the *Ward identities* [Wa 50] *for the generating functional of the Green functions* of QCD [Le 72, tH 71, tH 72, Sl 72, Ta 71]. These summarize all relationships between the Green functions of the theory resulting from gauge symmetry of the second kind.

If quark fields are neglected, for the sake of simplicity, the following equation results from a BRS variation of the fields in analogy to Eqns. (2.3.96), (2.3.99), (2.3.100):

$$\begin{aligned}\int\mathscr{D}[A^a_\mu]\mathscr{D}[u^a]\mathscr{D}[\bar{u}^a]\bigg(\int d^4x(j^\mu_a(x)\,\delta_s A^a_\mu(x)\\+(\delta_s\bar{u}^a(x))\omega_a(x)+\bar{\omega}_a(x)\,\delta_s u^a(x)\bigg)\exp\left(i\int d^4x\{\quad\}\right)=0\end{aligned} \tag{2.3.133}$$

with

$$\{\quad\} := \mathscr{L}_{\text{gauge}} + \mathscr{L}_{\text{fix}} + \mathscr{L}_{\text{ghost}} + j^{\mu}_a A^a_{\mu} + \bar{u}^a \omega_a + \bar{\omega}_a u^a. \qquad (2.3.133')$$

Using Eqn. (2.3.130) and replacing the fields by differentiation with respect to their sources, it follows that:

$$\int \mathrm{d}^4x \Big[j^{\mu}_b(x) \Big(\delta^b_c \partial^x_{\mu} - g f^{bcd} \frac{\delta}{\mathrm{i}\,\delta j^{\mu}_d(x)} \Big) \frac{\delta}{\mathrm{i}\,\delta \bar{\omega}_c(x)}$$

$$+ \frac{1}{\xi} \partial^{\mu} \frac{\delta}{\mathrm{i}\,\delta j^{\mu}_a(x)} \omega_a(x) + g \bar{\omega}_a(x) \tfrac{1}{2} f^{abc} \frac{\delta}{\mathrm{i}\,\delta \bar{\omega}_b(x)} \frac{\delta}{\mathrm{i}\,\delta \bar{\omega}_c(x)} \Big] T\{j; \bar{\omega}, \omega\} = 0 \qquad (2.3.134)$$

with

$$T\{j; \bar{\omega}, \omega\} = Z\{j; \bar{\omega}, \omega\} / Z\{0; 0, 0\},$$

$$Z\{j; \bar{\omega}, \omega\} = \int \mathscr{D}[A^a_{\mu}] \mathscr{D}[u^a] \mathscr{D}[\bar{u}^a] \exp\Big(\mathrm{i} \int \mathrm{d}^4x \{\quad\} \Big), \qquad (2.3.135)$$

(cf. Eqn. (2.3.133′)). Differentiation with respect to ω finally gives

$$\frac{\mathrm{i}}{\xi} \partial^{\mu}_y \frac{\delta T\{j\}}{\mathrm{i}\,\delta j^{\mu}_a(y)} - \int \mathrm{d}^4x\, j^{\mu}_b(x) \Big(\delta^b_c \partial^x_{\mu} - g f^{bcd} \frac{\delta}{\mathrm{i}\,\delta j^{\mu}_d(x)} \Big) T^{ca}\{j; x, y) = 0 \qquad (2.3.136)$$

for $\omega = \bar{\omega} = 0$. The last term in Eqn. (2.3.134) has a non-vanishing ghost number. Therefore it is zero for $\omega = \bar{\omega} = 0$ and does not contribute to Eqn. (2.3.136). The abbreviations $T\{j\} := T\{j; \bar{\omega} = 0, \omega = 0\}$ and

$$T^{ab}\{j; x, y) := \frac{\delta}{\mathrm{i}\,\delta \bar{\omega}_a(x)}\, \mathrm{i} \frac{\delta}{\delta \omega_b(y)} T\{j; \bar{\omega}, \omega\} \Big|_{\omega = \bar{\omega} = 0} \qquad (2.3.137)$$

have been used above. $T^{ab}\{j = 0; x, y)$ is the ghost field propagator. The more general ghost propagator in the presence of sources $j : T^{ab}\{j; x, y)$ can be calculated from the ghost field equation of motion. For this, the functional integral (2.3.135) is varied with respect to $\bar{u}^a(x)$:

$$\int \mathscr{D}[A^a_{\mu}] \mathscr{D}[u^a] \mathscr{D}[\bar{u}^a] \exp\Big(\mathrm{i} \int \mathrm{d}^4x \{\quad\} \Big)$$

$$\times \Big(\int \mathrm{d}^4x\, \delta \bar{u}^b(x) (\partial^{\mu} (\partial_{\mu} \delta_{bc} - g f_{bcd} A^d_{\mu}(x)) u^c(x) + \omega_b(x)) \Big) = 0. \qquad (2.3.138)$$

The equation

$$\Big[\partial^{\mu} \Big(\partial_{\mu} \delta_{bc} - g f_{bcd} \frac{\delta}{\mathrm{i}\,\delta j^{\mu}_d(x)} \Big) \frac{\delta}{\mathrm{i}\,\delta \bar{\omega}_c(x)} + \omega_b(x) \Big] T\{j; \bar{\omega}, \omega\} = 0$$

results just like in the derivation of Eqn. (2.3.134) and

$$\partial^{\mu}_x \Big(\partial^x_{\mu} \delta^b_c - g f_{bcd} \frac{\delta}{\mathrm{i}\,\delta j^{\mu}_d(x)} \Big) T^{ca}\{j; x, y)$$

$$\equiv \int \mathrm{d}^4z\, K^b_c \Big(x - z, \frac{\delta}{\mathrm{i}\,\delta j} \Big) T^{ca}\{j; z, y) = \mathrm{i}\, \delta^{ba} \delta^{(4)}(x - y) T\{j\}, \qquad (2.3.139)$$

by differentiating with respect to ω for $\omega = \bar{\omega} = 0$. The equations (2.3.136) and (2.3.139) are the desired Ward identities. They are the central equations for the proof that gauge field theories can be renormalized (Section 2.4). Their extension to theories with fermions is obvious. The following applies for the functional integral (2.3.129) of QCD:

$$\frac{i}{\xi}\partial_y^\mu \frac{\delta T\{j;\bar{\eta},\eta\}}{i\,\delta j_a^\mu(y)} - \int d^4x\left[j_b^\mu(x)\left(\partial_\mu\,\delta_c^b - gf^{bcd}\frac{\delta}{i\,\delta j_d^\mu(x)}\right)\right.$$
$$\left. + g\bar{\eta}(x)T_a\frac{\delta}{\delta\bar{\eta}(x)} - g\eta(x)T_a^t\frac{\delta}{\delta\eta(x)}\right]T^{ca}\{j;\bar{\eta},\eta;x,y) = 0 \quad (2.3.136')$$

and

$$\partial_x^\mu\left(\partial_\mu^x\,\delta_c^b - gf_{bcd}\frac{\delta}{i\,\delta j_d^\mu(x)}\right)T^{ca}\{j;\bar{\eta},\eta;x,y) = i\,\delta^{ba}\,\delta^{(4)}(x-y)T\{j;\bar{\eta},\eta\}$$
$$(2.3.139')$$

with $T\{j;\bar{\eta},\eta\} := T\{j;\bar{\eta},\eta;\bar{\omega}=0,\omega=0\}$ and $T^{ab}\{j;\bar{\eta},\eta;x,y)$ according to Eqn. (2.3.137). Identities for individual Green functions can be obtained by differentiating the functional differential equations with respect to the external sources and then putting them equal to zero (see Eqns. (2.3.50), (2.3.56)). A diagrammatic technique is the best way of providing a systematic evaluation [tH 73], in view of the many elementary fields and vertices in a non-Abelian gauge theory. However, no details are given here except by way of demonstrating how to evaluate Ward identities by means of a few simple examples.

(d) Applying $\partial_x^\nu\,\delta/i\,\delta j_b^\nu(x)$ to Eqn. (2.3.136′) and then putting $j=\eta=\bar{\eta}=0$ gives

$$\frac{i}{\xi}\partial_x^\nu\partial_y^\mu\frac{\delta^2 T\{j;\bar{\eta},\eta\}}{i\,\delta j_b^\nu(x)\,i\,\delta j_a^\mu(y)}\bigg|_{j=\eta=\bar{\eta}=0}$$
$$= \frac{1}{i}\partial_x^\nu\left(\partial_\nu^x\,\delta_c^b - gf^{bcd}\frac{\delta}{i\,\delta j_d^\nu(x)}\right)T^{ca}\{j;\bar{\eta},\eta;x,y)\bigg|_{j=\eta=\bar{\eta}=0}. \quad (2.3.140)$$

This is the Ward identity for the gluon propagator

$$iD'(x-y)_{\mu\nu}^{ab} := \frac{\delta^2 T\{j;\bar{\eta},\eta\}}{i\,\delta j_a^\mu(x)\,i\,\delta j_b^\nu(y)}\bigg|_{j=\eta=\bar{\eta}=0}. \quad (2.3.141)$$

If Eqn. (2.3.139′) is used with $j=\eta=\bar{\eta}=0$, then it follows from Eqn. (2.3.140) that:

$$\partial_x^\mu\,\partial_x^\nu D'(x-y)_{\mu\nu}^{ab} = \xi\,\delta^{ab}\,\delta^{(4)}(x-y)$$

or, in momentum space:

$$k^\mu k^\nu\tilde{D}'(k)_{\mu\nu}^{ab} = -\frac{\xi}{(2\pi)^4}\,\delta^{ab}. \quad (2.3.142)$$

This is the *Slavnov identity* [Sl 72, Ta 71], the Ward identity for the gluon propagator which requires the transversality of the gluon propagator in Landau gauge $\xi=0$. It is fulfilled in the tree graph approximation, i.e. for the free propagator. Eqn. (2.2.42) gives $iD'(k)^{ab}_{\mu\nu}$ in this approximation.

In the next section the important role played by Ward identities in the renormalization of Yang–Mills fields theories will be described. Thereby it will become clear that the propagator of the interacting gluon field must also satisfy the Slavnov identity (2.3.142) as well as its approximations to any order of the loop expansion (Section 2.3.2(e)).

(e) Since the derivation of Eqns. (2.3.136′) and (2.3.139′) obviously applies for any gauge group with structure constants f_{abc}, we finally consider *QED*. The Ward identities are considerably simpler in this case, because all structure constants vanish and the representation matrices T_a are trivial ($T_a=-1$, $g=-e$, see Section 2.2.2(c)). First of all

$$\partial^\mu_x\,\partial^x_\mu T^{ab}\{j;\bar\eta,\eta;x,y)=\mathrm{i}\,\delta^{ab}\,\delta^{(4)}(x-y)T\{j;\bar\eta,\eta\}$$

follows from Eqn. (2.3.139′). With the help of Eqn. (1.2.38′) for $M=0$, this equation can be integrated immediately to give:

$$T^{ab}\{j;\bar\eta,\eta;x,y)=-\mathrm{i}\,\delta^{ab}\,\Delta_F(x-y,M=0)T\{j;\bar\eta,\eta\}. \qquad (2.3.143)$$

The result is the ghost propagator in the presence of sources j, $\bar\eta$, and η. It factorizes into the ghost propagator and the generating functional of the 'ghost-free' theory. The reason for this is apparent from Eqn. (2.3.129‴): the ghost field in an Abelian gauge field theory ($f_{abc}\equiv 0$) is a free field; the functional $T\{j;\bar\eta,\eta;\ \bar\omega,\omega\}$ factorizes therefore into a ghost component $T\{\bar\omega,\omega\}$ and $T\{j;\bar\eta,\eta\}$ (see Eqn. (2.3.129) and Section 2.3.2(d)). Expressed in another way: no Faddeev–Popov ghost fields are necessary for the quantization of an Abelian gauge theory with linear gauge fixing.

If Eqn. (2.3.143) is inserted into Eqn. (2.3.136′) and if $f_{abc}\equiv 0$, $T_a\equiv -1$ and $g=-e$ is used, then the Ward identity

$$\frac{1}{\xi}\,\partial^\mu\,\frac{\delta T\{j;\bar\eta,\eta\}}{\mathrm{i}\,\delta j^\mu(x)}=\int \mathrm{d}^4x\left(\partial^\mu j_\mu(x)-e\bar\eta(x)\frac{\delta}{\delta\bar\eta(x)}+e\eta(x)\frac{\delta}{\delta\eta(x)}\right)\cdot$$
$$\times\Delta_F(x-y,0)T\{j;\bar\eta,\eta\} \qquad (2.3.144)$$

for the generating functional of the Green functions of QED is obtained. Differentiating with respect to j, gives for $j=\eta=\bar\eta=0$ the generalized transversality condition for the photon propagator

$$\partial^\mu_x D'(x-y)_{\mu\nu}=-\xi\,\partial^x_\nu\Delta_F(x-y,0)$$
$$\mathrm{i}D'(x-y)_{\mu\nu}:=\delta^2T/\mathrm{i}\,\delta j^\mu(x)\mathrm{i}\,\delta j^\nu(y)\big|_{j=\eta=\bar\eta=0} \qquad (2.3.145)$$

(see Eqn. (2.3.142)). Differentiation of Eqn. (2.3.144) with respect to η and $\bar\eta$ gives at the point $j=\eta=\bar\eta=0$:

$$\frac{1}{\xi}\,\partial^\mu_z\tau(x;y;z)_{\alpha;\beta;\mu}=e\mathrm{i}S'_F(x-y)_{\alpha\beta}[\Delta_F(y-z)-\Delta_F(x-z)] \qquad (2.3.146)$$

with the 3-point Green function

$$\tau(x; y; z)_{\alpha;\beta;\mu} = \langle T\psi_\alpha(x)\bar{\psi}_\beta(y)A_\mu(z)\rangle$$
$$= \frac{\delta^3 T\{j; \bar{\eta}, \eta\}}{\delta\bar{\eta}_\alpha(x)\, \delta\eta_\beta(y) \mathrm{i}\, \delta j^\mu(z)}\bigg|_{j=\eta=\bar{\eta}=0}$$

and the full electron propagator

$$\mathrm{i}S'_F(x-y)_{\alpha\beta} = \tau(x; y\mathrm{i})_{\alpha;\beta} = \langle T\psi_\alpha(x)\bar{\psi}_\beta(y)\rangle = \frac{\delta^2 T\{j; \bar{\eta}, \eta\}}{\delta\bar{\eta}_\alpha(x)\, \delta\eta_\beta(y)}\bigg|_{j=\eta=\bar{\eta}=0}.$$

The Green functions occurring in Eqn. (2.3.146) are connected. By amputating all field points according to Eqn. (2.3.33) the well-known identity for the electron-photon vertex proved by Y. Takahashi [Ta 57] is obtained (see Eqns. (2.3.33), (2.3.34)):

$$\begin{aligned} \partial_z^\mu \Gamma(x; y; z)_{\alpha;\beta;\mu} &\equiv \partial_z^\mu \tau(\underline{x}; \underline{y}; \underline{z})_{\alpha;\beta;\mu} \\ &= eS_F'^{-1}(x-y)_{\alpha\beta}(\delta^{(4)}(x-z) - \delta^{(4)}(y-z)). \end{aligned} \tag{2.3.147}$$

Its verification in the tree-graph approximation is carried out with the help of Eqn. (2.1.3) and Eqns. (2.3.92), (2.3.65), and (2.3.73) adapted for QED, or after Fourier transformation into the momentum space (Eqn. (2.3.26)) using Eqns. (1.2.37), (1.2.40). A systematic treatment of quantized gauge fields beyond the tree-graph approximation is the object of the next section.

2.4 Renormalization of quantized gauge field theories

The concept of renormalizability [RT] of a quantum field theory is right at the heart of the general discussion on the dynamical structure of strong and electroweak interaction; for only the renormalizability of a dynamics enables its non-conflicting understanding and therefore its evaluation beyond the level of simple collision theory represented by the tree graph approximation (Section 2.3.2). Moreover, the 'renormalization philosophy' has given rise to phenomenologically important concepts such as the 'running coupling constant' and 'dynamical dimensions' etc., (see Sections 2.5 and 2.7). Therefore, Section 2.4.1 is devoted to the main points of this renormalization programme and Section 2.4.2 illustrates it by calculating the gluon propagator in 1-loop approximation. Some simple combinatorial aspects together with dimensional considerations form the basis on which a decision as to whether or not a theory can be renormalized is taken. The complete proof that a gauge theory can be renormalized [Le 72, tH 72] was one of the most important stages in the development of QCD and the Glashow–Salam–Weinberg theory (GSW theory). Details cannot be entered into here but Section 2.4.3 gives an idea of how this proof is built up and the various stages reached *en route*. Reference is made to original work on the subject. However, knowledge of this last section is not essential for an initial grounding in QCD and GSW theory.

2.4.1 Divergences and renormalization

(a) The discussion of QED in Section 2.1.1 showed that internal radiative corrections (loop diagrams) modify the simple collision theory (tree graph approximation) within a field theory so much that the parameters of the Lagrangian function lose their original meaning as physical masses and directly measurable coupling constants. This fact is bad enough in itself but a further complication is presented by the fact that *the calculation of simple loop diagrams leads to divergent results.* Two examples of this are given below.

In QED, the internal radiative correction of the electron propagator is given in 1-loop approximation by the following graph

x_1 x_2

$$= \int \frac{d^4p}{(2\pi)^4} \frac{\exp[-ip(x_1-x_2)]i}{\gamma p - M + i\varepsilon} \Pi_2(p) \frac{i}{\gamma p - M + i\varepsilon} \tag{2.4.1}$$

with the expression (in Feynman gauge)

$$\Pi_2(p) = (ie)^2 \int \frac{d^4k}{(2\pi)^4} \frac{-i}{k^2+i\varepsilon} \gamma_\mu \frac{i}{\gamma \cdot (p-k) - M + i\varepsilon} \gamma^\mu \tag{2.4.2}$$

for the amputated Green function. A simple counting of the powers of k is the basis for the conjecture that this integral diverges linearly for large k values. In fact for symmetry reasons the leading divergent term $\sim \int d^4k k_\mu (k^2+i\varepsilon)^{-1} \cdot [(p-k)^2 - M^2 + i\varepsilon]^{-1}$ must be proportional to p_μ with a coefficient which therefore diverges only logarithmically.

In the case of the interaction of scalar particles mentioned in section (2.3.1(d)) the above graph gives for the amputated Green function:

$$\Pi_2(p) = (ig)^2 \int \frac{d^4k}{(2\pi)^4} \frac{i}{(k^2 - \mu^2 + i\varepsilon)} \cdot \frac{i}{(p-k)^2 - m^2 + i\varepsilon}, \tag{2.4.3}$$

an expression which is logarithmically divergent according to a superficial power counting in k. These simple examples should illustrate that when calculating Feynman graphs with loops by integrating with respect to loop momenta k, divergences occur with a power which is dependent on the composition of the loops made from vertices and internal lines.

(b) An initial idea of the structure of these divergences and from this a *criterion for renormalizability* can be obtained by a systematic *counting of the momentum powers* in the Feynman diagrams. In general there are l internal loop momenta in a general graph with l loops and L external lines with momenta $P_1, \ldots, P_L$ over which integration must be carried out. The integrand I is a product of vertices and propagators which depend on $P_1, \ldots, P_{L-1}$ and $k_1, \ldots, k_l$:

$$G(P_1, \ldots, P_{L-1}) = \int^\Lambda d^4k_1 \ldots d^4k_l I(P_1, \ldots P_{L-1}, k_1, \ldots, k_l) \tag{2.4.4}$$

(in our examples, l was always equal to 1). The possible divergent integration is presumed to be cut off at momentum Λ. Simple counting of the powers provides for a definition of the superficial degree of divergence of subintegrations. In this respect consider the increase of I when a partial set $S=\{k_i\}$ of m loop momenta becomes large, i.e. for $k_{i'}=\lambda r_{i'}$ with $k_{i'}\in S$, $k_i\notin S$; r_i, P_j fixed and in a general position (i.e. no partial sum must disappear). Then there is an upper bound for the power behaviour of I as a function of λ:

$$|I(\lambda)|\leqslant\lambda^{c(S)}\cdot\text{constant} \tag{2.4.5}$$

(in our examples, $l=1$, $S=\{k\}$ and $c=-3$ in Eqn. (2.4.2) and $c=-4$ in Eqn. (2.4.3)). The estimate:

$$\left|\int^{\Lambda}\mathrm{d}k_{i'_1}\ldots\mathrm{d}k_{i'_m}I\right|<\Lambda^{d_4(S)}\times[\text{logarithms of }\Lambda] \tag{2.4.6}$$

applies for integration with respect to momenta $k_{i'}\in S$. $d_4(S)=4\times m+c(S)$ is called the superficial degree of divergence [Dy 49] of the integration with respect to the momenta $k_{i'}\in S$ in four space-time dimensions. The integration is convergent when $d_4<0$. Note that this degree of divergence is lower when these same graphs are considered in lower space-time dimensions D: $d_D(S)=D\times m+c(S)$. The fact will be used later.

A graph G is referred to as primitively divergent [Dy 49] when only the final integration is the cause of divergence, i.e. when $d(S)<0$ for all $m(S)<l$. Thus, the divergence of an arbitrary divergent graph is based on the divergence of the integration over the loop momenta of primitive subgraphs. The structure of the primitively divergent graphs in a field theory, which are necessarily connected, can be deduced from combinatorial arguments. QED with graph rules described in Section 1.2.2(f) is taken as an example. If a graph has V photon–electron vertices and I_ψ, I_A (E_ψ, E_A) internal (external) electron and photon lines, respectively, then $d_D=D\cdot l-2I_A-I_\psi$, for each internal fermion line contributes one power of momentum to the denominator, each photon line two and the vertices do not alter the momentum behaviour. As two fermion lines and one photon line run into each vertex, apart from the loop equation (2.3.88):

$$l=I_A+I_\psi-V+1 \tag{2.4.7}$$

the following combinatorial equations are also valid:

$$E_\psi+2I_\psi=2V,\quad E_A+2I_A=V. \tag{2.4.8}$$

Thus, the following formula can be written for the degree of divergence for a graph in QED:

$$\begin{aligned}d_D&=D-\frac{D-1}{2}E_\psi-\frac{D-2}{2}E_A+(\tfrac{1}{2}D-2)V,\\ d_4&=4-\tfrac{3}{2}E_\psi-E_A.\end{aligned} \tag{2.4.9}$$

For the physical space–time dimension $D=4$, the number of external lines of primitively divergent graphs, i.e. graphs in which $d_4>0$, is limited. Symmetry principles such as charge conjugation, gauge invariance, etc. can make superficially divergent graphs convergent either individually or jointly. More detailed investigation shows that primitively divergent graphs with external lines in QED include the cases: (1) $E_\psi=2$, $E_A=0$; (2) $E_\psi=0$, $E_A=2$; and (3) $E_\psi=2$, $E_A=1$. The physical aspects of these types of graphs have already been discussed in Section 2.1.1. A field theory with a finite number of primitively divergent graphs is called renormalizable.

Weak interaction described in Section 1.3 may be also discussed in this context. Fermi theory assumes that only Dirac particles propagate and of these, each four interact with each other pointlike. This implies combinatorial equations similar to (2.4.7), (2.4.8): $l=I_\psi-V+1$, $E_\psi+2\cdot I_\psi=4\cdot V$, and therefore $d_4=4l-I_\psi=4+V-E_\psi/2$ is the degree of divergence. The number of external lines of primitively divergent graphs can increase *ad infinitum* with increasing order V, so Fermi theory is not renormalizable. Its high energy behaviour which violates the unitarity of the S-matrix cannot be simply corrected by higher-order graphs as these become increasingly undefined due to the increase of the number of primitively divergent graphs. This is the reason for extending the Fermi theory by massive, intermediate vector bosons $W^\pm$, Z^0 (Section 1.3.1), which then gives the same combinatorial equations as in QED (Eqns. (2.4.7–2.4.8)). The W, Z propagators from Eqns. (1.3.16) and (1.3.18) would give a unitary S-matrix (see Section 2.3.1(f)). However, their power behaviour is of the zeroth order in momentum k and thus $d_4=4+V-3E_\psi/2-2(E_W+E_Z)$ represents the superficial degree of divergence of primitively divergent graphs. The intermediate vector boson theory of weak interaction cannot be renormalized either.

The difference between interacting, massive vector particles as encountered so far in this discussion of weak interaction theory and QED is based on gauge invariance. By means of the latter, the photon propagator can be put into the simple form $\sim -g_{\mu\nu}/k^2$ (Eqn. (1.2.40)) which has the consequence that QED can be renormalized.

This seems to suggest that weak interaction theory might be renormalized by formulating it as a gauge field theory (Chapter 3), which is an extension of the intermediate vector boson hypothesis. The decisive step in this programme is the Higgs–Kibble mechanism (Section 3.2) which allows one to express massive vector fields as gauge fields. Within a class of gauges characterized by the gauge parameter ξ, the vector propagator can be written as

$$\frac{\mathrm{i}}{(2\pi)^4}\,\frac{-g_{\mu\nu}+k_\mu k_\nu(\xi-1)/(\xi k^2-M^2)}{k^2-M^2+\mathrm{i}\varepsilon}.$$

As in QED, the high energy behaviour is $\sim -g_{\mu\nu}/k^2$ for $\xi\neq 0$ and the theory can be renormalized (R-gauges). On the other hand, the propagator of Eqn. (2.3.20) is obtained for $\xi=0$. This one gives a unitary S-matrix (U-gauge).

Unitarity gets proven by the fact that the renormalization procedure (see Section 2.4.3) and the physical S-matrix are gauge-invariant [Le 72, Le 72a, tH 71, tH 72, Ty 72].

In the formulation of the Feynman rules of QCD (Section 2.2.3), the gauge freedom was expressed by the dependence of the gluon propagator on the gauge parameter ξ. Section 2.3.3 shows formally that physical, gauge-invariant quantities do not depend on ξ. This must also hold for correct perturbation theoretical calculations. For this the treatment of divergent graphs should not violate gauge invariance. If this is assumed, then primitively divergent graphs can be listed up for QCD too: let I_A, I_ψ, $I_G(E_A, E_\psi)$ be the numbers of the internal (external) lines of gluons, quarks, and ghosts and V_A^4, V_A^3, V_ψ, and V_G represent the 4-gluon, 3-gluon, quark–gluon and ghost–gluon vertices. Counting of powers then gives the expression $d_D = D \cdot l - 2 \cdot I_A - I_\psi - 2 \cdot I_G + V_A^3 + V_G$ for the superficial degree of divergence. The following combinatorial equations apply:

$$l = \mathbf{1} + I_A + I_\psi + I_G - V_A^4 - V_A^3 - V_\psi - V_G \quad \text{(loop rule)}$$

$$E_A + 2 \cdot I_A = 4 \cdot V_A^4 + 3 \cdot V_A^3 + V_G + V_\psi, \quad E_\psi + 2 \cdot I_\psi = 2 \cdot V_\psi, \quad 2 \cdot I_G = 2 \cdot V_G$$

(no external ghost lines!, end points of lines coincide at vertices). From this, the degree of divergence is derived as a function of the number of external lines and vertices:

$$\begin{aligned} d_D &= D + (D-4)V_A^4 + \left(\frac{D}{2} - 2\right)V_A^3 + \left(\frac{D}{2} - 2\right)V_G + \left(\frac{D}{2} - 2\right)V_\psi \\ &\quad - \frac{D-2}{2}E_A - \frac{D-1}{2}E_\psi, \qquad (2.4.10) \\ d_4 &= 4 - E_A - \tfrac{3}{2}E_\psi. \end{aligned}$$

There is only a finite number of types of primitively divergent graphs in four dimensions. QCD can be renormalized [tH 71].

(c) Now, how can the divergent graphs be treated in a non-Abelian gauge theory? A detailed or complete account of this aspect is obviously beyond the scope of this book [RT]. So we have restricted our survey to the *principles of the renormalization programme*, giving the headings and then illustrating the whole by means of a relatively simple example in the next section.

(i) As already mentioned several times, the renormalization of non-Abelian gauge field theories is determined by the principle of gauge invariance, indeed it could not be performed without it. Thus the first rule is that *gauge invariance must be observed.*

(ii) Feynman rules give mathematically incomplete expressions for primitively divergent graphs. Some sense must be made of these by relating them to convergent expressions. This is called the 'regularization' of a theory [Pa 49]. However, simply cutting off the momentum as was done for the purpose of calculating the degree of divergence of loop integrations, violates gauge invariance in complicated graphs and is therefore unacceptable. The trick for regularization is to consider gauge theories in the first instance in low space–time dimensions D, in which the primitively divergent graphs give finite results (see e.g. Eqn. (2.4.10)). Analytic continuation in D defines the divergent part of a primitively divergent graph as pole $1/(4-D)$. This method is known as *dimensional regularization* [tH 72a]. An alternative method is *lattice regularization* [Wi 74, Ko 75] which covers non-perturbative treatments of QCD as well as the perturbative one defined by Feynman graphs (see Section 2.6).

(iii) The basis of renormalization philosophy was the fact that the divergent parts of primitively divergent graphs belong to quantities which cannot be physically observed. Thus, in QED, the divergent part of the primitively divergent electron propagator ($E_\psi = 2$, $E_A = 0$) changes the mass M in the Lagrangian function $\mathscr{L}$ (Eqn. (2.1.17)) by ΔM to give the only observable, physical electron mass $M_e = M + \Delta M$. If the 'bare' mass M is substituted by $M_e - \Delta M$ in $\mathscr{L}$, this obviously provides a counter-term $\bar{\psi}\Delta M\psi$ which compensates the divergences by providing additional Feynman graphs of the form $\overset{\Delta M}{\longrightarrow\!\!\!\times\!\!\!\longrightarrow}$. This example illustrates the *subtraction* of divergences by introducing counter-terms into the Langrangian density.

A field theory is renormalizable when divergences extracted by the regularization process can be compensated in the Lagrangian density by a finite number of counter-terms. A typical counter-term will have the following form

$$\Lambda^j\left(\log\frac{\Lambda}{M}\right)^k M^n g^V \mathcal{O}_d(\psi, A_\mu, \partial, \ldots)$$

with $j, k, n \geqslant 0$ and $V > 0$. Λ is a momentum cut-off parameter ($\dim \Lambda = M^1$),* g a coupling constant ($\dim g = M^{[g]}$), V is the order of perturbation theory selected and $\mathcal{O}_d$ an operator constructed from the fields and their derivatives ($\dim \mathcal{O}_d = M^d$, see Table 2.2.) The dimension of the Lagrangian $\mathscr{L}$ implies:

$$M^4 = M^{j+n+V\cdot[g]+d}, \quad \text{i.e. } d \leqslant 4 - V\cdot[g].$$

If $[g]$ is negative, d can increase *ad infinitum* with increasing order of perturbation theory and the theory cannot be renormalized because the number of counter-terms and hence new coupling constants is unrestricted. $[g] \geqslant 0$ is required for renormalizability. A summary of earlier considerations in the light of the above might be worded as follows:

> a Lagrangian field theory can be renormalized when all coupling constants have non-negative mass dimensions and when all boson propagators behave as k^{-2} and all fermion propagators as k^{-1} for large momenta k.

* A dimensional parameter μ occurs also during dimensional regularization (see Section 2.4.2).

Table 2.2 Canonical mass dimensions of field operators and coupling constants in gauge theories. Each Langrangian density $\mathscr{L}$ has the dimension $\dim L = M^4$ in the natural system of units $\hbar = c = 1$, as the action $\int d^4x\mathscr{L}$ is dimensionless. The dimensions for the fields and coupling constants are derived from the kinetic energy and coupling terms

Dirac field	$\bar{\psi}M\psi$	$\rightarrow \dim \psi = M^{3/2}$
Scalar field	$\frac{1}{2}(\partial^\mu\phi)(\partial_\mu\phi)$	$\rightarrow \dim \phi = M^1$
Vector field	$-\frac{1}{4}F_{\mu\nu}F^{\mu\nu}$	$\rightarrow \dim F_{\mu\nu} = M^2, \quad \dim A_\mu = M^1$
Minimal coupling	$g\bar{\psi}\gamma_\mu\psi A^\mu$	$\rightarrow \dim g = M^0$
3-gauge boson coupling	$gA_\mu A^\mu\ \partial_\nu A^\nu$	$\rightarrow \dim g = M^0$
4-gauge boson coupling	$g^2(A_\mu A^\mu)^2$	$\rightarrow \dim g = M^0$
4-fermion coupling	$G_F(\bar{\psi}\gamma\psi)^2$	$\rightarrow \dim G_F = M^{-2}$
Yukawa coupling	$g\bar{\psi}\psi \cdot \phi$	$\rightarrow \dim g = M^0$
Coupling of scalars	$g^*A_\mu A^\mu\phi$	$\rightarrow \dim g^* = M^1$
	$gA_\mu(\partial^\mu\phi)\cdot\phi$	$\rightarrow \dim g = M^0$
	$g^2\phi^2\cdot A_\mu A^\mu$	$\rightarrow \dim g = M^0$

Counter-terms to the Lagrangian function must be Poincaré-invariant operators with dimension M^4. In the case of QCD* the requirement of local gauge invariance implies that all types of possible counter-terms are already contained in the non-renormalized Lagrangian function, so that the *subtraction* of divergences can be carried out *by multiplicative renormalization*, e.g.

$$-\tfrac{1}{4}F_{\mu\nu}F^{\mu\nu} \rightarrow -\tfrac{1}{4}Z_A F_{\mu\nu}F^{\mu\nu} = -\tfrac{1}{4}F_{\mu\nu}F^{\mu\nu} + (Z_A - 1)(-\tfrac{1}{4}F_{\mu\nu}F^{\mu\nu}).$$

All divergent parts of primitively divergent graphs can be absorbed in multiplicative renormalization constants Z_i of fields and coupling constants, which become infinite in the limiting case $\Lambda \rightarrow \infty$ (Eqn. (2.4.6)) or with dimensional regularization for $\varepsilon = 4 - D \rightarrow 0$ (cf. Eqn. (2.4.29)):

$$\begin{aligned} A_\mu &= [Z_A(\varepsilon)]^{1/2} A^R_\mu, \\ \psi &= [Z_q(\varepsilon)]^{1/2}\psi^R, \quad g = \frac{Z(\varepsilon)}{Z_q(\varepsilon)\cdot[Z_A(\varepsilon)]^{1/2}}\, g_R. \\ u &= [Z_u(\varepsilon)]^{1/2}u^R. \end{aligned} \tag{2.4.11}$$

Renormalized fields A^R, ψ^R, u^R, and renormalized coupling constants g^R are finite for $\varepsilon \rightarrow 0$.

(iv) The condition which requires renormalization constants Z_i to absorb the divergences fixes these only up to finite parts—these are determined by *renormalization conditions* for vertex functions of renormalized fields:

$$\varphi_{n_A,n_q}(k_i; p_j) \quad \text{and} \quad \Gamma_{n_A,n_q}(k_i; p_j), \qquad i = 1, \ldots, n_A, j = 1, \ldots, n_q$$

designates the renormalized connected Green functions and renormalized vertex functions in momentum space (see Section 2.3.1). n_A represents the

* For theories with Higgs fields (see Section 3.2).

number of gluon fields and n_q the number of ψ plus $\bar{\psi}$ quark fields (n_q is, therefore, always even). Based on Eqn. (2.4.11), the following formula can be written for non-renormalized Green functions:

$$\varphi_{n_A,n_q}(k_i\,;\,p_j) = \lim_{\varepsilon\to 0}\,[Z_A(\varepsilon)]^{-n_A/2}[Z_q(\varepsilon)]^{-n_q/2}\varphi^{(u)}_{n_A,n_q}(k_i\,;\,p_j\,;\,\varepsilon) \quad (2.4.12)$$

and for the propagators in particular

$$\begin{aligned}\varphi_{2,0}(k_1,\,k_2\mathrm{i}) &= \lim_{\varepsilon\to 0}\,[Z_A(\varepsilon)]^{-1}\varphi^{(u)}_{2,0}(k_1,\,k_2;\,\varepsilon)\\ \varphi_{0,2}(\mathrm{i}p_1,\,p_2) &= \lim_{\varepsilon\to 0}\,[Z_q(\varepsilon)]^{-1}\varphi^{(u)}_{0,2}(\mathrm{i}p_1,\,p_2;\,\varepsilon).\end{aligned} \quad (2.4.12')$$

To obtain the vertex functions Γ_{n_A,n_q}, the one-particle irreducible component of φ_{n_A,n_q} must be amputated with n_A renormalized propagators $\varphi_{2,0}$ and n_q renormalized propagators $\varphi_{0,2}$ (see Section 2.3.1). Thus, the following equation applies:

$$\Gamma_{n_A,n_q}(k_i\,;\,p_j) = \lim_{\varepsilon\to 0}\,[Z_A(\varepsilon)]^{n_A/2}[Z_q(\varepsilon)]^{n_q/2}\Gamma^{(u)}_{n_A,n_q}(k_i\,;\,p_j\,;\,\varepsilon). \quad (2.4.13)$$

The renormalization conditions compare the vertex functions which can be directly interpreted physically at chosen points in momentum space with physically measurable quantities. Because of Eqn. (2.4.13) this fixes the finite parts of the renormalization constants $Z(\varepsilon)$. In QED, where three renormalization constants must be determined the following conditions are usually imposed

$$\Gamma_{2,0}(k,\,-k\,;)^{\mu\nu} \to \mathrm{i}(-g^{\mu\nu}k^2) \qquad \text{for } k^2\to 0, \quad (2.4.14)$$

$$\Gamma_{0,2}(\mathrm{i}p,\,p)_{\alpha\beta} \to \mathrm{i}(\gamma p - M)_{\alpha\beta} \qquad \text{for } p^2\to M^2, \quad (2.4.14')$$

$$\Gamma_{1,2}(0;\,p,\,p)^{\mu}_{\alpha\beta} \to -\mathrm{i}e(\gamma^{\mu})_{\alpha\beta} \qquad \text{for } p^2\to M^2, \quad (2.4.14'')$$

in which M is the physical electron mass and e the electron charge ($e^2/4\pi = 1/137$). The renormalization point, i.e. the momenta at which the theory is fitted to physical parameters, is chosen in Eqn. (2.4.14″) for example such that a photon with momentum transfer zero is scattered on a real, physical electron ($p^2 = M^2$). Then the Thomson cross-section gets its usual form. In general, however, the renormalization point is chosen according to convention (see Section 2.5). In QCD, spacelike momenta $p^2 = -\mathcal{M}^2$ are chosen in order to avoid possible infrared divergences [Ge 54]:

$$\tilde{\Gamma}_{2,0}(k,\,-k\,;)^{\mu\nu}_{ab}\,|_{k^2=-\mathcal{M}^2} = \mathrm{i}(-g^{\mu\nu}k^2 + k^{\mu}k^{\nu})\,\delta_{ab} \quad \text{for } \xi = 0 \quad (2.4.15)$$

where $\tilde{\Gamma}_{2,0}$ is the transverse part of $\Gamma_{2,0}$

$$\Gamma_{0,2}(;\,p,\,p)\,|_{p^2=-\mathcal{M}^2} = \mathrm{i}(\gamma p - M(\mu^2)) \quad (2.4.15')$$

where $M(\mu^2)$ is a running mass

$$\Gamma_{1,2}(p;\,p,\,p)^{\mu}_{a}\,|_{p^2=-\mathcal{M}^2} = -\mathrm{i}g\frac{\lambda_a}{2}\gamma^{\mu}. \quad (2.4.15'')$$

2.4.2 Example: calculation of the gluon propagator in the 1-loop approximation

Calculation of the gluon propagator in 1-loop approximation [Po 73, Gr 73b] is carried out in detail to group together the various aspects of the renormalization programme described in the previous section.

The following graphs contribute to the gluon propagator in 1-loop approximation:

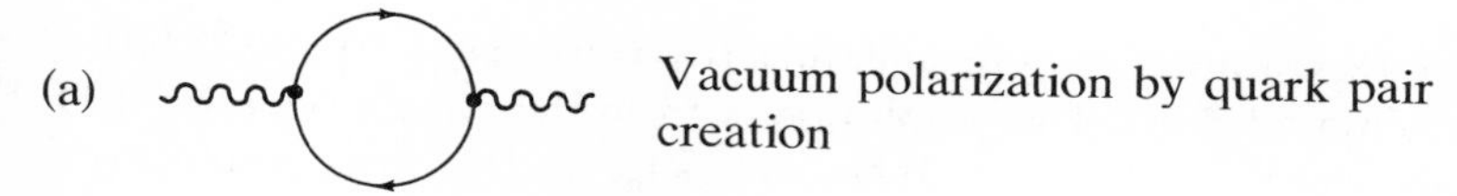

(a) Vacuum polarization by quark pair creation

(b) Vacuum polarization by gluon pair creation

(c) direct gluon self-interaction based on the 4-gluon coupling

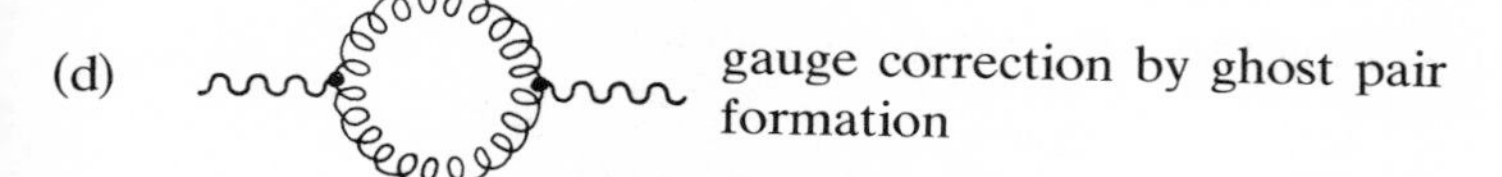

(d) gauge correction by ghost pair formation

The 'tadpole graphs'

disappear after symmetrical integration with respect to the loop momentum or after dimensional regularization (see below), because the quark–gluon, gluon–gluon and ghost–gluon coupling are traceless.

We start by calculating graph (a). In accordance with the Feynman rules described in Section 2.2.3, the following formula can be written for the Landau gauge ($\xi = 0$):

$$D^{(a)}(x-x')^{cd}_{\mu\nu} = \delta^{cd}\frac{g^2}{(2\pi)^4}\int d^4k e^{ik(x-x')}\frac{-g_{\mu\rho}+k_\mu k_\rho/k^2}{k^2+i0}\Pi^{(a)}(k)^{\rho\sigma}$$

$$\times\frac{-g_{\sigma\nu}+k_\sigma k_\nu/k^2}{k^2+i0} \quad (2.4.16)$$

with

$$\Pi^{(a)}(k)_{\rho\sigma} = \frac{N_F}{2} \mathrm{Tr} \int \frac{d^4q}{(2\pi)^4} \frac{1}{\gamma_\mu (q+\frac{1}{2}k)^\mu - M + i0} \gamma_\rho \frac{1}{\gamma_\nu (q-\frac{1}{2}k)^\nu - M + i0} \gamma_\sigma$$
$$= 2N_F \int \frac{d^4q}{(2\pi)^4} \frac{-g_{\rho\sigma}[(q+k)^2 - \frac{1}{4}k^2 - M^2] + 2(q_\rho q_\sigma - \frac{1}{4}k_\rho k_\sigma)}{[(q+\frac{1}{2}k)^2 - M^2 + i0][(q-\frac{1}{2}k)^2 - M^2 + i0]}. \tag{2.4.17}$$

The factor $N_F/2$ is obtained by calculating the trace $\mathrm{Tr}_F 1 \times \mathrm{Tr}_C(\lambda^c \lambda^d/4) = N_F \delta^{cd}/2$ (see Eqn. (1.2.5)). The graph differs from the corresponding QED graph by this factor (Section 2.1.1(2)). $\Pi^{(a)}(k)_{\rho\sigma}$ is the free propagator-amputated contribution of graph (a) to the 2-point Green function; M is the quark mass which is put equal to zero for the sake of simplicity.

The integration with respect to the loop momentum q in Eqn. (2.4.17) diverges quadratically in four dimensions. It must be *regularized* by the *dimensional* method, for gauge invariance reasons. For this, the integral (2.4.17) is considered in D dimensions:

$$\int \frac{d^4q}{(2\pi)^4} \to \int \frac{d^Dq}{(2\pi)^D}.$$

In order to evaluate this integral for arbitrary D it is rewritten as a Gaussian integral, by representing the propagators as follows:

$$i(p^2 - M^2 + i0)^{-1} = \int_0^\infty dz \exp[iz(p^2 - M^2 + i0)].$$

The dimension of g^2 in D dimensions is $\sim[\text{mass } \mu]^{4-D}$. Hence:

$$\Pi^{(a)}(k)_{\rho\sigma} = -2N_F \int_0^\infty dz_1 \int_0^\infty dz_2 \mu^{4-D} \int \frac{d^Dq}{(2\pi)^D}$$
$$\times P(q,k)_{\rho\sigma} \exp\{i[(z_1+z_2)(q^2 + \tfrac{1}{4}k^2) - (z_1 - z_2) q \cdot k']\}|_{k'=k}. \tag{2.4.18}$$

The polynomial in the numerator of the integrand in Eqn. (2.4.18) has been abbreviated by $P(q,k)_{\rho\sigma}$. In the bilinear term of the exponent k has been replaced by k' in order to substitute as usual q^μ by $-i(z_1 - z_2)^{-1} \partial/\partial k'_\mu$ in the above polynomial. Thus the q-integration becomes Gaussian, defined for D dimensions by

$$\mu^{4-D} \int \frac{d^Dq}{(2\pi)^D} \exp(iq^2 a) = \frac{i}{(4\pi)^{D/2}} \exp(-\tfrac{1}{4} i\pi D) \frac{(\mu^2 a)^{(4-D)/2}}{a^2}, \quad a > 0.$$

This is carried out, a Gaussian function in k' is obtained and the differential operator $P(-i(z_1 - z_2)^{-1} \partial/\partial k', k)$ is applied. Then k' is put equal to k and the result is:

$$\Pi^{(a)}(k)_{\rho\sigma} = \mathrm{i}\,\frac{N_F}{2}\,\frac{\exp[+\mathrm{i}\frac{1}{4}\pi(4-D)]}{4\pi^2}\int_0^\infty \mathrm{d}z_1 \int_0^\infty \mathrm{d}z_2 \exp\left(\mathrm{i}\,\frac{z_1 z_2}{z_1+z_2}\,k^2\right)$$
$$\times \frac{(\tilde{\mu}^2(z_1+z_2))^{\frac{1}{2}(4-D)}}{(z_1+z_2)^2}\left(\frac{g_{\rho\sigma}(\frac{1}{2}D-1)}{\mathrm{i}(z_1+z_2)} + \frac{z_1 z_2}{(z_1+z_2)^2}(g_{\rho\sigma}k^2 - 2k_\rho k_\sigma)\right),$$
$$\tilde{\mu} = 2\mu\sqrt{\pi} \qquad (2.4.19)$$

Substituting $z_1 = \lambda(1-z)$, $z_2 = \lambda z$, $0 \leqslant \lambda < \infty$, $0 \leqslant z \leqslant 1$, the integral over λ is the Euler integral for the Γ function, the z integral is elementary; the result is given below:

$$\Pi^{(a)}(k)_{\rho\sigma} = \mathrm{i}\,\frac{N_F}{4\pi^2}\,(g_{\rho\sigma}k^2 - k_\rho k_\sigma)\left(\frac{-k^2-\mathrm{i}0}{\tilde{\mu}^2}\right)^{-(4-d)/2}$$
$$\times \Gamma\left(\frac{4-D}{2}\right)\frac{[\Gamma(2-\frac{1}{2}(4-D))]^2}{\Gamma(4-(4-D))}$$
$$= \mathrm{i}\,\frac{N_F}{16\pi^2}\,\tfrac{2}{3}(-g_{\rho\sigma}k^2 + k_\rho k_\sigma)\left(-\frac{2}{\varepsilon} - \frac{5}{3} + \log(-k^2-\mathrm{i}0)/\tilde{\kappa}^2\right) + O(\varepsilon),$$
$$\Gamma(\varepsilon) = \frac{1}{\varepsilon} + \Gamma'(1) + O(\varepsilon), \quad \varepsilon = 4-D, \quad \tilde{\kappa} = \mathrm{e}^{\Gamma'(1)/2}\tilde{\mu}. \qquad (2.4.20)$$

The poles of the Γ function at $D = 4, 2, \ldots$ express the logarithmic, quadratic ... divergence of the vacuum polarization graph in the corresponding space-time dimensions. The pole at $D = 4$ defines the infinite term of this graph in dimensional regularization, which together with the corresponding terms from the other graphs must be subtracted by multiplicative renormalization.

The other graphs must be calculated in a corresponding manner. The vacuum polarization by gluon pair creation (graph (b)) is particularly interesting. It results from the self-interaction of coloured gluons and is not present in an Abelian gauge theory with neutral gauge bosons like QED. The contribution made by this graph to $\Pi^{(2)}(k)_{\rho\sigma}$ is:

$$\Pi^{(b)}(k)_{\rho\sigma} = -2C_2\int \frac{\mathrm{d}^4 q}{(2\pi)^4}\,\frac{P(q,k)_{\rho\sigma}}{\{[(q+\frac{1}{2}k)^2+\mathrm{i}0][(q-\frac{1}{2}k)^2+\mathrm{i}0]\}^2} \qquad (2.4.21)$$

with

$$P(q,k)_{\rho\sigma} = (g_{\rho\mu}k_\nu - g_{\mu\nu}q_\sigma - g_{\nu\rho}k_\mu)$$
$$\times[g^{\nu\nu'}(q-\tfrac{1}{2}k)^2 - (q-\tfrac{1}{2}k)^\nu (q-\tfrac{1}{2}k)^{\nu'}]$$
$$\times(g_{\sigma\mu'}k_{\nu'} - g_{\mu'\nu'}q_\sigma - g_{\nu'\sigma}k_{\mu'})[g^{\mu\mu'}(q+\tfrac{1}{2}k)^2 - (q+\tfrac{1}{2}k)^\mu (q+\tfrac{1}{2}k)^{\mu'}],$$
$$C_2\,\delta_{cd} = \sum_{a,b} f_{abc}f_{abd} = 3\,\delta_{cd} \quad \text{(see Eqn. (1.2.8))}.$$

This graph is also quadratically divergent. Dimensional regularization is caried out by converting Eqn. (2.4.21) into a Gaussian integral with respect to $d^D q$ by means of the formula $\alpha^{-2}=\int_0^\infty dz z e^{-\alpha z}$. The subsequent steps proceed as described above and give the following result:

$$\Pi^{(b)}(k)_{\rho\sigma}=\frac{\mathrm{i}}{16\pi^2}\left(-\frac{C_2}{12}\right)\Big[(-16g_{\rho\sigma}k^2+15k_\rho k_\sigma)$$
$$+(-25g_{\rho\sigma}k^2+28k_\rho k_\sigma)\left(-\frac{2}{\varepsilon}-\frac{5}{3}+\log\frac{-k^2-\mathrm{i}0}{\tilde{\kappa}^2}\right)\Big]+O(\varepsilon). \tag{2.4.22}$$

Graph (c) gives an indefinite contribution which is independent of the propagator momentum k. This can be set zero after dimensional regularization [Le 74].

The gauge correction by ghost-pair formation (graph (d)) leads to:

$$\Pi^{(d)}(k)_{\rho\sigma}=-C_2\int\frac{d^4q}{(2\pi)^4}\frac{(q+\frac{1}{2}k)_\rho(q-\frac{1}{2}k)_\sigma}{[(q+\frac{1}{2}k)^2+\mathrm{i}0][(q-\frac{1}{2}k)^2+\mathrm{i}0]}$$
$$=\frac{\mathrm{i}}{16\pi^2}\left(\frac{-C_2}{12}\right)\Big[g_{\rho\sigma}k^2+(-g_{\rho\sigma}k^2-2k_\rho k_\sigma)$$
$$\times\left(-\frac{2}{\varepsilon}-\frac{5}{3}+\log\frac{-k^2-\mathrm{i}0}{\tilde{\kappa}^2}\right)\Big]+O(\varepsilon). \tag{2.4.23}$$

Thus, the results of the calculations can be added:

$$\Pi^{(2)}(k)_{\rho\sigma}=\Pi^{(a)}(k)_{\rho\sigma}+\Pi^{(b)}(k)_{\rho\sigma}+\Pi^{(c)}(k)_{\rho\sigma}+\Pi^{(d)}(k)_{\rho\sigma}$$
$$=\frac{-\mathrm{i}}{16\pi^2}(\tfrac{13}{6}C_2-\tfrac{2}{3}N_F)\log\frac{-k^2-\mathrm{i}0}{\tilde{\kappa}^2}(-g_{\rho\sigma}k^2+k_\rho k_\sigma)$$
$$+\frac{\mathrm{i}}{16\pi^2}(\tfrac{13}{6}C_2-\tfrac{2}{3}N_F)\frac{2}{\varepsilon}(-g_{\rho\sigma}k^2+k_\rho k_\sigma)$$
$$+\frac{\mathrm{i}}{16\pi^2}(\tfrac{15}{4}C_2-\tfrac{10}{9}N_F)(-g_{\rho\sigma}k^2+k_\rho k_\sigma)$$
$$+O(\varepsilon),\quad \varepsilon=4-D. \tag{2.4.24}$$

The free gluon propagator (2.2.42) is used to get the regularized, non-renormalized gluon propagator in 1-loop approximation according to Eqn. (2.4.16):

$$\tilde{D}^{(2)}(k)^{cd}_{\mu\nu}=\tilde{D}(k)^{cd}_{\mu\nu}-\mathrm{i}(2\pi)^4\tilde{D}(k)^{ce}_{\mu\rho}\Pi^{(2)}(k)^{\rho\sigma}_{ef}\tilde{D}(k)^{fd}_{\sigma\nu}$$
$$=\frac{-1}{(2\pi)^4}\left(g_{\mu\nu}-\frac{k_\mu k_\nu}{k^2}\right)\frac{\delta^{cd}}{k^2+\mathrm{i}0}(1-\Pi(k^2)). \tag{2.4.25}$$

This introduces the 'vacuum polarization'

$$\Pi^{(2)}(k)^{\rho\sigma}_{ab}:=\Pi^{(2)}(k)^{\rho\sigma}\delta_{ab}=\mathrm{i}(-g_{\rho\sigma}k^2+k_\rho k_\sigma)\Pi(k^2)\,\delta_{ab} \tag{2.4.26}$$

and the 'self-energy':

$$\Pi(k^2) := \frac{g^2}{16\pi^2}\left(\tfrac{13}{6}C_2 - \tfrac{2}{3}N_F\right)\left(\log\frac{-k^2 - \mathrm{i}0}{\kappa^2} - \frac{2}{\varepsilon}\right) + O(g^4) \qquad (2.4.27)$$

Moreover, the last term in Eqn. (2.4.24) was combined with the first by a redefinition $\tilde{\kappa} \to \kappa$. The result provides the basis for a discussion of various, typical aspects of the renormalization programme for non-Abelian gauge theories.

The general solution of the Slavnov identity (2.3.142) can be represented in the form

$$\tilde{D}'(k)^{ab}_{\mu\nu} = \frac{-1}{(2\pi)^4}\left[\left(g_{\mu\nu} - \frac{k_\mu k_\nu}{k^2}\right)\frac{1}{1+\Pi(k^2)} + \xi\frac{k_\mu k_\nu}{k^2}\right]\frac{\delta^{ab}}{k^2 + \mathrm{i}0} \qquad (2.4.28)$$

with the undetermined function $1/(1+\Pi(k^2))$. If this is expanded into a geometric series, then the symbolic result is $\tilde{D}' = \tilde{D} + \tilde{D}\Pi\tilde{D} + \tilde{D}\Pi\tilde{D}\Pi\tilde{D} + \ldots = D(1+\Pi D')$ and hence the vacuum polarization:

$$\Pi(k)^{\rho\sigma}_{ab} = [\tilde{D}(k)^{-1} - \tilde{D}'(k)^{-1}]^{\rho\sigma}_{ab} = (2\pi)^4(g^{\rho\sigma}k^2 - k^\rho k^\sigma)\,\delta_{ab}.$$

The Slavnov identity requires the transversality of the vacuum polarization (independent of the gauge). A comparison between Eqns. (2.4.28) and (2.4.25) shows that our calculation gives the regularized 1-loop approximation for the vacuum polarization in Landau gauge and the result as shown in Eqns. (2.4.24) and (2.4.26) is transversal. Thus, this example illustrates that:

(1) The dimensional regularization process is gauge-invariant; the transversality of the vacuum polarization is a result of the Slavnov identity derived from the gauge invariance of the theory.
(2) The contribution made by the Faddeev–Popov ghost fields $\Pi^{(d)}(k)_{\mu\nu}$ is important for the transversality of $\Pi^{(2)}(k)_{\mu\nu}$. Neither the gluon contribution (b), nor the ghost contribution (d) are themselves transversal. Only the sum of $\Pi^{(b)}$ and $\Pi^{(d)}$ gives a transversal pole term and a transversal k-dependent contribution to the dressed propagator. This underlines once again the role played by ghost fields for the gauge invariance of QCD.

The divergent part of the gluon propagator in four dimensions was separated as a pole in the self-energy (2.4.27) with help of the dimensional regularization. The transverse part of the vertex function is according to (2.4.28):

$$\tilde{\Gamma}_{2,0}(k, -k;)^{\mu\nu}_{ab} = \frac{\mathrm{i}}{(2\pi)^4}(\tilde{D}^{-1})(k)^{\mu\nu}_{ab}\big|_{\mathrm{trans}} = \mathrm{i}(-g^{\mu\nu}k^2 + k^\mu k^\nu)\,\delta_{ab}(1+\Pi(k^2)), \qquad (2.4.28')$$

so that the pole in $\varepsilon = 4 - D$ also occurs here. Following Eqn. (2.4.13), this pole is subtracted by multiplicative renormalization with a renormalization constant

$$Z_A(\varepsilon)=1-\frac{g^2}{16\pi^2}(\tfrac{13}{6}C_2-\tfrac{2}{3}N_F)\left(-\frac{2}{\varepsilon}+\log\frac{\mathcal{M}^2}{\kappa^2}\right)+O(g^4) \qquad (2.4.29)$$

which diverges at $\varepsilon=0$. Its finite part is fixed by the renormalization condition (2.4.15). The results from Eqns. (2.4.13), (2.4.28′), (2.4.29), and (2.4.27) give the transversal component of the renormalized vertex function $\tilde{\Gamma}_{2,0}(k,-k;)^{\mu\nu}_{ab}$ in 1-loop approximation and in Landau gauge:

$$\tilde{\Gamma}^{(2)}_{2,0}(k,-k;)^{\mu\nu}_{ab}=\mathrm{i}(-g^{\mu\nu}k^2+k^\mu k^\nu)\cdot\left(1+\frac{g^2}{16\pi^2}(\tfrac{13}{6}C_2-\tfrac{2}{3}N_F)\log\frac{-k^2-\mathrm{i}0}{\mathcal{M}^2}\right)\delta_{ab}. \qquad (2.4.30)$$

This result plays a very important role in proving the asymptotic freedom of QCD (see Section 2.5.2, p. 149).

2.4.3 Remarks on the proof of renormalizability of quantized gauge theories

The last section used the example of the gluon propagator in 1-loop approximation to sketch out the plan for the perturbative evaluation of a quantized gauge field theory:

(1) gauge invariant (dimensional) regularization;
(2) subtraction of divergences extracted by regularization by introducing counter-terms into the Lagrangian density (equivalent to multiplicative renormalization of the gauge field in the example); and
(3) adjustment of the renormalized (finite) parameters of the theory to their measured values (renormalization conditions).

In order to prove the renormalizability of a theory one has to show that:

(1) the procedure can be extended to arbitrary orders of the loop expansion without having to continually introduce new types of counter-terms with indefinite coupling constants; and
(2) the finite, physical S-matrix which can then be calculated is unitary and causal.

The most important stages of the proof of renormalizability are shown in the highly simplified, logical plan shown in Fig. 2.10.

Some of the points which may help to clarify the role of gauge invariance are described briefly below.

(a) The quantized and dimensionally regularized gauge theory as described in Section 2.3.3 forms the basis of this discussion. The gauge chosen by $\mathcal{L}_{\text{fix}}$ is assumed to be renormalizable as defined in Section 2.4.1 (**R**-gauge). The gauges (2.3.129″) have this property. The generating functional of Green functions is finite in $D=4-\varepsilon$ dimensions, $\varepsilon\neq 0$ and satisfies the Ward identities (Eqns. (2.3.136′) and (2.3.139′)) due to the gauge invariance

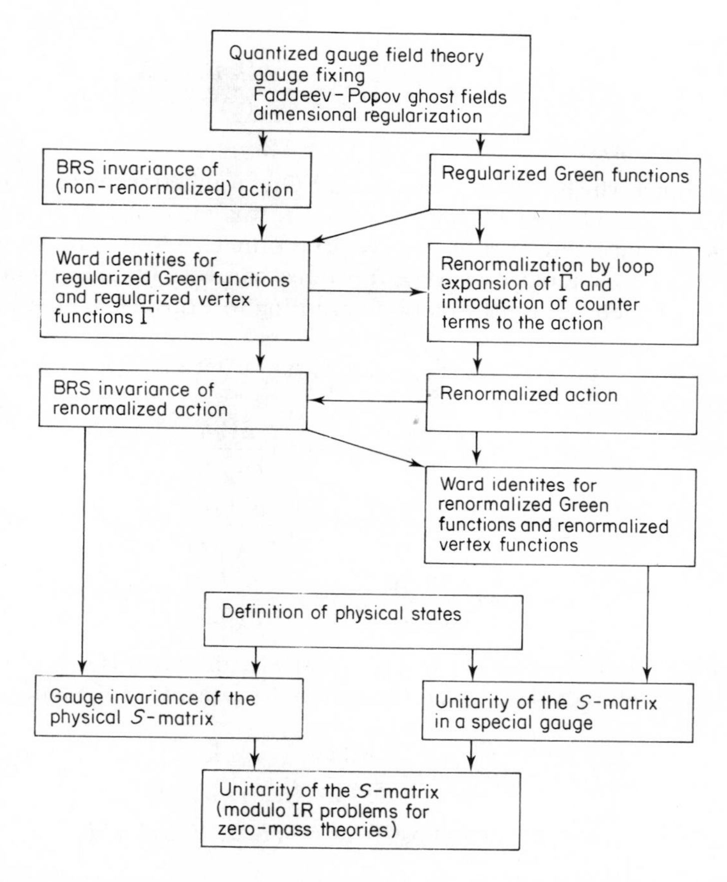

Figure 2.10 Simplified, logical plan for the proof of renormalizability of Yang–Mills field theories

of the dimensional regularization procedure. These equations must be transformed into an identity for the generating functional Γ of vertex functions, to the so-called *Lee identity* for the purposes of the present discussion. Only a pure Yang–Mills theory is considered. Eqn. (2.3.136) is rewritten for the functional $\Phi\{j; \bar{\omega}, \omega\}$ of the connected Green functions as an initial step. From $T\{j; \bar{\omega}, \omega\} = \exp(\Phi\{j; \bar{\omega}, \omega\}/\hbar)$ (Eqn. (2.3.66)) follows

$$\frac{\mathrm{i}}{\xi}\partial^{\mu}_{y}\frac{\delta\Phi\{j\}/\hbar}{\mathrm{i}\,\delta j^{\mu}_{a}(y)} - \int \mathrm{d}^4x j^{\nu}_{b}(x)\left(\delta^{b}_{c}\,\partial^{x}_{\nu} - gf^{bcd}\frac{\delta\Phi\{j\}/\hbar}{\mathrm{i}\,\delta j^{\nu}_{d}(x)} - gf^{bcd}\frac{\delta}{\mathrm{i}\,\delta j^{\nu}_{d}(x)}\right)\hbar G^{ca}\{j; x, y) = 0 \tag{2.4.31}$$

In this

$$\Phi\{j\} := \Phi\{j; \bar{\omega} = 0, \omega = 0\},$$

and

$$G^{ca}\{j; x, y)=\frac{1}{\hbar}T^{ca}\{j; x, y)e^{-\Phi\{j\}/\hbar}=\left.\frac{\delta^2\Phi\{j; \bar{\omega}, \omega\}/\hbar^2}{\delta\bar{\omega}_c(x)\,\delta\omega_a(y)}\right|_{\omega=\bar{\omega}=0} \tag{2.4.32}$$

is the functional which generates connected Green functions with two ghost field points x, c and y, a. Planck's constant $\hbar$ was introduced for counting loops in accordance with Section 2.3.2(e) (see Eqn. (2.3.89)). An identity for the generating functional of vertex functions can be obtained from Eqn. (2.4.31) by a Legendre transformation according to Eqns. (2.3.69)–(2.3.71):

$$\frac{1}{\hbar}\Gamma\{a; \chi, \bar{\chi}\}=-\mathrm{i}\int d^4x(j_a^\mu(x)a_\mu^a(x)+\bar{\omega}_a(x)\chi^a(x)+\bar{\chi}^a(x)\omega_a(x))+\frac{1}{\hbar}\Phi\{j; \bar{\omega}, \omega\}, \tag{2.4.33}$$

$$a_\mu^a(x)=\frac{\delta\Phi/\hbar}{\mathrm{i}\,\delta j_a^\mu(x)}, \quad j_a^\mu(x)=\mathrm{i}\frac{\delta\Gamma/\hbar}{\delta a_\mu^a(x)},$$
$$\chi^a(x)=\frac{\delta^L\Phi/\hbar}{\mathrm{i}\,\delta\bar{\omega}_a(x)}, \quad \bar{\omega}_a(x)=\frac{\delta^L\Gamma/\hbar}{\mathrm{i}\,\delta\chi^a(x)}, \tag{2.4.33'}$$
$$\bar{\chi}^a(x)=\mathrm{i}\frac{\delta^L\Phi/\hbar}{\delta\omega_a(x)}, \quad \omega_a(x)=\mathrm{i}\frac{\delta^L\Gamma/\hbar}{\delta\bar{\chi}^a(x)}.$$

To transform the term $\delta G^{ca}/\delta j$ in Eqn. (2.4.31), the term $(G^{-1})_{ab}\{a; x, y)$, must be considered. This is given in analogy to Eqns. (2.3.72) and (2.3.73) by:

$$(G^{-1})_{ab}\{a; x, y)=\left.\frac{\delta^2\Gamma\{a; \chi, \bar{\chi}\}}{\delta\bar{\chi}^a(x)\,\delta\chi^b(y)}\right|_{\chi=\bar{\chi}=0}. \tag{2.4.34}$$

It generates vertices with two ghost field points x, a and y, b.

$$(G^{-1})_{ab}\{a; x, y)\,|_{a(x)=0}$$

is the inverse ghost propagator. By differentiating the equation

$$\int d^4z G^{ab}\{j; x, z)(G^{-1})_{bc}\{a; z, y)=\delta_c^a\,\delta^{(4)}(x-y)$$

with respect to $j_d^\mu(x)$, the result is

$$gf^{bcd}\frac{\delta G^{ca}\{j; x, y)}{\mathrm{i}\,\delta j_d^\mu(x)}=-\hbar\int d^4z\gamma_\mu^{cb}\{a; z, x)G^{ca}\{j; z, y) \tag{2.4.35}$$

with

$$\gamma_\mu^{ab}\{a; x, y)=gf^{bcd}\int d^4z\, d^4z' X_{\mu\nu}^{de}(y, z)G^{cf}\{j; y, z')\frac{\delta(G^{-1})_{fa}\{a; z', x)}{\delta a_\nu^e(z)}. \tag{2.4.36}$$

In agreement with Eqn. (2.3.75),

$$\frac{\delta}{\delta j_d^\mu(x)}=\mathrm{i}\hbar\int d^4z X_{\mu\nu}^{de}(x, z)\frac{\delta}{\delta a_\nu^e(z)} \tag{2.4.37}$$

is used in Eqn. (2.4.36). $X^{de}_{\mu\nu}(x, z) = \delta^2\Phi\{j\}/\hbar^2 \mathrm{i}\, \delta j^{\mu}_{d}(x) \mathrm{i}\, \delta j^{\nu}_{e}(z)$ is the gauge field propagator in the presence of the source j. Substituting Eqns. (2.4.33′) and (2.4.35) in Eqn. (2.4.31) gives:

$$\int \mathrm{d}^4y \left(L^{ab}_{\mu}\{a; x, y\} \frac{\delta\Gamma\{a\}}{\delta a^{b}_{\mu}(y)} - \frac{1}{\xi} (\partial^{\mu} a^{b}_{\mu}(y))(G^{-1})_b{}^a\{a; y, x\} \right) = 0 \quad (2.4.38)$$

with

$$L^{ab}_{\mu}\{a; x, y) = (\delta^{ab}\, \partial^{x}_{\mu} + g f^{abc} a^{c}_{\mu}(x))\, \delta^{(4)}(x-y) + \hbar\gamma^{ab}_{\mu}\{a; x, y). \quad (2.4.39)$$

The ghost field equation of motion (2.3.139) is transformed in a similar way:

$$(G^{-1})^{ba}\{a; y, x) = \frac{1}{\mathrm{i}}\, \partial^{\mu}_{y} L^{ab}_{\mu}\{a; x, y) \quad (2.4.40)$$

and this is then substituted in Eqn. (2.4.37). The result is the *Lee identity for the generating functional of the vertices* [Le 73, Be 74, Kl 75]:

$$\int \mathrm{d}^4y L^{ab}_{\mu}\{a; x, y) \frac{\delta}{\delta a^{b}_{\mu}(y)} \left(\Gamma\{a\} - \mathrm{i} \int \mathrm{d}^4z\, \mathscr{L}_{\text{fix}}(a(z)) \right) = 0 \quad (2.4.41)$$

with

$$\mathscr{L}_{\text{fix}}(a(z)) = -\frac{1}{2\xi} (\partial^{\mu} a^{a}_{\mu}(z))^2. \quad (2.3.129'')$$

The physical significance of the Lee identity is apparent from the following observations:

(i) In the tree graph approximation ($\hbar = 0$), it is

$$\Gamma\{a\} = \mathrm{i} \int \mathrm{d}^4z (\mathscr{L}(a(z)) + \mathscr{L}_{\text{fix}}(a(z))),$$

(with $\chi = \bar{\chi} = 0$) according to Eqn. (2.3.92), and the operator $L^{ab}_{\mu}\{a; x; y)$ is reduced in this approximation to

$$\begin{aligned} L^{ab}_{\mu}\{a; x, y)|_{\hbar=0} &= (\delta^{ab}\, \partial^{x}_{\mu} + g f^{abc} a^{c}_{\mu}(x))\, \delta^{(4)}(x-y) \\ &= \frac{\delta a^{a}_{\mu}(y)}{\delta\theta^{b}(x)} \end{aligned} \quad (2.4.42)$$

(see Eqn. (2.2.7)). Eqn. (2.4.41) expresses the gauge invariance of the action without the gauge fixing term (compare Eqn. (2.3.101)):

$$\frac{\delta}{\delta\theta^{b}(x)} \left(\int \mathrm{d}^4z (\mathscr{L}(a(z)) \right) = \int \mathrm{d}^4y \frac{\delta a^{a}_{\mu}(y)}{\delta\theta^{b}(x)} \frac{\delta}{\delta a^{a}_{\mu}(y)} \left(\int \mathrm{d}^4z \mathscr{L} \right) = 0. \quad (2.4.43)$$

For higher orders of the loop expansion the Lee identity is a generalization of this assertion of gauge invariance. The transformation law described by $L^{ab}_{\mu}|_{\hbar=0}$ in Eqn. (2.4.42) is modified by the term $\hbar\gamma^{ab}_{\mu}\{a; x, y)$ (Eqn. (2.4.36)) for non-Abelian gauge groups. As a result, not only fields and coupling constants but also the gauge group and the Ward identities based on it must all be renormalized [Le 72, Le 73, Sl 72, Ta 71, tH 72].

(ii) $\gamma^{ab}_{\mu}\{a; x, y)$ can be represented by structural graphs (Eqn. (2.3.34)) in the following way:

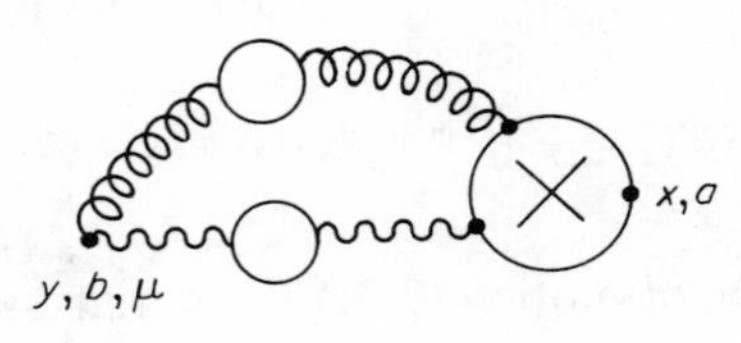

Thus, the reaction of the Feddeev–Popov ghosts with the gauge fields is the reason for this gauge group renormalization. (*Note*: this statement depends on the choice of the gauge.)

(iii) We have shown that a gauge-invariant action leads to the Lee identity for the generating functional of the vertices. However, the reverse is also true [tH 72]: if an action satisfies the lee identity, then it is invariant under a group of gauge transformations.

(b) The introduction of suitable gauge fixing terms and associated ghost fields formed the basis of a set of Feynman rules for gauge theories, and these belong in the formal sense defined by counting the degrees of divergence (Section 2.4.1) to a renormalizable theory. The generating functional of vertices Γ is finite in $D=4-\varepsilon$ dimensions. Addition of a finite number of counter terms to the Lagrangian density gives the renormalized functional of the vertex functions Γ^R, which remains finite for $\varepsilon \to 0$. The most important part in confirming the *renormalizability of a gauge theory* in a physical consistent way is the proof that the subtraction terms can be chosen such that Γ^R like Γ satisfies Lee identities of the type described in Eqn. (2.4.41). This is a decisive factor, as it is the only way of showing that the introduction of non-physical degrees of freedom through $\mathcal{L}_{\text{ghost}}$ and $\mathcal{L}_{\text{fix}}$ (Eqn. (2.3.129)) does not violate the unitarity of the S-matrix strictly required by physics.

The proof is by induction with respect to the number of loops l in the perturbation expansion. Further details can be found in refs. [Ab 73, Zi 74, Ma 78]. The reason for the expansion in powers of the loop parameter $\hbar$ (see Section 2.3.2(e)) rather than in powers of the coupling constant g is that the loop expansion preserves any symmetry of the Lagrangian density in any order. This is the case because the loop parameter multiplies the whole Lagrangian density (Eqn. (2.3.87)). In contrast to this, splitting up the Yang–Mills Lagrangian density into a free (g-independent) part and an interaction part would not be gauge-invariant, as non-Abelian gauge transformations explicitly contain the coupling constant (Eqns. (2.2.4) and (2.2.5)). Thus the consequences of gauge invariance cannot be discussed separately for any order of perturbation theory with respect to g.

In the order $\hbar^0$, the renormalized functional Γ^R agrees with renormalized action (Eqn. (2.3.92)); this satisfies the (renormalized) Lee identity (2.4.41) and is thus invariant under a group of gauge transformations (the renormalized gauge group)—see (iii) above. For the simplest Green functions, the renormalized action is a result of multiplicative renormalization according to Eqn. (2.4.11). The renormalization of Green functions for composite local field operators is discussed in refs. [Jo 76] and [Di 74].

(c) The covariant quantization of a gauge field theory in a renormalizable gauge, as described in Section 2.3.3, generally results in unphysical degrees of freedom:

(1) (zero mass) gauge fields produce negative norm states (Eqn. (2.3.21)) and
(2) fictitious ghost degrees of freedom provide the formal gauge covariance of the theory (Section 2.3.3(a), (b)).

The physical S-matrix transforms physical initial states into physical final states of possible reactions. Thus a *definition of physical states* must be made before any further investigation of the S-matrix is carried out [Be 74, Co 75]. Quite apart from the question as to whether such states can actually be dynamically realized, the definition of physical states is phrased as follows:

> Physical degrees of freedom comprise all degrees of freedom invariant under the renormalized gauge group.

For the purpose of their characterization, the Lee identity is taken for the inverse gauge-field propagator, i.e. for the 2-point vertex function $(\tilde{D}'^{-1})(k)$ of gluons, photons etc.. This is obtained by differentiation of Eqn. (2.4.41) with respect to $a^c_\nu(z)$ at the point $a \equiv 0$. This leads to:

$$L^{ab}_\mu(k)\left((\tilde{D}'^{-1})(k)^{\mu\nu}_{bc} + \frac{(2\pi)^4}{\xi}\,\delta_{bc}k^\mu k^\nu\right) = 0. \tag{2.4.44}$$

in momentum space. $L^{ab}_\mu(k)$ is the Fourier-transform of:

$$L^{ab}_\mu\{a; x, y)|_{a\equiv 0} = \delta^{ab}\,\partial^x_\mu\,\delta^{(4)}(x-y) + \hbar\gamma^{ab}_\mu\{a; x, y)|_{a\equiv 0} \tag{2.4.45}$$

(Eqns. (2.4.39) and (2.4.36)). According to Eqn. (2.4.44), the gauge-independent part in the gauge-field propagator is characterized by the fact that it is annihilated by L^{ab}_μ, whereas the gauge-dependent part must be compensated by the second term containing the gauge parameter ξ. If $\varepsilon^a_\mu(k, r)$ and $\bar{\varepsilon}^a_\mu(k, s)$ are the eigenvectors of $(\tilde{D}')^{-1}$:

$$\begin{aligned}(\tilde{D}'^{-1})(k)^{bc}_{\mu\nu}\varepsilon^\nu_c(k, r) &= \Gamma^{(r)}\varepsilon^b_\mu(k, r),\\ (\tilde{D}'^{-1})(k)^{bc}_{\mu\nu}\bar{\varepsilon}^\nu_c(k, s) &= \bar{\Gamma}^{(s)}\bar{\varepsilon}^b_\mu(k, s),\end{aligned} \tag{2.4.46}$$

with:

$$L^\mu_{ab}(k)\varepsilon^b_\mu(k, r) = 0 \quad \text{for all } r, \tag{2.4.47}$$

$$L^\mu_{ab}(k)\bar{\varepsilon}^b(k, s) \neq 0 \quad \text{for all } s, \tag{2.4.47'}$$

then the vectors $\varepsilon_\mu^a(k, r)$ describe the physical degrees of freedom and the vectors $\bar{\varepsilon}_\mu^a(k, s)$ the unphysical ones.

These facts are well-known and especially transparent in QED, in which the ghost fields decouple from the other degrees of freedom (Eqns. (2.3.143)) and $\gamma_\mu\{a; x, y\} \equiv 0$. Eqns. (2.4.47) can be simplified to $k^\mu \cdot \varepsilon_\mu(k, r) = 0$ and $k^\mu \cdot \bar{\varepsilon}_\mu(k, s) \neq 0$, $k^\mu = (|\mathbf{k}|, \mathbf{k})$. For the tree graph approximation, the photon propagator in the Feynman gauge is given by Eqn. (1.2.40) and its inverse is proportional to $g_{\mu\nu}$. The physical photon states are described by

$$\varepsilon^\mu(k, 1) = (0, \hat{\mathbf{k}}_\perp^{(1)}),\ \varepsilon^\mu(k, 2) = (0, \hat{\mathbf{k}}_\perp^{(2)}),\ \hat{\mathbf{k}}_\perp^{(i)} \cdot \mathbf{k} = 0,\ \hat{\mathbf{k}}_\perp^{(j)} = \delta^{ij}, \quad (2.4.48)$$

and the unphysical ones by

$$\bar{\varepsilon}^\mu(k, 0) = (1, \mathbf{0}),\ \bar{\varepsilon}^\mu(k, 3) = (0, \mathbf{k}/|\mathbf{k}|) \quad (2.4.48')$$

Thus, the time-like excitation $\bar{\varepsilon}^\mu(k, 0)$, which introduces negative norm states (Eqn. (2.3.21)) is characterized as unphysical. According to Eqn. (2.4.48), physical photons are transversal.

As a result of Eqn. (2.4.47), $L_\mu^{ab}(k)$ can be expanded according to:

$$L_\mu^{ab}(k) = \sum_s \bar{L}^a(k, s)\bar{\varepsilon}_\mu^b(k, s). \quad (2.4.49)$$

Eqn. (2.4.40) then shows that the inverse ghost propagator is proportional to $k^\mu \cdot \sum_s \bar{L}^a(k, s) \cdot \bar{\varepsilon}_\mu^b(k, s)$ for $a \equiv 0$. This means: *all Faddeev–Popov ghost degrees of freedom are unphysical* (n degrees of freedom for an n-dimensional gauge group).

If the linear gauge fixing always adopted so far is written in the form $C^a[\mathscr{A}_\mu(x)] = \int d^4y F_\mu^{ab}(x, y) A_b^\mu(y)$, then the following applies:

$$F_\mu^{ab}(k) Q^{(ph)}(k)_{bc}^{\mu\nu} = 0. \quad (2.4.50)$$

$$Q^{(ph)}(k)_{ab}^{\mu\nu} := \sum_r \varepsilon_a^\mu(k, r)\varepsilon_b^\nu(k, r) \quad (2.4.51)$$

is the projector onto the physical degrees of freedom which are not affected by gauge fixing according to Eqn. (2.4.50). For the Lorentz gauge (Eqn. (2.3.124)), $F_\mu^{ab}(k)$ is proportional to $\delta^{ab}k_\mu$ and this statement is trivial for the tree graph approximation of QED (see Eqn. (2.4.48), (2.4.48')). For a non-Abelian gauge theory, definition (2.4.47) of the physical degrees of freedom, also contains the ghost propagator and the proof of Eqn. (2.4.50) establishes amongst other things that the 'masses' (i.e. zeros of the inverse propagators) of the Faddeev–Popov ghosts and of the unphysical gauge field degrees of freedom are degenerate. For massless gauge fields this is also true for the physical gauge field degrees of freedom. This is an extra difficulty in addition to the infrared problem.

(d) The next step in the renormalization programme is to provide the proof of the *unitarity* of the *S-matrix* [tH 71, Ta 71, Co 75], i.e. to confirm

Eqn. (2.3.49). The Cutkosky cutting rule (Eqn. (2.3.47)), which can be represented graphically in the following form:

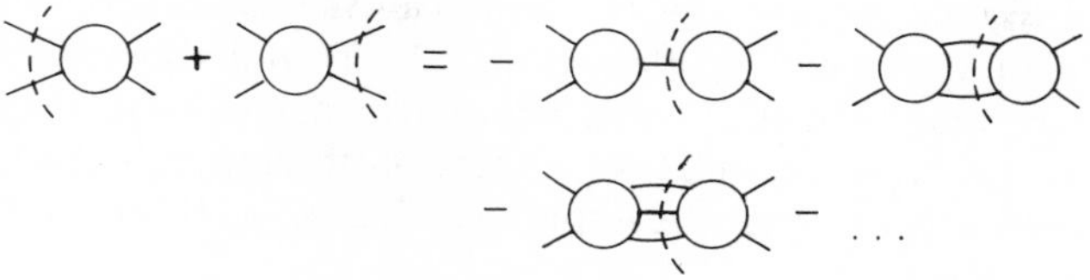

(2.4.52)

(see Section 2.3.1(g)) is the starting point. 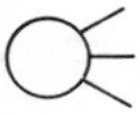 represents the amputated Green functions and 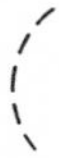 indicates that *all* degrees of freedom, unphysical ones included, must be considered in the cutted propagator. If the physical degrees of freedom only are cutted, in accordance with the definition Eqn. (2.4.47), then this is represented by  . By refining the graph rules in this way, the unitarity condition (2.3.49) can now be represented by:

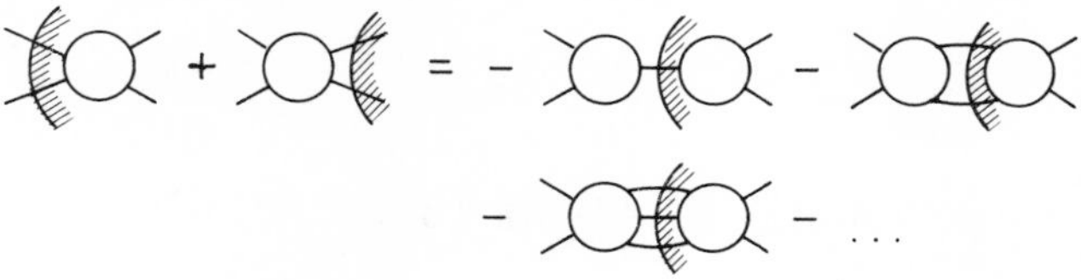

(2.4.53)

Cuts in external lines separate incoming and outgoing particles but have no meaning otherwise. After multiplying with the wave functions of the physical states (Eqn. (2.3.38)), can be replaced by in the external lines. From now on, this will be assumed and no external lines will appear in the graphs. Then, both graph equations ((2.4.52) and (2.4.53)) are exactly equivalent provided that unphysical degrees of freedom do not contribute to the summation with respect to the intermediate states, or that these contributions cancel each other:

(2.4.54)

The proof of unitarity reduced to Eqn. (2.4.54) can be made with the help of Ward identities. This idea can be sketched by using a pure Yang–Mills theory in Feynman gauge for one- and two-particle intermediate states, by way of example. The generalization to multiparticle intermediate states means that either higher Ward identities must be considered or an inductive proof with respect to the number of particles in the intermediate state must be supplied. Theories involving charged particles in which there is a minimal coupling with the gauge field can be treated in the same way.

As a result of theorem (1) in Section 2.3.1(f) and the renormalization condition (Eqn. (2.4.14)), the exact propagator agrees with its tree graph approximation, i.e. with the free propagator, on the mass shell $k^2=0$ of the gauge field. The projector onto the unphysical degrees of freedom in Feynman gauge is then given by Eqn. (2.4.48′):

$$\begin{aligned} Q^{(un)}(k)^{\mu\nu}_{ab} &= -\bar{\varepsilon}^{\mu}_{a}(k,0)\bar{\varepsilon}^{\nu}_{b}(k,0)+\bar{\varepsilon}^{\mu}_{a}(k,3)\bar{\varepsilon}^{\nu}_{b}(k,3) \\ &= \frac{\delta_{ab}}{2\,|\mathbf{k}|^2}(k^\mu\hat{k}^\nu+\hat{k}^\mu k^\nu) \end{aligned} \tag{2.4.55}$$

with

$$k^\mu = (|\mathbf{k}|, \mathbf{k}), \quad \hat{k}^\mu = (-|\mathbf{k}|, \mathbf{k}),$$

or graphically:

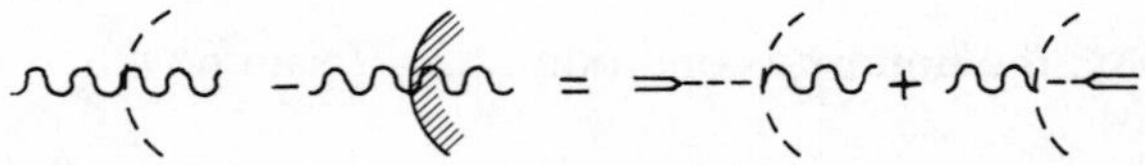

(2.4.55′)

with

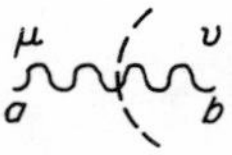

$\mu\,a \;\;\nu\,b$: $=\tilde{\Delta}^-(k)^{\mu\nu}_{ab}=\dfrac{\delta_{ab}}{(2\pi)^3}\,\theta(k^0)\,\delta(k^2)(-g^{\mu\nu})$ (Eqn. (2.3.43)),

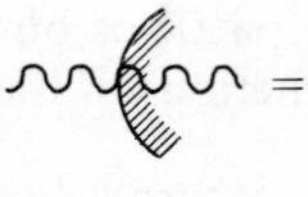

$$=\frac{\delta_{ab}}{(2\pi)^3}\,\theta(k^0)\,\delta(k^2)(-g^{\mu\nu}-(k^\mu\hat{k}^\nu+\hat{k}^\mu k^\nu)/2\,|\mathbf{k}|^2),$$

(2.4.55″)

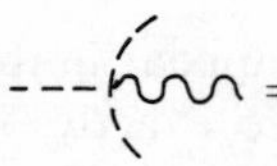

$$=\frac{\delta_{ab}}{(2\pi)^3}\,\theta(k^0)\,\delta(k^2)(-i\hat{k}^\nu)/2\,|\mathbf{k}|^2,$$

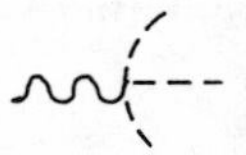

$$=\frac{\delta_{ab}}{(2\pi)^3}\,\theta(k^0)\,\delta(k^2)i\hat{k}^\mu/2\,|\mathbf{k}|^2,$$

═ at the left or right of the cut means multiplication with $+ik^\mu$ or $-ik^\nu$ respectively.

The unphysical components of gauge-field states described by (2.4.55) give no contribution to the imaginary part of the scattering amplitude. This is shown by differentiating the Ward identity (2.3.136) with respect to the ource and then putting this equal to zero. The result for $\xi = 1$

$$\delta_y^{\mu} \frac{\delta^{n+1} T\{j\}}{i\,\delta j_a^{\mu}(y) i\,\delta j_{a_1}^{\nu_1}(x_1) \ldots i\,\delta j_{a_n}^{\nu_n}(x_n)}\bigg|_{j=0}$$

$$= \sum_i \frac{1}{i} \left(\delta_c^{a_i} \partial_{\nu_i}^{x_i} - g f^{a_i c d} \frac{\delta}{i\,\delta j_d^{\nu_i}(x_i)} \right)$$

$$\times \frac{\delta^{n-1} T^{ca}\{j; x_i, y)}{i\,\delta j_{a_1}^{\nu_1}(x_1) \ldots i\,\delta j_{a_{i-1}}^{\nu_{i-1}}(x_{i-1}) i\,\delta j_{a_{i+1}}^{\nu_{i+1}}(x_{i+1}) \ldots i\,\delta j_{a_n}^{\nu_n}(x_n)}\bigg|_{j=0} \qquad (2.4.56)$$

s the explicit Ward identity for the (full) Green functions. Taking the lefinition of the physical degrees of freedom Eqns. (2.4.47), (2.4.39), 2.4.35), and (2.4.36) into account, the right-hand side vanishes when all ield points $x_1, \ldots, x_n$ describe physical excitations. Then, current conserva-ion holds, which leads to the transversality condition

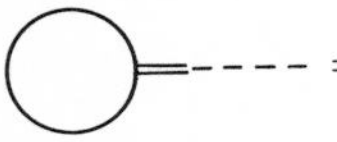

$$= 0 \qquad (2.4.57)$$

physical, i.e. transversal gauge field lines are not shown) and thus the inphysical contributions of the gauge field propagator in the one-particle ntermediate state disappear (see Eqn. (2.4.55′)):

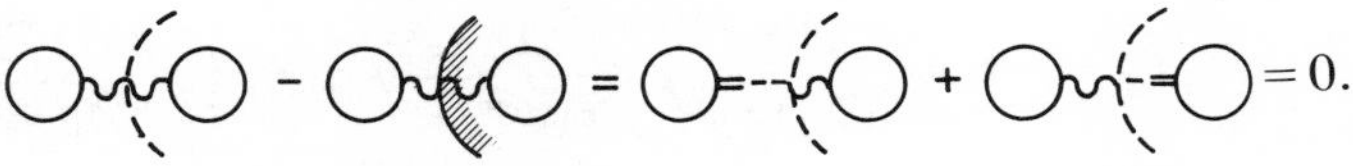

$$- \quad = \quad + \quad = 0.$$

'he unphysical Faddeev–Popov ghosts which do not contribute to the ;-matrix as single intermediate state as a result of the ghost number onservation must be considered in two-particle intermediate states. It ollows from Eqn. (2.4.56) that when all lines satisfy their mass shell onditions $k^2 = 0$ and all field points describe transversal gauge fields up to y ınd x_1, then the Ward identity* is [tH 71, Ta 71]:

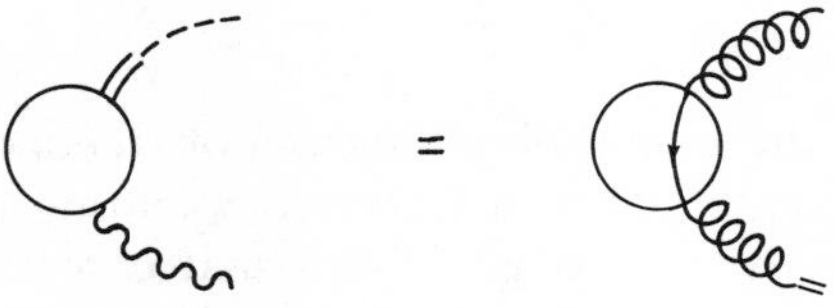

$$= \qquad (2.4.58)$$

* In view of the ghost number conservation at each vertex, the ghost line can be followed hrough any Feynman diagram summarized in ○, hence 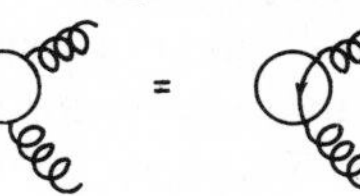 =

From Eqns. (2.4.58), (2.4.55), and (2.4.57), the following pattern emerges:

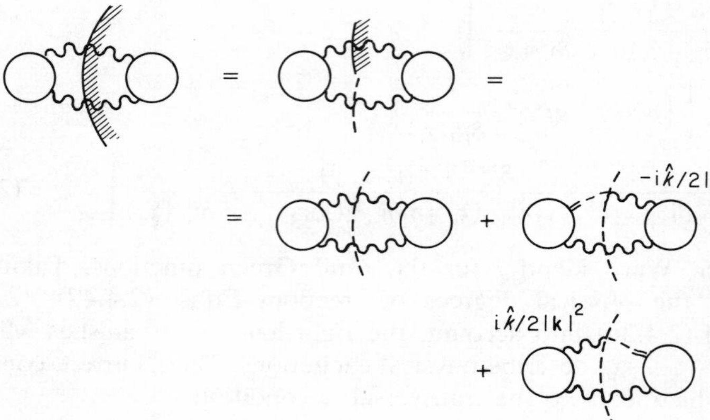

$= \qquad =$

$= \qquad + \quad -i\hat{k}/2|\mathbf{k}|^2 \quad +$ (2.4.59)

$+ \quad i\hat{k}/2|\mathbf{k}|^2$

and then with $k \cdot \hat{k} = -2\,|\mathbf{k}|^2$

$ik \quad -i\hat{k}/2|\mathbf{k}|^2 \quad = \quad -i\hat{k}/2|\mathbf{k}|^2 \quad ik \quad = \quad - \quad -i\hat{k}/2|\mathbf{k}|^2 \quad -ik$

$$= - \qquad \frac{-i\hat{k}}{2|\mathbf{k}|^2}(-ik) = -$$

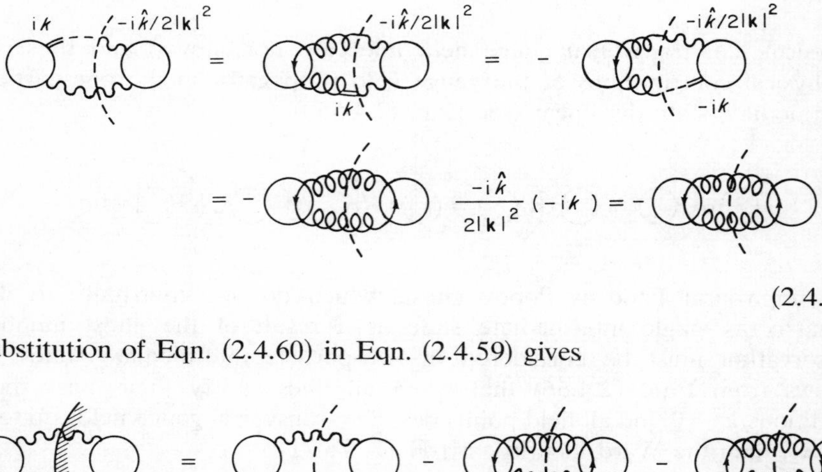

(2.4.60)

Substitution of Eqn. (2.4.60) in Eqn. (2.4.59) gives

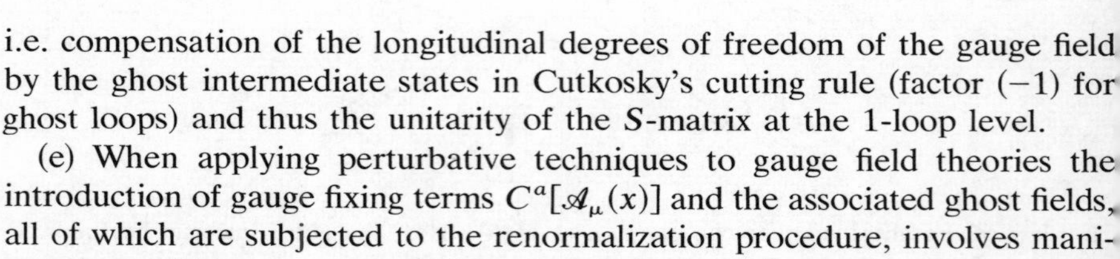

$= \qquad - \qquad -$

i.e. compensation of the longitudinal degrees of freedom of the gauge field by the ghost intermediate states in Cutkosky's cutting rule (factor (-1) for ghost loops) and thus the unitarity of the S-matrix at the 1-loop level.

(e) When applying perturbative techniques to gauge field theories the introduction of gauge fixing terms $C^a[\mathcal{A}_\mu(x)]$ and the associated ghost fields, all of which are subjected to the renormalization procedure, involves manipulations which are not explicitly gauge-invariant. Gauge invariance of the physical results must therefore be shown by confirming the *gauge independence of the S-matrix*. This can be done by replacing the gauge fixing functional $C^a[\mathscr{A}_\mu(x)]$ (Eqn. (2.3.129″)), by a neighbouring functional and then showing that the S-matrix remains unchanged for variations $C^a \to C^a + \delta C^a$ [Co 75].

For massive gauge fields, which obtain their mass by a dynamical mechanism, which has yet to be explained, without destroying the gauge symmetry (Section 3.2), the gauge independence of the S-matrix supplies a more simple proof of the unitarity of the renormalized S-matrix [tH 72]. In this case, the renormalization procedure gives a unitary, renormalized S-matrix, provided that:

(1) a gauge $\mathscr{L}^R_{\text{fix}}$ exists, which is renormalizable in the sense of power counting according to Section 2.4.1 (R-gauge);
(2) a gauge $\mathscr{L}^U_{\text{fix}}$ exists, in which the ghost fields are free fields and the propagators of the massive vector particles take the form $i(2\pi)^{-4}(-g_{\mu\nu} + k_\mu k_\nu/M^2)/(k^2 - M^2 + i0)$ (Eqn. (2.3.20)). (We know from discussions in Section 2.3.1(c) that this propagator contains only physical polarization states and therefore gives a unitary S-matrix—U-gauge);
(3) a continuous set of gauge functions $\mathscr{L}^\lambda_{\text{fix}}$ interpolates between $\mathscr{L}^R_{\text{fix}}$ and $\mathscr{L}^U_{\text{fix}}$, e.g. $\mathscr{L}^{\lambda=0}_{\text{fix}} = \mathscr{L}^R_{\text{fix}}$, $\mathscr{L}^{\lambda=1}_{\text{fix}} = \mathscr{L}^U_{\text{fix}}$, so that the ghost propagator is defined for all λ (the R-gauges in Section 2.4.1 with $\lambda = \xi$ form an example);
(4) that no Alder–Bell–Jackiw anomalies occur.

This latter condition, the anomaly-freedom, is the last point in the discussions of this section.

(f) So far, our explanation of the renormalization programme has blindly assumed that each gauge field theory is suitable for dimensional and therefore gauge-invariant regularization. However, this is only correct if no coupling of gauge fields to axial spinor currents occurs in the perturbation series and this certainly does not apply to any field theory of the weak interaction. The problem is that in $D \neq 4$ dimensions there is no analogy to the Dirac matrix γ_5, or equivalently, to the antisymmetrical tensor $\varepsilon_{\mu\nu\rho\sigma}$ ($\gamma_5 = (i/4!)\varepsilon_{\mu\nu\rho\sigma}\gamma^\mu\gamma^\nu\gamma^\rho\gamma^\sigma$). As a result, an axial current conserved from a chiral symmetry of the Langrangian density (Section 1.2.3(a)) is no longer divergence-free in D dimensions: $\partial^\mu j^5_\mu = (D-4)\mathscr{O}$ ($\mathscr{O} \equiv$ local operator). If matrix elements of physical states are considered and $\langle\,|\mathscr{O}|\,\rangle$ is calculated beyond the tree graph approximation, then divergences, as shown in Section 2.4.1, generally occur which can then be separated as poles in $\varepsilon = 4 - D$: $\langle\,|\mathscr{O}|\,\rangle = A/(4-D) +$ regular terms. The axial current of interacting fields is not conserved, even when changing to four space–time dimensions: $\partial^\mu\langle\,|j^5_\mu|\,\rangle = A$, an anomaly A occurs which is known as a γ_5 anomaly or an Adler–Bell–Jackiw type anomaly [Ad 69a, b, Be 69, Ad 70, Ja 72a]. As a result of the close connection between current conservation, symmetry of the Lagrangian density and Ward identities (see Eqns. (2.3.98)–(2.3.103)) each anomaly of a current coupled to a gauge field destroys the gauge symmetry and hence the Ward identities which are decisive in the proof of renormalizability. Anomalies of currents which are only connected to global symmetries, e.g. the chiral $U(1)$ current [tH 76], generally do not affect the renormalizability.

Anomaly-free theories can be classified with the help of the following observations:

(i) anomalies occur only as a result of 1-loop diagram contributions [Ad 69b, Gr 72];
(ii) 1-loop diagrams with no spinor lines can undergo gauge-invariant regularization and thus do not give any anomalies [tH 72a].

Thus, it is sufficient to restrict the discussion of anomalies to the diagrams represented in Fig. 2.11. All other anomalies in fermion 1-loop diagrams (*AAA* triangular diagram, *VVVA* and *VAAA* square diagram, *VVVVA*, *VVAAA*, and *AAAAA* pentagon diagram) are related to these [Ba 69, We 71, Av 72].

For a theory with internal symmetry, the γ_5 anomaly in the triangular graph in Fig. 2.11 is proportional to

$$\mathrm{Tr}[(\underline{\lambda}^a\underline{\lambda}^b+\underline{\lambda}^b\underline{\lambda}^a)\underline{\lambda}^c] \qquad (2.4.61)$$

$(\underline{\lambda}^a)_{ij}$, $(\underline{\lambda}^b)_{ij}$ and $(\underline{\lambda}^c)_{ij}$ are the coupling matrices in the fermion gauge field vertices *a*, *b*, and *c*. The trace should be carried out over all possible degrees of freedom of the fermions, as γ_5 anomalies are independent of the fermion masses.

A theory is certain to be anomaly-free when the trace in Eqn. (2.4.61) is equal to zero. This always applies for some 'safe' internal symmetry groups. Thus,

$$\mathrm{Tr}[(\underline{\lambda}^a\underline{\lambda}^b+\underline{\lambda}^b\underline{\lambda}^a)\underline{\lambda}^c]=0 \qquad (2.4.62)$$

applies for all representations of the simple Lie groups:

$$\begin{array}{lll} SU(2)\simeq SO(3), & & G(2), \\ SO(N) & \text{for } N\geqslant 5, N\neq 6, & F(4), \\ Sp(2N) & \text{for } N\geqslant 3 & E(6), E(7), \text{ and } E(8), \end{array} \qquad (2.4.63)$$

and for products of these groups, e.g. $SO(4)\simeq SU(2)\times SU(2)$ [Ge 72, Gü 76]. On the other hand, theories with factors of unitary symmetry groups $SU(N)$, $N\geqslant 3$ or with Abelian factors $U(1)$ are not 'safe' from anomalies. However, even in this instance:

(1) γ_5 anomalies can be avoided by choosing representations of these groups which satisfy Eqn. (2.4.62) for fermions; or
(2) γ_5 anomalies can be compensated, i.e. anomalies of left- and right-handed coupled fermions cancelling each other (see Sections 1.3.1 and 1.3.2) (vectorlike theories).

Finally, let us consider the theory of electromagnetic and weak phenomena

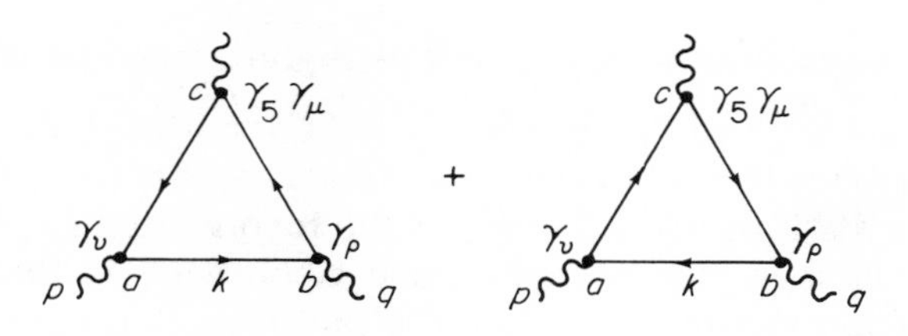

'igure 2.11 Fermion triangular graphs with γ_5 anomaly (*VVA* coupling); the external lines are vector boson lines

rom the point of view of anomaly-freedom. The internal symmetry group is he group $SU(2)_W \times U(1)$ in which left-handed leptons and quarks are ransformed as doublets and the right-handed ones as singlets (Section .3.1(c), (d)). As a result of the properties of the Pauli-matrices (Eqn. 1.2.3)), Eqn. (2.4.62) is not fulfilled only when the external lines in the liagrams in Fig. 2.11 represent two identical intermediate vector bosons Section 1.3.1(f)) and a photon or one vector boson and two photons. In ach of these cases, the condition for anomaly-freedom [Bo 72, Gr 72] is

$$\mathrm{Tr}\, Q_L = 0 \tag{2.4.64}$$

vhere Q_L is the matrix of electrical charges of left-handed fermions. The *implest* compensation scheme for the anomalies is when the number of eft-handed lepton and quark doublets is equal, since $Q = -1$ for lepton loublets $(l, \nu_l)_L$ and $Q = 3 \times (\frac{2}{3} - \frac{1}{3}) = +1$ for quark doublets $(u, d)_L$, $c, s)_L, \ldots$ (see Sections 1.3.1(c) and 1.3.2(a)). The fact that three leptons, *e*, ι, and τ, are now known is used as an argument in favour of the existence of he flavour degree of freedom known as 'top' which has not yet been liscovered. However, as γ_5 anomalies are independent of mass, no predic-ion for the *t*-quark mass can be made. But a further, totally unexpected rgument in favour of the colour degree of freedom emerges; the sum of the harges in a left-handed quark doublet is only equal to $+1$ for quarks vith three colours and this is the only way that the -1 charge of the orresponding lepton doublet can be neutralized.

The renormalization programme solves the ultraviolet problem of quan-ized gauge field theories, i.e. it eliminates divergences which occur in 'eynman integrals at large values of the loop momenta. However, field heories with zero-mass particles are subject to infrared divergences to vhich very little attention has been paid in this chapter. An acceptable olution to the infrared problem for QED is based on the work of F. Bloch nd A. Nordsieck [Bl 37]. However, for the case of QCD where the gauge osons of the non-renormalized theory also have zero mass (Eqns. (2.2.38), 2.2.42)) but carry a charge themselves, in contrast to the photon, the nfrared problem is still unsolved and is tied up with the question of quark onfinement.

2.5 Renormalization group and asymptotic freedom of QCD

QCD is a quantum field theory which can be renormalized. The following section investigates the behaviour of its Green functions for large externa momenta. The basis of this discussion is the renormalizability of QCD. The main result will show that the quark and gluon fields in this limit can be regarded as approximately free. QCD is said to be an asymptotically free theory. This QCD property has its basis in the self-interaction of the gluons present in non-Abelian gauge theories. The asymptotic freedom of QCD helps to explain the success of the parton model in interpreting experiment on deep inelastic lepton scattering, etc. (Section 1.4). Section 2.7 deals with the improvements which QCD brings to experimental analysis, as compared to the parton model.

2.5.1 Renormalization group equation

Several conclusions can now be drawn on the behaviour of Green function within a renormalizable quantum field theory, for large Euclidean values o the momenta $p_i^2 \to -\infty$. It is assumed that the mass terms in the Lagrangian density $\mathscr{L}$ plays no significant role and this has been confirmed by perturba tion theory [We 60, Sy 70]. The infrared singularities which occur in zero mass theories cause no problems if the analysis is based on general Euclidean momenta [Sy 71]. Hence, the basic considerations of the renor malization group techniques are illustrated in zero-mass QCD [St 53 Ce 54].

The renormalized Green function of the zero-mass theory depend on the renormalized coupling constant g and the renormalization point paramete $\mathcal{M}$, which is used in the formulation of renormalization conditions. However in the original Lagrangian function there is only one parameter, the non renormalized coupling constant g_0. Hence there must be a relation, though somewhat hidden, between g and $\mathcal{M}$ and this is given by the renormalization group equation for Green functions [RG].

(a) The *renormalization group equation* is derived for *renormalized verte functions.* Calculation of Green functions from the Feynman rules has been seen to result in divergent expressions. However, following the renormaliza tion programme for gauge theories (Section 2.4), from expressions which are made finite by regularization techniques divergent factors can be isolated such that the rest—i.e. the renormalized functions—remain finite in the physical limit of the cut-off parameter. This results in Eqn. (2.4.13) for the vertex functions:

$$\Gamma_{n_A,n_q}(k_i;p_j)=\lim_{\varepsilon\to 0}[Z_A(\varepsilon)]^{n_A/2}[Z_q(\varepsilon)]^{n_q/2}\Gamma^{(u)}_{n_A,n_q}(k_i;p_j,\varepsilon). \qquad (2.5.1)$$

The finite parts of the renormalized vertex functions Γ_{n_A,n_q} are determined from the renormalization conditions (2.4.15). The vertex functions depend on the gauge (i.e. on the gauge parameter ξ). However, this dependence no longer exists when calculating S-matrix elements, as the Green functions fulfil the Ward identities. The Landau gauge $\xi = 0$ is used in the following description. The renormalization group equation (2.5.3) is then valid in its given form for vertex functions in Landau gauge. An additional term $\delta(\xi, g)\,\partial/\partial\xi$ has to be incorporated in (2.5.3) for other gauges [Gr 73].

The unrenormalized coupling constant, i.e. the one in the Lagrangian function $\mathscr{L}$, is designated as g_0 (or $g_0\mu^{\varepsilon/2}$, $\varepsilon = 4 - D$ for dimensional regularization (see Eqn. (2.4.18))). The renormalized coupling constant g defined by the renormalization condition in Eqn. (2.4.15″) is a function of the unrenormalized coupling constant g_0, cut-off parameter ε and renormalization mass $\mathcal{M}$. The renormalized vertex functions are regarded as functions of the renormalized coupling constant:

$$\Gamma_{n_A, n_q} = \Gamma_{n_A, n_q}(k_i, p_j; g; \mathcal{M}), \quad g = g(g_0\mu^{\varepsilon/2}, \varepsilon, \mathcal{M}). \tag{2.5.2}$$

They depend only indirectly on the unrenormalized coupling constant.

The fact that a parameter which was originally not present has been introduced into the theory by the renormalization point $\mathcal{M}$ is expressed by the fact that the unrenormalized vertex function is independent of $\mathcal{M}$. Thus, the following equations apply:

$$\begin{aligned}
0 &= \mathcal{M}\frac{d}{d\mathcal{M}}\Gamma^{(u)}_{n_A, n_q}(k_i, p_j; g_0\mu^{\varepsilon/2}) \\
&= \mathcal{M}\frac{d}{d\mathcal{M}}[Z_A^{-n_A/2}(g_0\mu^{\varepsilon/2}, \varepsilon, \mathcal{M})Z_q^{-n_q/2}(g_0\mu^{\varepsilon/2}, \varepsilon, \mathcal{M}) \\
&\quad \times \Gamma_{n_A, n_q}(k_i, p_j; g(g_0\mu^{\varepsilon/2}, \varepsilon, \mathcal{M}); \mathcal{M})] \\
&= Z_A^{-n_A/2}Z_q^{-n_q/2} \\
&\quad \times \mathcal{M}\left(\frac{n_A}{2}Z_A^{-1}\frac{\partial Z_A}{\partial\mathcal{M}} - \frac{n_q}{2}Z_q^{-1}\frac{\partial Z_q}{\partial\mathcal{M}} + \frac{\partial g}{\partial\mathcal{M}}\frac{\partial}{\partial g} + \frac{\partial}{\partial\mathcal{M}}\right)\Gamma_{n_A, n_q}(k_i, p_j; g; \mathcal{M})
\end{aligned}$$

or:

$$\left(\mathcal{M}\frac{\partial}{\partial\mathcal{M}} + \beta(g)\frac{\partial}{\partial g} - n_A\gamma_A(g) - n_q\gamma_q(g)\right)\Gamma_{n_A, n_q}(k_i, p_j; g; \mathcal{M}) = 0 \tag{2.5.3}$$

with the RG function (also Gell–Mann–Low function or Callan–Symanzik function)

$$\beta(g) = \lim_{\varepsilon \to 0}\mathcal{M}\frac{\partial}{\partial\mathcal{M}}g(g_0\mu^{\varepsilon/2}, \varepsilon, \mathcal{M}) \tag{2.5.4}$$

and the anomalous dimension

$$\gamma_i(g) = \tfrac{1}{2}\lim_{\varepsilon \to 0}\mathcal{M}Z_i^{-1}\frac{\partial Z_i}{\partial\mathcal{M}}(g_0\mu^{\varepsilon/2}, \varepsilon, \mathcal{M}), \qquad i = A, q. \tag{2.5.5}$$

Functions $\beta(g)$, $\gamma_i(g)$ are the same for all Green's functions and thus are a characteristic of the theory.

An elementary dimensional consideration can be used to reshape Eqn. (2.5.3) somewhat. From the usual mass dimensions for quark fields: dim $\psi = M^{3/2}$, and gluon fields: dim $A = M^1$, given in Table 2.2, it follows dim $\tau_{n_A,n_q}(k_i, p_j) = M^{4-3n_A-5/2n_q}$ (the δ-function for energy momentum conservation is split off). As a result dim $\Gamma_{n_A n_q} = M^{4-n_A-3/2n_q}$ and Euler's differential equation for homogeneous functions gives:

$$\left[\lambda\frac{\partial}{\partial\lambda}+\mathcal{M}\frac{\partial}{\partial\mathcal{M}}+\tfrac{3}{2}n_q+n_A-4\right]\Gamma_{n_A,n_q}(\lambda k_i, \lambda p_j; g; \mathcal{M})=0. \qquad (2.5.6)$$

Thus, if $\mathcal{M}(\partial/\partial\mathcal{M})$ in Eqn. 2.5.3 is replaced by the corresponding expression from Eqn. 2.5.6, then the Gell–Mann–Low or homogenous Callan–Symanzik equation is obtained [Ca 70, Sy 70]:

$$\left[\lambda\frac{\partial}{\partial\lambda}-\beta(g)\frac{\partial}{\partial g}+(\tfrac{3}{2}+\gamma_q)n_q+(1+\gamma_A)n_A-4\right]\Gamma_{n_A,n_q}(\lambda k_i, \lambda p_j; j; \mathcal{M})=0 \qquad (2.5.7)$$

for zero mass theory. The anomalous dimensions $\gamma(g)$ describe a deviation in vertex function dimensions from natural ones caused by the dynamic processes involved. This becomes clear when the GL equation is solved (2.5.7).

(b) The GL equation (2.5.7) is a linear, partial differential equation of the first order which can be worked out by the method of characteristic curves (Co 68). Combined with the initial condition of $\bar{g}(1, g) = g$, the solution of the non-trivial equation of the characteristic curve

$$\frac{d\bar{g}(\lambda, g)}{d(\log\lambda)} = \beta(\bar{g}) \qquad (2.5.8)$$

defines the *running coupling constant* $\bar{g}$. This is given by solving

$$\int_1^\lambda \frac{d\lambda'}{\lambda'} = \log\lambda = \int_g^{\bar{g}} dg'\beta(g')^{-1} \qquad (2.5.9)$$

for $\bar{g}(\lambda, g)$ and hence the explicit solution to the GL equation can be expressed as

$$\Gamma_{n_A,n_q}(\lambda k_i, \lambda p_j; g; \mathcal{M})$$
$$=\Gamma_{n_A,n_q}(k_i, p_j; \bar{g}(\lambda, g); \mathcal{M})\lambda^{4-n_A-3n_q/2}\exp\left(-\int_g^{\bar{g}} dg'(n_A\gamma_A(g') + n_q\gamma_q(g'))\beta(g')^{-1}\right). \qquad (2.5.10)$$

The solution of the GL equation gives a scale change equation which states that a change in the momenta by a factor λ can be expressed by a change in the coupling constant and a multiplicative factor. This factor differs from that of homogeneous functions with canonical dimension ($\lambda^{4-n_A-3n_q/2}$) by the contribution coming from the dynamical dimensions γ_A and γ_q, averaged with by $\beta(g)^{-1}$. For the sake of simplicity, this result has been derived for a zero mass theory. If λ is large, Eqn. (2.5.10) provides the leading term for a theory with quark masses [We 73a, Sy 70].

The running coupling constants $\bar{g}$ and the anomalous dimensions $\gamma(g)$ thus determine the behaviour of the vertex functions for high values of λ. If $\beta(g)$ and $\gamma(g)$ are calculated in 1-loop approximation, then the GL equation provides the asymptotic summation of the leading logarithms in all orders of g and thus improves the perturbation series in the asymptotic region.

(c) The structure of zero mass QCD (or zero mass field theories in general) is given essentially by the limits of the variable coupling constant $\bar{g}(\lambda, g)$ for large or small λ. These limit values are determined by the *fixed points* of the differential equation (2.5.8) for the running coupling constant, i.e. by the values of g, for which $\beta(g)=0$. $\beta(g)$ is expanded around g^* to determine the behaviour of $\bar{g}$ around a fixed point g^*:

$$\beta(g)=(g-g^*)^r\beta', \quad r\geq 1.$$

Eqn. (2.5.9) then gives the following formula for the behaviour of the running coupling constant around the fixed point:

$$\bar{g}-g^*=(g-g^*)\lambda^{\beta'}, \quad \text{for } r=1 \tag{2.5.11}$$

$$(\bar{g}-g^*)^{r-1}=\frac{(g-g^*)^{r-1}}{1-\beta'(r-1)(g-g^*)^{r-1}\log\lambda} \quad \text{for } r=2, 3, \ldots. \tag{2.5.11'}$$

For a simple zero ($r=1$), the following formula applies:

$$\bar{g}(\lambda, g)\to g^* \quad \text{for} \begin{cases} \lambda\to\infty, & \text{when } \beta'<0, \quad (2.5.12) \\ \lambda\to 0, & \text{when } \beta'>0. \quad (2.5.12') \end{cases}$$

g^* is an infrared stable fixed point for $\beta'>0$ and an ultraviolet stable fixed point for $\beta'<0$. The vertex functions are homogenous functions of the order $4-(\frac{3}{2}+\gamma_q(g^*))n_q-(1+\gamma_A(g^*))n_A$ at the stable fixed point, in agreement with the scale change equation (2.5.10). Around an infrared (ultraviolet) unstable fixed point for large (small) λ, $\bar{g}(\lambda, g)$ moves away from the value of the coupling constants g^* at the fixed point. Thus the shape of the RG function $\beta(g)$ can detemine the general properties of a field theory. This is illustrated by Fig. 2.12.

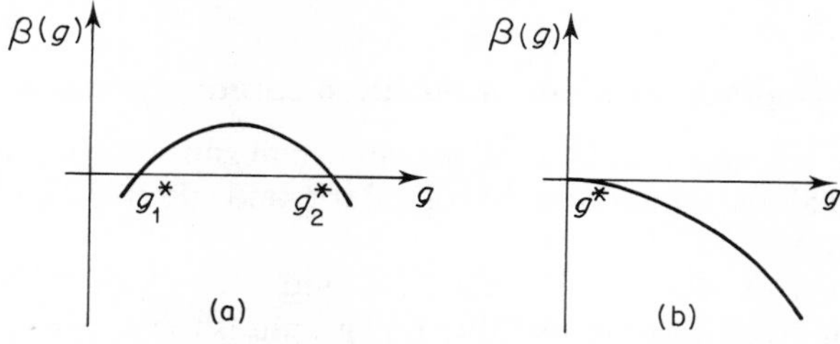

Figure 2.12 Examples of fixed points for the RG function $\beta(g)$

Theory (a) has an ultraviolet stable fixed point in g_2^* and an infrared one in g_1^*. If the renormalized coupling constant g is between g_1^* and g_2^*, then the theory approaches the corresponding limit theories with homogenous vertex functions, for large and small distances, respectively.

Theory (b) has an ultraviolet stable fixed point, at $g^*=0$ and this type of theory is called asymptotically free. Thus, for small distances, it has the scale behaviour of a free theory. However, this is modified by logarithmic terms for $r>1$ as Eqn. (2.5.11) shows. It is clear from Eqn. (2.5.12) that as $\lambda\to\infty$, so does $\bar{g}(\lambda, g)\to\infty$. Theory (b) is infrared-unstable. At this point, it must be stressed that leaving out mass terms in the Lagrangian function is only insignificant as regards the high-energy behaviour. For small momenta, mass terms can have a considerable effect on the behaviour of Green functions.

The aim of the following section is to show that QCD is asymptotically free. Section 2.7 deals with physical consequences from the corresponding scaling behaviour. The renormalization group techniques must be modified with methods used in statistical mechanics [Wi 75] for the analysis of the infrared behaviour of QCD. This is dealt with later in the chapter on the confinement problem.

(d) Within a renormalization group concept the consideration of physical quantities is of fundamental significance. As emphasized many times, QCD with zero mass quarks contains no parameter on which an energy scale can be based. Such a quantity is introduced only by the renormalization point $\mathcal{M}$. So, if a measurable physical quantity ('Phys') e.g. a mass is calculated in QCD, this will depend on g and $\mathcal{M}$: Phys = Phys(g, $\mathcal{M}$), but only in a combination which is invariant under renormalization [Gr 74, Ca 78]:

$$\left(\mathcal{M}\frac{\partial}{\partial\mathcal{M}}+\beta(g)\frac{\partial}{\partial g}\right)\text{Phys}(g,\mathcal{M})=0; \tag{2.5.13}$$

as a change in $\mathcal{M}$ means only a change in the meaning of g. As a mass M must be proportional to $\mathcal{M}$ for dimensional reasons, it must have in QCD with massless quarks the *renormalization-invariant* form

$$M=\mathcal{M}\exp\left(-\int^{g}\frac{dg'}{\beta(g')}\right). \tag{2.5.14}$$

The special value of M is determined by the integration constant.

2.5.2 The asymptotic freedom of quantum chromodynamics

A calculation will now be made of the renormalization group function $\beta(g)$ and the anomalous dimensions γ_i of quark and gluon fields in QCD, in 1-loop approximation

(a) Three vertex functions $\Gamma_{2,0}$, $\Gamma_{0,2}$, and $\Gamma_{1,2}$ derived in the above approximation from the Green function graphs shown below are used to determine the β *function* and *anomalous dimensions*:

(α) Gluon propagator:

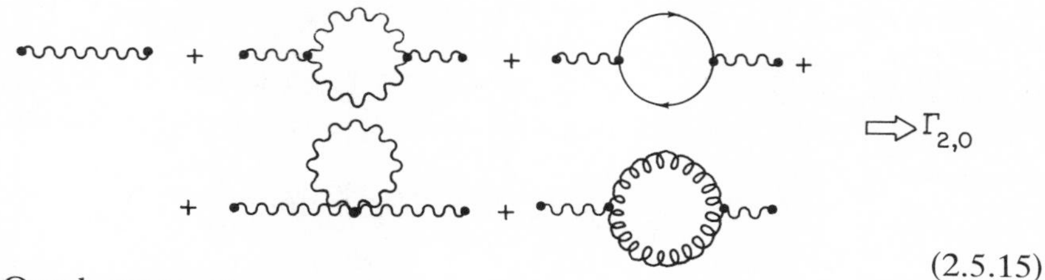

(2.5.15)

(β) Quark propagator

(γ) Quark–gluon vertex:

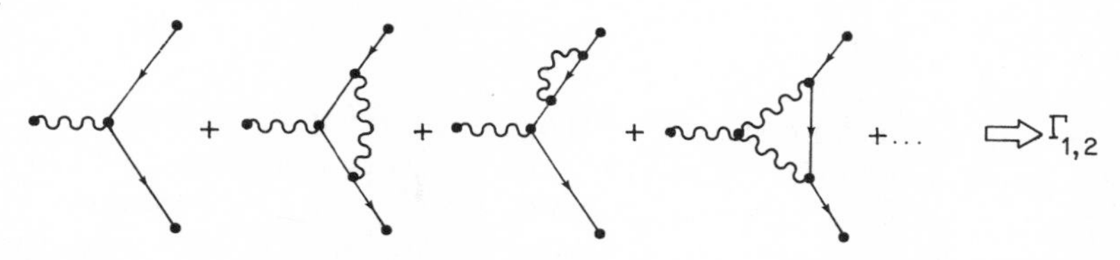

Section 2.4.2 gave details on the calculation of (α)—the gluon propagator in Landau gauge—as an example. The result (Eqns. (2.4.29), (2.4.30)) was the transversal part of the renormalized vertex function $\Gamma_{2,0}$ and the regularized renormalization constant

$$Z_A(g, \mathcal{M}) = 1 - \frac{g^2}{16\pi^2}\left(\tfrac{13}{6}C_2(G) - \tfrac{2}{3}N_F\right)\left(-\frac{2}{\varepsilon} + \log\frac{\mathcal{M}^2}{\kappa^2}\right) + O(g^4). \tag{2.5.16}$$

The self-energy term of the second order in the quark propagator for zero mass quarks disappears in Landau gauge. So, in this order, the following formula holds:

$$Z_q(g, \mathcal{M}) = 1 + O(g^4). \tag{2.5.17}$$

A similar calculation to that of the gluon propagator gives the renormalized quark–gluon vertex function:

$$\Gamma_{1,2}(p, p, p; g; \mathcal{M})^\mu_a = -ig\frac{\lambda_a}{2}\gamma^\mu\left[1 - \tfrac{3}{4}C_2(G)\frac{g^2}{16\pi^2}\log\left(\frac{-p^2 - i\varepsilon}{\mathcal{M}^2}\right)\right],$$

and the associated renormalization constant:

$$Z(g, \mathcal{M}) = 1 + \frac{g^2}{16\pi^2}\tfrac{3}{4}C_2(G)\left(-\frac{2}{\varepsilon} + \log\frac{\mathcal{M}^2}{\kappa^2}\right) + O(g^4). \tag{2.5.18}$$

The relation between the unrenormalized and the renormalized coupling constants g_0 and g can be seen from Eqns. (2.4.11) and (2.5.16)–(2.5.18):

$$g = g_0\left[1 - \frac{g_0^2}{16\pi^2}\left(\tfrac{11}{6}C_2(G) - \tfrac{1}{3}N_F\right)\left(-\frac{2}{\varepsilon} + \log\frac{\mathcal{M}^2}{\kappa^2}\right)\right] + O(g_0^5). \quad (2.5.19)$$

By substituting the expressions (2.5.16), (2.5.17), and (2.5.19) into Eqns. (2.5.4) and (2.5.5), the following formula in g^3 order emerges for the RG function of QCD:

$$\begin{aligned}\beta(g) &= -\frac{g^3}{16\pi^2}\left(\tfrac{11}{3}C_2(G) - \tfrac{2}{3}N_F\right) = -\beta_0\frac{g^3}{16\pi^2}\\ &= -\frac{g^3}{16\pi^2}\left(11 - \tfrac{2}{3}N_F\right) \quad \text{for } G = SU(3)_C \end{aligned} \quad (2.5.20)$$

and

$$\begin{aligned}\gamma_A(g) &= -\frac{g^2}{16\pi^2}\left(\tfrac{13}{6}C_2(G) - \tfrac{2}{3}N_F\right) = \gamma_A^0\frac{g^2}{16\pi^2}\\ &= -\frac{g^2}{16\pi^2}\left(\tfrac{13}{2} - \tfrac{2}{3}N_F\right) \quad \text{for } G = SU(3)_C \end{aligned} \quad (2.5.21)$$

$$\gamma_q(g) = 0$$

for the anomalous dimensions of the gluon and quark fields in Landau gauge. According to the criterion for the $\beta(g)$ function formulated in Eqn. (2.5.12), $g = 0$ is an ultraviolet stable fixed point. Non-Abelian gauge theories are asymptotically free provided the number N_F of fermions is not too large.

The gluon–gluon coupling which gives the term proportional to $C_2(G)$ and thus the negative sign to the RG function, as shown in graph (b) in Section 2.4.2, is responsible for the asymptotic freedom of QCD. It is $C_2(G) = 0$ in QED, i.e. the β-function, has a positive sign. In agreement with Eqn. (2.5.11), this means that the effective coupling constant increases as distance becomes smaller, i.e. decreases as distance becomes larger. Within the realm of QED, this is known as screening of the unrenormalized ('bare') charge [La 55, Be 76, No 78]. Vacuum polarization by virtual electron–positron pairs can screen the 'bare' charge, as the latter attracts particles with the opposite charge and repels those with the same. So, from a given distance, the bare charge is seen surrounded by a cloud of charges with opposite sign. The physical electrical charge measured in a sphere with this radius is smaller than the bare 'point' charge. This represents a special case of the general quantum mechanical rule, that fluctuations of the ground state into physical intermediate states lead to a reduction of the energy of the undisturbed ground state—in this case the field energy of a charge [No 78]. Hence, vacuum polarization by charged transversal, i.e. physical gluon pairs (Eqn. (2.4.48)) also causes colour charge screening. However, this argument does not apply to non-physical intermediate states such as a pair made up of a transversal and a longitudinal (Coulomb-type) gluon. In fact, this intermediate state makes a contribution of opposite sign which is greater by a factor of 12. The different sign is explained by the fact that the interaction between equal gluons which is mediated by transversal gluons is attractive, (Biot–Savart law), whereas the Coulomb force between equal charges is repulsive. Hence the repulsive Coulomb force between charged gluons causes the antiscreening effect which results in asymptotic freedom. However, this intuitive interpretation of the calculation leading to asymptotic freedom is somewhat dependent on the gauge.

Almost all known renormalizable theories which are not non-Abelian gauge theories, are not asymptotically free. For the gauge group $SU(3)_C$ in QCD, the sign of the β-function changes for $N_F > 33/2$. If the number of quarks is below 33/2, then QCD is asymptotically free. Meanwhile, parameters for the renormalization group equation of QCD have also been calculated in 2-loop approximation (Eqn. (2.5.27)). $\beta(g) < 0$ remains, provided that quarks are not too numerous.

In addition to fermion fields, gauge theories for the electroweak interaction and the unified strong and electroweak interaction also contain scalar fields. Their gauge-invariant coupling with the gauge bosons makes a positive contribution to the RG function which is similar in structure to that of the fermions and which destroys asymptotic freedom when the number of fermions and scalars is too high. The theories described require scalar field self-interaction in addition, which then means that they are no longer asymptotically free [Co 73a].

(b) The GL equation for the vertex functions in QCD can now be solved with the help of Eqns. (2.5.20) and (2.5.21). First of all, Eqns. (2.5.20) and (2.5.11′) are used to calculate the QCD *running coupling constant* for $r = 3$

$$(\bar{g}(\lambda, g))^2 = \frac{g^2}{1 + 2\beta_0(g^2/16\pi^2)\log\lambda} = \frac{g^2}{1 + (g^2/8\pi^2)(11 - \frac{2}{3}N_F)\log\lambda}. \tag{2.5.22}$$

The strong, fine structure constant $\bar{\alpha}_s = \bar{g}^2/4\pi$ is often used, for which (2.5.22) reads

$$\bar{\alpha}_s(\lambda) = \frac{\alpha_s}{1 + (\alpha_s/12\pi)(33 - 2N_F)\log\lambda^2}.$$

$\alpha_s = g^2/4\pi$ is the strong coupling constant which is linked to renormalization mass $\mathcal{M}$ by the renormalization condition (2.4.15″). Using the fact that $\lambda^2 = Q^2/\mathcal{M}^2$ is the scale factor with which the four-momentum transfer at the vertex is modified, the following formula can be written:

$$\bar{\alpha}_s(Q^2) = \frac{12\pi}{(33 - 2N_F)\log(Q^2/\Lambda^2)} \tag{2.5.23}$$

with

$$\Lambda = \mathcal{M}\exp\left(-\frac{6\pi}{(33 - 2N_F)\alpha_s}\right) \approx \mathcal{M}\exp\left(-Pf'\int_0^g \frac{dg'}{\beta(g')}\right). \tag{2.5.24}$$

The expression (2.5.23) shows most clearly the dependence of the running coupling constant $\bar{\alpha}_s(Q^2)$ on momentum. The asymptotic scale parameter (ASP) Λ is a renormalization group invariant quantity as described in Eqn. (2.5.14). A prescription (Pf') which gives the integral a finite value is required for a general determination of the integration constant of the integral in Eqn. (2.5.24), which is divergent when $g = 0$. This can be done by splitting off the divergent first- and second-order contributions and fixing them individually. Eqn. (2.5.24) expresses the value for the first order. The general prescription is given below (Eqns. (2.5.28), (2.5.29)), together with the second order expression of Eqn. (2.5.23).

Λ is the only, independent, free parameter in zero mass QCD which has a dimension. It determines the coupling strength for strong interactions (Eqn. (2.5.23)). For $Q^2 \gg \Lambda^2$, $\bar{\alpha}_s(Q^2)$ is small and the perturbative calculations are automatically consistent. However, the next order must be considered if Λ is to be determined experimentally (see below). Current values are in the region of $\Lambda \approx 200$–400 MeV (see Section 2.7.2(g)).

The introduction of a dimensional parameter into a dimensionless field theory by the renormalization procedure is also known as dimensional transmutation [Co 73a]. According to this scheme the dimensional parameter—which is Λ in our particular case—replaces the dimensionless coupling constant.

If the renormalized coupling constant $g(\mathcal{M}^2)$ is replaced by the running coupling constant $\bar{g}(Q^2)$ in the simple quark–gluon vertex, then it is obvious from Eqn. (2.5.22) that this quark–gluon interaction is described by summing up contributions of arbitrary order in α_s. These $(\log Q^2/\Lambda^2)^n$ contributions come from graphs formed from the elementary vertex by iterative introduction of radiative corrections of the type given in Eqn. (2.5.15). In this sense, application of the renormalization group leads beyond perturbation theory of finite order. Methods for partial summation of logarithmic contributions to the perturbation series will be described in Section 2.7.3.

(c) The integral:

$$\int_g^{\bar{g}} dg'(n_A\gamma_A(g') + n_q\gamma_q(g'))\beta(g')^{-1} = \int_1^{\lambda} \frac{d\lambda'}{\lambda'} [n_A\gamma_A(\bar{g}(\lambda')) + n_q\gamma_q(\bar{g}(\lambda'))].$$

is required for an explicit solution to the GL Eqn. (2.5.10). With Eqn. (2.5.21) and (2.5.22), this becomes:

$$n_A\gamma_A^0 \frac{g^2}{16\pi^2}\int_1^{\lambda}\frac{d\lambda'}{\lambda'}\frac{1}{1+2\beta_0(g^2/16\pi^2)\log\lambda'} = n_A\frac{\gamma_A^0}{2\beta_0}\log\left(1+2\beta_0\frac{g^2}{16\pi^2}\log\lambda\right).$$

Finally, this then gives:

$$\begin{aligned}\Gamma_{n_A,n_q}(\lambda k_i, \lambda p_j; g; \mathcal{M}) &= \Gamma_{n_A,n_q}(k_i, p_j; \bar{g}; \mathcal{M})\lambda^{4-n_A-3/2n_q} \\ &\quad \times \left(1+2\beta_0\frac{g^2}{16\pi^2}\log\lambda\right)^{-n_A\gamma_A^0/2\beta_0} \\ &= \Gamma_{n_A,n_q}(k_i, p_j; \bar{g}; \mathcal{M})\lambda^{4-n_A-3n_q/2}\left(\frac{\bar{g}^2}{g^2}\right)^{n_A\gamma_A^0/2\beta_0} \qquad (2.5.25)\end{aligned}$$

for the scaling behaviour of vertex functions in QCD in 1-loop approximation.

According to this, the vertex functions in QCD show an asymptotic scaling behaviour in which the dynamical dimensions are in the form of $(\log \lambda)^{-n_A \gamma_A^0/2\beta_0}$. The reason for this is that the RG function in QCD has a triple zero ($r=3$) when $g=0$ (see Section 2.7.2(d) for details on calculation of the dynamical dimensions of composite local operators in QCD).

(d) The previous section outlined asymptotic freedom under the most simple conditions, viz.:

(1) a zero mass theory;
(2) the lowest order in perturbation theory for the calculation of $\beta(g)$ etc.;
(3) a fixed gauge, the Landau gauge; and
(4) a renormalization scheme as described in Section 2.4.1.

However, there are many results available from literature which go beyond this scope. To start with, it has been shown that consistent use of methods associated with asymptotic freedom for interpreting experimental results does require consideration of higher orders.

Calculation of the expansion coefficients for the RG function $\beta(g)$:

$$\beta(g) = -\beta_0 \frac{g^3}{16\pi^2} - \beta_1 \frac{g^5}{(16\pi^2)^2} + \ldots = -g\left[\beta_0 \frac{\alpha_s}{4\pi} + \beta_1\left(\frac{\alpha_s}{4\pi}\right)^2 + \ldots\right] \tag{2.5.26}$$

in 2-loop approximation, results in [Ca 74, Ja 74]:

$$\beta_0 = \tfrac{11}{3}C_2 - \tfrac{2}{3}N_F \overset{SU(3)}{=} 11 - \tfrac{2}{3}N_F,$$

$$\beta_1 = \tfrac{34}{3}(C_2)^2 - \tfrac{10}{3}C_2 N_F - 2C_2^F N_F \overset{SU(3)}{=} 102 - \tfrac{38}{3}N_F \tag{2.5.27}$$

in which C_2 or C_2^F is the value of the Casimir operator of the gauge group for the adjoint or fundamental representation as given by Eqns. (1.2.7) and (1.2.8). Hence, the ASP in this approximation becomes:

$$\Lambda^2 = \mathcal{M}^2 \exp\left(-2Pf' \int_0^g \frac{dg'}{\beta(g')}\right) = \mathcal{M}^2 \exp\left[-\frac{16\pi^2}{\beta_0 g^2} - \frac{\beta_1}{\beta_0^2} \log\left(\frac{\beta_0}{16\pi^2} g^2\right)\right] \tag{2.5.28}$$

and in reverse, the running coupling constant up to terms of the order $(\log \log Q^2/\Lambda^2)^2 (\log Q^2/\Lambda^2)^{-3}$ is:

$$\frac{\bar{g}^2(Q^2)}{4\pi} = \frac{4\pi}{\beta_0 \log Q^2/\Lambda^2} - \frac{4\pi\beta_1}{\beta_0^3} \frac{\log\log Q^2/\Lambda^2}{(\log Q^2/\Lambda^2)^2}. \tag{2.5.29}$$

The indefiniteness associated with the evaluation of the singular integral $Pf' \int dg'$ (Eqn. (2.5.28)) is cured by the condition that no terms of the form $\sim (\log Q^2/\Lambda^2)^{-2}$ occur in the solution for the running coupling constant

(2.5.29) in the second order. The first-order ASP cannot be fixed until this second-order-based convention has been established [Ba 78, Ba 79a].

The fundamental significance of second order for determining the ASP can be further clarified by the following discussion [Ba 78]. If the 'magnitude' of a running coupling constant is defined on the basis of the first order:

$$\bar{\alpha}_s(Q^2) = \frac{4\pi}{\beta_0 \log Q^2/\Lambda^2} \tag{2.5.30}$$

by two different asymptotic scale parameters Λ and Λ', then the two coupling constants differ by a second-order term:

$$\begin{aligned}\bar{\alpha}'_s(Q^2) &= \frac{4\pi}{\beta_0 \log Q^2/\Lambda^2} = \frac{4\pi}{\beta_0(\log Q^2/\Lambda^2 + \log \Lambda^2/\Lambda'^2)} \\ &= \bar{\alpha}_s(Q^2) + \rho\bar{\alpha}_s^2(Q^2) + \ldots \end{aligned} \tag{2.5.31}$$

with

$$\rho = \frac{\beta_0}{4\pi} \log \frac{\Lambda'^2}{\Lambda^2},$$

i.e. different values of Λ lead to the same running coupling constant in the first order. However, in the second-order equation (2.5.29), the substitution $\Lambda \to \Lambda'$ leads to terms in the form $\sim(\log Q^2/\Lambda^2)^{-2}$ and thus violate our convention.

Now, if Λ in formula (2.5.30) is determined by comparison with experiment then a second-order contribution to a specific effect can be formally represented by a first-order formula with modified $\Lambda \to \Lambda_{\text{eff}}$. QCD corrections (Section 2.7.3) for the normalized cross-section R for e^+e^- pair annihilation can be taken as an example (Eqn. (1.4.28)):

$$\begin{aligned} R &= \sum_i Q_i^2[1 + \bar{\alpha}_s(Q^2)/\pi + F_3\bar{\alpha}_s^2(Q^2) + O(\bar{\alpha}^3)] \\ &= \sum_i Q_i^2[1 + \bar{\alpha}_s^{\text{eff}}(Q^2)/\pi + O(\bar{\alpha}^3)] \end{aligned} \tag{2.5.32}$$

with

$$\Lambda_{\text{eff}}^2 = \Lambda^2 \exp\left(\frac{4\pi^2}{\beta_0} F_3\right).$$

Thus a Λ_{eff} which differs from Λ by a factor which varies from process to process and is independent of the value of the running coupling constant, is determined experimentally.

The various definitions of the running coupling constant in asymptotically free theories often lead to different formulae in the second order. The importance of controlling this problem has been emphasized above. Therefore a systematic review of the transformation of coupling constants seems in order. Quantum field theory gives a set of independent, dimensionless running coupling constants, $g_i(\mu)$ which satisfy differential equations of the form:

$$\mu \frac{d}{d\mu} g_i = \beta_i(g_j) \tag{2.5.33}$$

(i.e. $\mu^2 = Q^2$ in conjunction with Eqn. (2.5.23)). It is useful to extend the discussion to several coupling constants [Sy 73] and indeed this would be necessary when considering the GL equation with general gauge [Gr 73]. If another renormalization procedure leads to another set of coupling constants $\tilde{g}_i(\tilde{\mu})$ which satisfy Eqn. (2.5.33) with $\tilde{\beta}_i(\tilde{g}_j)$, then within the theory, there must exist a transformation formula between these coupling constants, which for dimensional reasons must be of the form:

$$g_i(\mu) = F_i\left(\tilde{g}(\tilde{\mu}), \frac{\tilde{\mu}}{\mu}\right). \tag{2.5.34}$$

The simple calculations:

$$0 = \tilde{\mu} \frac{d}{d\tilde{\mu}} g_i(\mu) = \frac{\partial g_i}{\partial \tilde{g}_j} \tilde{\mu} \frac{d\tilde{g}_j}{d\tilde{\mu}} + \frac{\tilde{\mu}}{\mu} \frac{\partial F_i}{\partial(\tilde{\mu}/\mu)}$$

and

$$\mu \frac{d}{d\mu} g_i(\mu) = -\frac{\tilde{\mu}}{\mu} \frac{\partial F_i}{\partial(\tilde{\mu}/\mu)}$$

give the transformation formula for the RG functions β_i:

$$\beta_i(g) = \sum_j \frac{\partial g_i}{\partial \tilde{g}_j} \tilde{\beta}_j(\tilde{g}), \tag{2.5.35}$$

A direct result of this is that, by the transformation (2.5.34), fixed points are transformed into fixed points. A further result of Eqn. (2.5.35) is that in perturbative transformations of QCD coupling constants

$$\begin{aligned} &g = \tilde{g} + \rho \frac{\tilde{g}^3}{4\pi} + \ldots, \quad \tilde{g} = g - \rho \frac{g^3}{4\pi} + \ldots, \\ &\frac{dg}{d\tilde{g}} = 1 + 3\rho \frac{g^2}{4\pi} \end{aligned} \tag{2.5.36}$$

the first two coefficients β_0 and β_1 of $\beta(g)$ in Eqn. (2.5.26), remain invariant [Sy 73]:

$$\beta_0 = \tilde{\beta}_0, \quad \beta_1 = \tilde{\beta}_1. \tag{2.5.37}$$

Namely, if the expressions in Eqns. (2.5.26) and (2.5.36) are substituted in Eqn. (2.5.35), comparison of coefficients in g turns Eqn. (2.5.37) into:

$$\begin{aligned} \beta_0 &= -\left(\beta_0 \frac{g^3}{(4\pi)^2} + \beta_1 \frac{g^5}{(4\pi)^4} + \ldots\right) = \frac{\partial g}{\partial \tilde{g}} \tilde{\beta}(\tilde{g}) \\ &= -\left(1 + 3\rho \frac{g^2}{4\pi} + \ldots\right)\left[\tilde{\beta}_0\left(\frac{g^3}{(4\pi)^2} - 3\rho \frac{g^5}{(4\pi)^4} + \ldots\right) + \tilde{\beta}_1\left(\frac{g^5}{(4\pi)^4} + \ldots\right)\right] \\ &\quad -\left(\tilde{\beta}_0 \frac{g^3}{(4\pi)^2} + \tilde{\beta}_1 \frac{g^5}{(4\pi)^4} + \ldots\right). \end{aligned}$$

Perturbation theoretical transformations between coupling constants as in Eqn. (2.5.35) occur when a comparison of various renormalization procedures is made. So far, we have used the dimensional regularization method and the renormalization conditions in Eqn. (2.4.15). However, for both practical and historical reasons, various other methods are described in the literature. In 't Hooft's [tH 73a] minimal subtraction process (MS), only the singular terms, e.g. the term $\sim 1/\varepsilon$ in Z_A, Eqn. (2.4.29), are subtracted. This simplifies higher-order calculations. However, as the transition to a dimensionless coupling constant in the form $g^2 \to g^2\mu^\varepsilon$, (Eqn. (2.4.18)), is not unique, even the MS method can be generalized [Co 74]. For $f(\varepsilon, g^2) = 1 - \frac{1}{2}\varepsilon(\log 4\pi - \gamma)$, $g^2 \to g^2\mu^\varepsilon f(\varepsilon, g^2)$, $f(0, g^2) = 1$ gives the modified minimal subtraction procedure ($\overline{\text{MS}}$). The determination of finite renormalization terms by Eqn. (2.4.15)—known as the momentum subtraction procedure (MOM), can of course be done by fixing the 3-gluon or 4-gluon vertex normalization instead of that of the quark–gluon vertex at symmetrical momenta. All these various ways of renormalization give the same relations between physical quantities. In perturbation theory, the relevant, renormalized coupling constants are linked by transformations as in Eqn. (2.5.35). These leave the $\beta(g)$ functions invariant up to the second order and hence the transition to another scheme can be compensated up to this order by rescaling the ASP: $\Lambda_{\text{MOM}} = k\Lambda_{\text{MS}}$. A list of conversion factors can be found in the literature [Ce 79]. ($\Lambda_{\text{MOM}} \approx 2 \cdot \Lambda_{\overline{\text{MS}}}$. All information given in this book is based on the renormalization process as defined by Eqn. (2.4.15).

To sum up: the asymptotic scale parameter Λ in QCD can be regarded as a fundamental physical constant which defines the energy scale for an evaluation of QCD based on perturbative calculations, but only if:

(1) Λ is accurately defined as a renormalization group invariant by a second-order calculation at the very least (Eqns. (2.5.8), (2.5.9));
(2) Λ is related to a specific renormalization process;
(3) Λ is determined experimentally by comparison with a second-order calculation.

The investigation of the dependence on the gauge parameter ξ of the various quantities tied up with the RG show, as above, for the $\beta(g)$ function, that these are independent up to the second order. Conversely, dynamical dimensions are a function of the gauge parameter even in the first order:

$$\begin{aligned} \gamma_q &= \frac{g^2}{(4\pi)^2}\, \tfrac{4}{3}\xi \\ \gamma_A &= -\frac{g^2}{(4\pi)^2}\left[\left(\frac{13}{6} - \frac{\xi}{2}\right)3 - \tfrac{2}{3}N_F\right]. \end{aligned} \tag{2.5.38}$$

Λ changes with the gauge.

(e) The RG equations allow the renormalization procedure of a quantum field theory to be considered from a different angle [Gl 76]. The assumption is made that the singular expressions for Green functions can be made finite by any regularization procedure with a cut-off parameter $1/\varepsilon$. Physical expressions then result after subtraction or infinite renormalization in the limiting case $\varepsilon \to 0$. Section 2.6.3 will show that the lattice approximation technique is a suitable regularization process for dealing with the confinement problem. In this instance ε is the lattice constant.

A *running non-renormalized coupling constant* $g_0(\varepsilon)$ is now considered instead of the running renormalized one $\bar{g}$ and its dependence on the renormalization mass $\mathcal{M}$[Wi 75, Sy 77, Sy 79]:

$$\begin{aligned} \frac{1}{\varepsilon}\frac{\mathrm{d}}{\mathrm{d}(1/\varepsilon)}\, g_0(\varepsilon) &= -\varepsilon\frac{\mathrm{d}}{\mathrm{d}\varepsilon}\, g_0(\varepsilon) = \tilde{\beta}(g_0(\varepsilon)) \\ &= -\beta_0\frac{g_0^3}{16\pi^2} - \beta_1\frac{g_0^5}{(16\pi^2)^2} + \ldots, \end{aligned} \tag{2.5.39}$$

This is a function of the cut-off parameter $1/\varepsilon$. For finite $1/\varepsilon$, the renormalization procedure relates renormalized and non-renormalized coupling constants by a perturbation expansion (Eqn. (2.5.19)). Thus, it follows from Eqn. (2.5.37) that the lowest-order coefficients of the $\tilde{\beta}$ function agree with the β_i given in Eqn. (2.5.27) for QCD. This variation of the non-renormalized coupling constants with the cut-off parameter gives a finite limit for vertex functions after renormalization: $\Gamma(p_i) = Z(\varepsilon)\Gamma^{(u)}(p, g_0(\varepsilon))$ in agreement with Eqn. (2.5.1).

2.6 Quark confinement

In QED, the basic field quanta describe the observed physical particles, e.g. photons, electrons, muons, etc.. In contrast, particles with properties belonging to the basic field quanta in QCD have not yet been discovered. There is no experimental confirmation of

(1) free *quarks*, i.e. particles with non-integer flavour charges;
(2) free *gluons*, i.e. zero mass, strongly interacting, flavour-neutral vector particles; and/or
(3) *open colour charges*, i.e. multiplets of particles with various colour charges (Section 1.2.1(e)).

Together, these facts are known as the problem of 'quark confinement' and 'colour screening'. This is the central problem of QCD and its solution will change the status of a quark field concept with no free quarks from speculation to actual theory. Quantitative application of QCD to problems associated with the hadron spectrum, quark fragmentation in the parton model, etc. is tied up with quark confinement but unfortunately this is not yet fully understood. However, some introduction into its treatment should be given, firstly because the subject is of great physical significance and secondly it provides methods which could open up new and important aspects of gauge field theories.

Quark confinement is a non-perturbative problem. Perturbation theory as derived from the path integral formulation of field theory (Section 2.4.2) assumes that field quanta in first approximation propagate like free particles and experience weak interaction at the next approximation. Quarks do not propagate as free particles over great distances. Moreover, the discussion on asymptotic freedom showed that the effective interaction in QCD increases with increasing distance. This increasing strength of the interaction must be responsible for quark confinement in the QCD theory. Dimensional transmutation is a more formal argument in favour of quark confinement as a non-perturbative phenomenon. If the quark mass is omitted when calculating a hadron mass M_h, then the result is $M_h = c\mathcal{M}\exp(-16\pi^2/\beta_0 g^2 + \ldots)$, according to Eqns. (2.5.14) and (2.5.28). M_h cannot be expanded in a perturbation series in powers of g. A phenomenological analysis based on the approximate validity of chiral invariance, favours small masses of u and d quarks [Le 74b]. Hence the statement from zero mass QCD is relevant. Non-perturbative QCD, i.e. non-perturbative evaluation of the path integral representation (2.3.129) demands new methods. The connection between field theory and statistical mechanics referred to in Section 2.3.2(g), allows that the strong coupling, i.e. the high-temperature expansion can be successfully applied to the confinement problem (Section 2.6.3). Effects which are related to non-perturbative solutions of the classical Yang–Mills equations may play an important part in the transition from the asymptotic freedom region of small distances to the quark confinement region for larger distances (Section 2.6.4). Before describing these recent theoretical developments a more precise definition of the problem of quark confinement and colour screening and the associated physical concepts is given in a general frame in Sections 2.6.1 and 2.6.2. We would like to point out once again that we are dealing here with an unsolved and somewhat controversial problem. [QC]

2.6.1 The Wilson criterion

The most simple physical explanation of quark confinement can be given by describing quark interaction by means of an attractive potential which strongly increases as distances become larger—the 'confinement potential'. A similar type of model for binding $c\bar{c}$ quarks in the charmonium system was discussed in Section 1.1.3. But does the concept of a confinement potential have any significance in QCD? This is the quetion which is considered now.

(a) When *defining a potential* within the realm of QCD, the relativistic nature of the interaction together with gauge invariance gives rise to several problems which must first of all be dealt with. The finite propagation velocity of the interaction in a relativistic field theory means that the interaction within this theory—and this also includes QCD—is always linked with the creation of virtual and real particles. These type of effects are not described by a potential. Therefore, the first step in ensuring that the confinement potential concept can be used for formulating the confinement problem is to ascribe an infinite mass to quarks in QCD. This 'freezes' all kinetic degrees of freedom and there is no quark pair creation. The quarks are merely static sources of colour charges which are characteristic for an $SU(3)_C$ triplet. The finiteness of the gluon interaction propagation velocity is not significant for static sources. In this approximation, the existence of a confinement potential between such static 3–$\bar{3}$ colour charges is a property belonging to the realm of pure gluon dynamics. When considering the existence of a confinement potential in pure gauge theory as evidence of confinement, it should always be remembered that this potential refers to the 'interaction' of static quarks. The physical confinement problem is only fully solved if it can be shown that the existence of a confinement potential in a pure gauge theory means that no free quarks can be produced even after coupling gluons with low mass 'dynamical' quarks.

K. Wilson attempted to define a potential between a quark q and an antiquark $\bar{q}$ in a colour singlet state, by considering the quantum mechanical vacuum expectation value $W(\mathscr{C})$ for a classical path $\mathscr{C}$ of a heavy external quark–antiquark pair between x and y:

$$\begin{aligned} W(\mathscr{C}) &= \left\langle \mathrm{Tr}\, P \exp\left(-\mathrm{i}g \int_{\mathscr{C}} \mathrm{d}s_\mu A_\mu^a(s) \frac{\lambda_a}{2}\right)\right\rangle \\ &= \frac{1}{Z} \int \mathscr{D}[A_\mu]\, \mathrm{Tr}\, U(\mathscr{C}) \exp(S_E\{A\}). \end{aligned} \tag{2.6.1}$$

$$U(\mathscr{C}) = P \exp\left(-\mathrm{i}g \int_{\mathscr{C}} \mathrm{d}s_\mu A_\mu^a(s)\lambda_a/2\right)$$

$U(\mathscr{C})$ represents the parallel displacement described by the gauge field of the q and $\bar{q}$ colour vectors along $\mathscr{C}$ as in Eqn. (2.2.20). As the antiquark $\bar{q}$ has the opposite colour charge to quark q, the colour displacement is along a closed line (see Fig. 2.13). q and $\bar{q}$ are considered in a colour singlet state; the trace formation of $U(\mathscr{C})$ gives the average over singlet wavefunctions. The quantum mechanical expectation value of $U(\mathscr{C})$ can be calculated from the path integral formula (2.3.82). This has been written down symbolically with the aid of the Euclidean action $S_E\{A\}$—in the second line of Eqn. (2.6.1); see

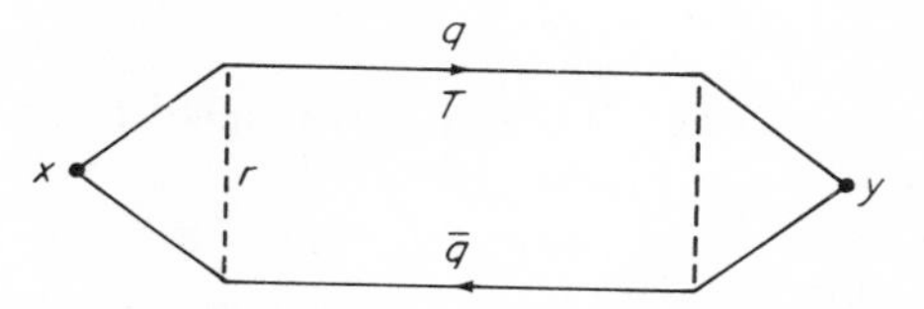

Figure 2.13 'Wilson loop' for defining a gauge-invariant static quark–antiquark potential

Eqns. (2.3.106), (2.3.107)—complications arising from gauge fixing have thus been suppressed (see Eqn (2.3.11)).

The gauge invariance of $W(\mathscr{C})$ follows directly from the transformation of colour displacement (Eqn. (2.2.19)) under gauge transformations:

$$\operatorname{Tr} U(\mathscr{C}) \to \operatorname{Tr}(g(x)U(\mathscr{C})g^{-1}(x)) = \operatorname{Tr} U(\mathscr{C}). \tag{2.6.2}$$

If $\mathscr{C}$ is a planar, closed curve like Fig. 2.13, then

$$V(r) = -\lim_{T\to\infty} \frac{1}{T} \log W(\mathscr{C}) \tag{2.6.3}$$

can be interpreted as a static potential $V(r)$ between a quark–antiquark pair in a colour singlet state. This gauge-invariant definition of a potential will be justified in the following Sections ((b), (c)) and in Section 2.6.3 (Eqn. (2.6.80)).

(b) Eqns. (2.6.1) and (2.6.3) are considered now for *QED* in order to give a better idea of the details of the gauge-invariant *potential definition* by *K. Wilson.* For this purpose, the pure electromagnetic Lagranian function $\mathscr{L}_{\text{gauge}} = -\frac{1}{4}F_{\mu\nu}F^{\mu\nu}$ (Eqns. (2.1.3), (2.1.4)) is supplemented by a gauge fixing term $\mathscr{L}_{\text{fix}} = -\frac{1}{2}(\partial^\mu A_\mu)^2$ (Eqns. (2.3.124), (2.3.125)). The Euclidean action:

$$\begin{aligned} S_E\{A\} &= -\int \mathrm{d}^4x[\tfrac{1}{4}(\partial_\mu A_\nu - \partial_\nu A_\mu)(\partial_\mu A_\nu - \partial_\nu A_\mu) + \tfrac{1}{2}(\partial_\mu A_\mu)^2] \\ &= -\tfrac{1}{2}\int \mathrm{d}^4x(\partial_\mu A_\nu(x))(\partial_\mu A_\nu(x)). \end{aligned} \tag{2.6.4}$$

is obtained after continuation to Euclidean time $x_4 = \mathrm{i}x_0$, $A_4 = \mathrm{i}A_0$. Summation with respect to equal indices in Euclidean metric, means in this case $A_\mu A_\mu = A_1^2 + A_2^2 + A_3^2 + A_4^2$. Hence, the Wilson integral (2.6.1) becomes

$$W(\mathscr{C}) = \frac{1}{Z}\int \mathscr{D}[A_\mu] \exp\left(ie\int_{\mathscr{C}} \mathrm{d}s_\mu A_\mu(s)\right) \exp\left(-\frac{1}{2}\int \mathrm{d}^4x(\partial_\mu A_\nu)(\partial_\mu A_\nu)\right). \tag{2.6.5}$$

This is a Gaussian integral as in Eqn. (2.3.58), with $\int \mathrm{d}^4x J_\mu(x)A_\mu(x) = e\int_{\mathscr{C}} \mathrm{d}s_\mu A_\mu(s)$ which can then be evaluated like the examples (2.3.76)–(2.3.78). The result is:

$$W(\mathscr{C}) = \exp\left(-\frac{e^2}{2}\int_{\mathscr{C}} \mathrm{d}s_\mu \int_{\mathscr{C}} \mathrm{d}t_\nu S_{\mu\nu}(s-t)\right). \tag{2.6.6}$$

$S_{\mu\nu}(x)$ is the Euclidean Green function, also called the Schwinger function, in Feynman gauge:

$$S_{\mu\nu}(x) = \frac{1}{4\pi^2}\delta_{\mu\nu}\frac{1}{x^2} = \frac{\delta_{\mu\nu}}{(2\pi)^4}\int \frac{\mathrm{e}^{\mathrm{i}xp}}{p^2}\mathrm{d}^4p, \quad -\partial_\rho\,\partial_\rho S_{\mu\nu}(x-y) = \delta_{\mu\nu}\,\delta^{(4)}(x-y). \tag{2.6.7}$$

The relationship between the 2-point Green function $\tau_{\mu\nu}(x, y) = -\mathrm{i}g_{\mu\nu}\,\Delta_F(x-y)$ (Eqns. (1.2.40), (2.3.23)), and the Schwinger function $S_{\mu\nu}(x, y) = \delta_{\mu\nu}S(x, y)$ is

$$S(x, y) = \mathrm{i}\Delta_F(x-y)\big|^{x_0=-\mathrm{i}x_4}_{y_0=-\mathrm{i}y_4}, \quad \tilde{S}(p) = \mathrm{i}(\mathrm{i}\widetilde{\Delta}_F(p))\big|_{p_0=-\mathrm{i}p_4}, \tag{2.6.8}$$

in which the Green function is continued analytically to Euclidean points. There is a similar relationship between n-point functions [Sy 66].

A rectangle with side lengths r and T (see Fig. 2.13) was chsoen for $\mathscr{C}$ in order to evaluate the Wilson integral (2.6.6) even further. Hence, with Eqn. (2.6.7), we now have:

$$\begin{aligned}\int_{\mathscr{C}} \mathrm{d}s_\mu \int_{\mathscr{C}} \mathrm{d}t_\nu S_{\mu\nu}(s-t) = &-\frac{1}{4\pi^2}\left(2\int_0^T\int_0^T \frac{\mathrm{d}t'\,\mathrm{d}t''}{r^2+(t'-t'')^2}\right.\\ &\left.+2\int_0^r\int_0^r \frac{\mathrm{d}r'\,\mathrm{d}r''}{T^2+(r'-r'')^2}\right)\\ &+\frac{1}{4\pi^2}\left(2\int_0^T\int_0^T \frac{\mathrm{d}t'\,\mathrm{d}t''}{(t'-t'')^2}+2\int_0^r\int_0^r \frac{\mathrm{d}r'\,\mathrm{d}r''}{(r'-r'')^2}\right).\end{aligned} \tag{2.6.9}$$

Because of the factor $\delta_{\mu\nu}$ in $S_{\mu\nu}$, only integrations betwen parallel sides give a contribution. Those integrals in which t and s run over opposite (same) sides are shown in the first two (third) lines. Elementary integration gives:

$$\int_0^T\int_0^T \frac{\mathrm{d}t'\,\mathrm{d}t''}{r^2+(t'-t'')^2} = \frac{2T}{r}\arctan\frac{T}{r} - \log\left(1+\frac{T^2}{r^2}\right). \tag{2.6.10}$$

Integrals over the same sides diverge. Systematic discussion of these divergences would involve gauge-invariant regularization as in Section 2.4.1(c), but a more simple regularization is sufficient for the following consideration. This omits the integral around the singularity:

$$\operatorname{reg}\int_0^T\int_0^T \frac{\mathrm{d}t'\,\mathrm{d}t''}{(t'-t'')^2} = 2\int_0^{T-\varepsilon}\mathrm{d}t'\int_{t'+\varepsilon}^T \mathrm{d}t\frac{1}{(t-t')^2} = \frac{2}{\varepsilon}(T-\varepsilon) + 2\log\frac{\varepsilon}{T}. \tag{2.6.11}$$

Addition of all the contributions gives:

$$\begin{aligned}W(\mathscr{C}) = \exp\Bigg[\frac{e^2}{2\pi^2}&\left(\frac{T}{r}\arctan\frac{T}{r}+\frac{r}{T}\arctan\frac{r}{T}-\tfrac{1}{2}\log\left(1+\frac{T^2}{r^2}\right)\left(1+\frac{r^2}{T^2}\right)\right)\\ &-2\,\Delta m(T+r) - \frac{e^2}{2\pi^2}\log\frac{\varepsilon^2}{rT}\Bigg].\end{aligned} \tag{2.6.12}$$

for the Wilson integral. Hence, the em potential between opposite charges according to Eqn. (2.6.3) is

$$V(r) = -\lim_{T\to\infty}\frac{1}{T}\log W(\mathscr{C}) = -\frac{e^2}{4\pi}\frac{1}{r} + 2\,\Delta m, \tag{2.6.13}$$

the known Coulomb potential. The constant term $2\,\Delta m \approx e^2/2\pi^2\varepsilon$ represents the self-energy of a regularized point charge. It is subtracted during the course of the renormalization process. The logarithmic divergent term $\sim\log(\varepsilon^2/r)T$ does not occur for paths $\mathscr{C}$ with continuous tangent [Do 80]. When transferring to another gauge, a term $((1-\xi)/16\pi^2)\,\partial_\mu\,\partial_\nu \log x^2/\Delta$ is added to $S_{\mu\nu}$. A path integral along closed loops with respect to such gradient terms vanishes after regularization. The Wilson integral is independent of gauge fixing.

(c) For calculating the potential for small distances in QCD the relationship between the *Wilson integral* and the *Feynman graphs* must be known [Fi 77a, Ap 77, Br 79]. QED in Minkowski space–time is considered as an example. The photon propagator $D_F(x)_{\mu\nu}$ is best considered in Coulomb gauge for the interaction of static sources. To this end, the Feynman gauge is converted to Coulomb gauge $A^C_\mu(x) = A_\mu(x) - \partial_\mu(-\Delta)^{-1}(\boldsymbol{\partial}\mathbf{A})(x)$, $\Delta = \boldsymbol{\partial}^2$, and the photon propagator

$$\tilde{D}^C_{00}(p) = \frac{1}{(2\pi)^4}\frac{1}{\mathbf{p}^2}, \quad D^C_{00}(x) = \frac{1}{4\pi}\frac{1}{|\mathbf{x}|}\,\delta(x_0)$$

$$\tilde{D}^C_{ik}(p) = \frac{1}{(2\pi)^4}\left(\delta_{ik} - \frac{p_i p_k}{|\mathbf{p}^2|}\right)\frac{1}{p^2 + \mathrm{i}\varepsilon}, \tag{2.6.14}$$

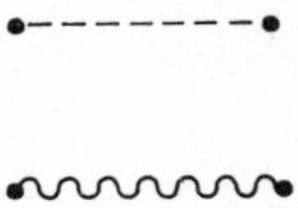

graphically

$$\tilde{D}^C_{0i}(p) = 0 \quad i, k = 1, 2, 3.$$

is obtained.

The 00 component describes an instantaneous interaction. The propagator of a locally fixed, massive particle is obtained according to Eqns. (1.2.26) and (1.2.38), as a Green function of the differential equation $(\mathrm{i}\,\partial/\partial t - M)G(x-x') = \delta^{(4)}(x-x')$. It is

$$G(x-x') = -\mathrm{i}\theta(t-t')\exp[-\mathrm{i}M(t-t')]\,\delta(\mathbf{x}-\mathbf{x}'). \tag{2.6.15}$$

The Feynman rules are completed by specification of the photon–particle vertex. A static particle represents only charge density and no spatial current density. Therefore only the Coulomb component A_0 is coupled with the static charge with strength $\mp e$ for particles or antiparticles. If the Feynman rules in configuration space (Eqn. (2.3.42)), are adapted in this way to the case in question, then the following formula emerges for Green functions in lowest, non-trivial order

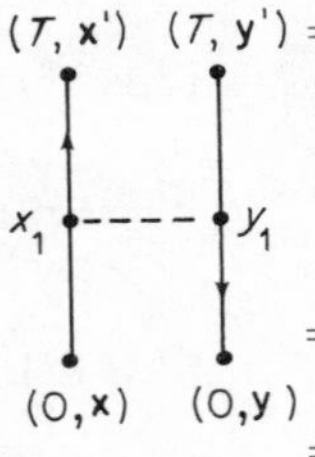

$$\begin{aligned} &= \int \mathrm{d}^4x_1 \int \mathrm{d}^4y_1\,\theta(T-x_1^0)\theta(x_1^0)\theta(T-y_1^0)\theta(y_1^0)\mathrm{e}^{-2\mathrm{i}MT} \\ &\quad \times \mathrm{i}\frac{e^2}{4\pi}\frac{\delta(x_1^0-y_1^0)}{|\mathbf{x}_1-\mathbf{y}_1|}\,\delta(\mathbf{x}'-\mathbf{x}_1)\,\delta(\mathbf{x}_1-\mathbf{x})\,\delta(\mathbf{y}'-\mathbf{y}_1)\,\delta(\mathbf{y}_1-\mathbf{y}) \\ &= -\frac{e^2}{4\pi}\frac{-\mathrm{i}T}{|\mathbf{x}-\mathbf{y}|}\,\delta(\mathbf{x}-\mathbf{x}')\,\delta(\mathbf{y}-\mathbf{y}')\mathrm{e}^{-2\mathrm{i}MT} \\ &= -\mathrm{i}V(\mathbf{x}-\mathbf{y})T\,\delta(\mathbf{x}-\mathbf{x}')\,\delta(\mathbf{y}-\mathbf{y}')\mathrm{e}^{-2\mathrm{i}MT}. \end{aligned} \tag{2.6.16}$$

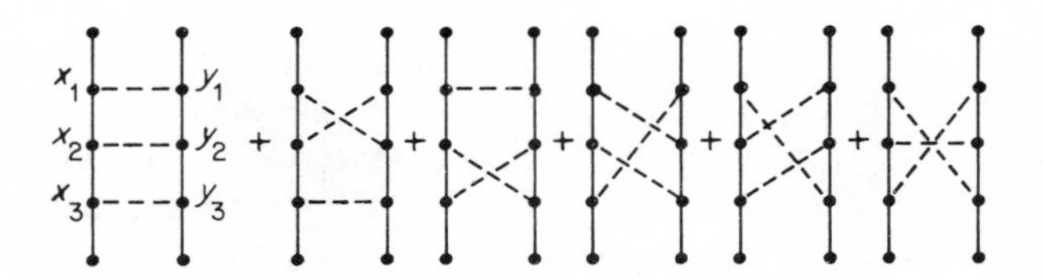

Figure 2.14 Contributions of Feynman diagrams with $n=3$ photon lines to the Wilson integral

Evaluation of a graph with an arbitrary exchange of n-photon lines, e.g. $n=3$ corresponding to $2n$th($=6$th) order (Fig. 2.14), can be made following the same simple method. Only integration over the time intervals has to be considered in more detail because of the $\theta(t_i - t_k)$ functions of the charge propagators. If all points x_i and y_j of the first graph in Fig. 2.14 are permuted independently, then all other graphs are obtained ($n!$) times, as permutation of y_j alone produces all other graphs. In the sum over all permutations of x_i and y_j time ordering can be disregarded, and an n-fold integral over the time interval $(0, T)$ is obtained. This can be evaluated to:

$$\frac{1}{n!}(-\mathrm{i}V(\mathbf{x}-\mathbf{y})T)^n\, \delta(\mathbf{x}-\mathbf{x}')\, \delta(\mathbf{y}-\mathbf{y}')\mathrm{e}^{-2\mathrm{i}MT}. \tag{2.6.17}$$

The direct result of this is:

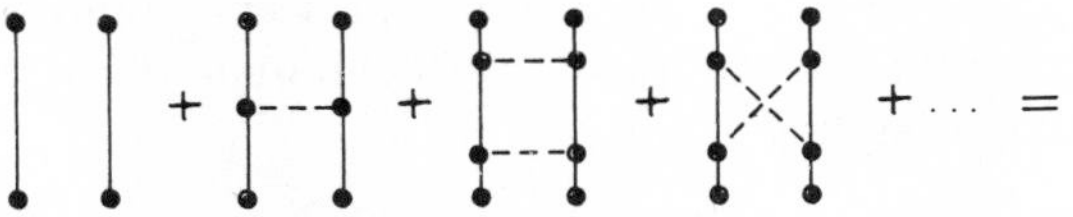

$$\begin{aligned} &= \exp(-\mathrm{i}V(\mathbf{x}-\mathbf{y})T - 2\mathrm{i}MT)\, \delta(\mathbf{x}-\mathbf{x}')\, \delta(\mathbf{y}-\mathbf{y}') \\ &= W'(\mathscr{C})\mathrm{e}^{-2\mathrm{i}MT}\, \delta(\mathbf{x}-\mathbf{x}')\, \delta(\mathbf{y}-\mathbf{y}'). \end{aligned} \tag{2.6.18}$$

This sum of exchange graphs corresponds to the essential part $W'(\mathscr{C})$ of the Wilson integral. So, based on the consideration that in the expansion:

$$\left\langle T\exp\left(-\mathrm{i}e\int_{\mathscr{C}} \mathrm{d}s^{\mu}A_{\mu}\right)\right\rangle = \sum_n \frac{(-\mathrm{i})^n e^n}{n!}\int_{\mathscr{C}}\dots\int_{\mathscr{C}} \mathrm{d}s_1^{\mu_1}\dots \mathrm{d}s_n^{\mu_n}\langle TA_{\mu_1}(s)\dots A_{\mu_n}(s_n)\rangle$$

the Green functions of free QED factorize according to Eqn. (2.3.16):

$$W(\mathscr{C}) = \sum_n \frac{(-1)^n \mathrm{e}^{2n}}{(2n)!}\int_{\mathscr{C}}\dots\int_{\mathscr{C}} \mathrm{d}s_1^{\mu_1}\dots \mathrm{d}s_{2n}^{\mu_{2n}} \sum_{\text{partitions}} D_F(s_i - s_k)_{\mu_i\mu_k} \tag{2.6.19}$$

and restricting integration to particle lines along the time axis

$$\int_{\mathscr{C}} \mathrm{d}s = \int_{\mathbf{s}=\mathbf{x}} \mathrm{d}s^0 - \int_{\mathbf{s}=\mathbf{y}} \mathrm{d}s^0,$$

then the part of the graph in which the photon propagator does not connect points on the same particle line in Coulomb gauge gives directly $W'(\mathscr{C})$ of Eqn. (2.6.18). Lines which do connect points along the same particle line are self-energy terms which when added up lead to renormalization of the physical mass M. Of course the rseult of Eqn. (2.6.18) agrees with the calculation performed in the Euclidean region in Eqns. (2.6.12) and (2.6.13), if T is replaced by $-\mathrm{i}T$.

In QCD, a perturbative representation of the Wilson integral has problems because of two reasons. The colour charges $Q^a = \frac{1}{2}\lambda^a$ of static quarks have quantum character which is expressed by the non-commutativity of the λ matrices. The self-interaction of gluons gives contributions which are more complex than simple gluon exchange. Thus, the fact that the general sum of exchange graphs (2.6.18), can be added up to give the exponential function of the potential can be shown only for the most simple gluon graphs (2.6.20).

$$\sim V(r) = -\frac{4}{3}\frac{\alpha_s(\mathscr{M})}{r}\left(1+\frac{11}{8\pi^2}\alpha_s(\mathscr{M})\log \mathscr{M}r+\dots\right) \qquad (2.6.20)$$

Graphs of the type shown in Fig. 2.15 behave like $T \log T$ when T is large.

The problems thus arising can be overcome by an infinite re-arrangement of the perturbation series [Ap 77]. Further details are given in the literature.

Finally, it must be stressed once again, that the confinement problem has to be worked out by a non-perturbative technique valid for large distances and for the strong coupling regime.

(d) A potential interpretation of the Wilson integral (2.6.3) gives the Wilson criterion for static quark confinement directly. If the Euclidean expectation value $W(\mathscr{C})$ for similarly shaped paths $\mathscr{C}$, decreases exponentially with the minimal surface $Ar(\mathscr{C})$ enclosed by $\mathscr{C}$:

$$W(\mathscr{C}) \sim \text{constant} \times \exp(-\kappa Ar(\mathscr{C})) \quad \text{for large paths } \mathscr{C} \qquad (2.6.21)$$

then this is regarded as the criterion for confinement; because a confinement potential

$$V(r) = \kappa r \quad \text{for large } r \qquad (2.6.22)$$

follows according to Eqn. (2.6.3) for rectangular paths $\mathscr{C}$, κ is called the

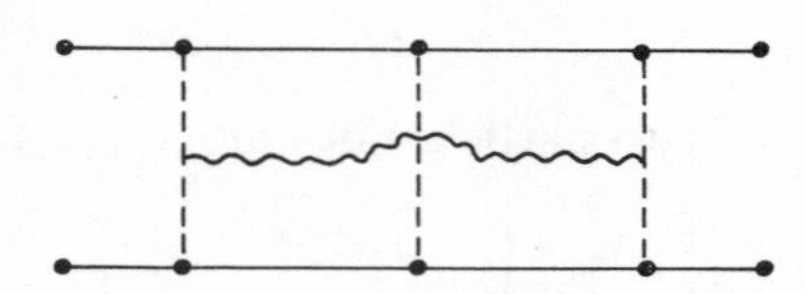

Figure 2.15 A Feynman graph which contributes to the Wilson integral $\sim T \log T$

'string constant', the reasons for which are described in the following section. κ has the dimension $(\text{mass})^2$ and is, therefore, a renormalization group-invariant quantity of the form (2.5.14) which cannot be calculated by perturbation theory.

The expectation value of the Wilson integral

$$W(\mathscr{C}) = \left\langle \operatorname{Tr} P \exp\left(\mathrm{i} \int_{\mathscr{C}} \mathrm{d}s_\mu \mathscr{A}_\mu\right)\right\rangle$$

plays the role of the 2-point Green function for gluons in QCD in many ways. Eqn. (2.6.19) shows this for free QED. The advantage of $W(\mathscr{C})$ is that it is gauge-invariant and therefore independent of the gauge fixing. In this sense, the Wilson criterion describes the asymptotic behaviour of $W(\mathscr{C})$ in pure QCD (with no quark coupling), which can be compared with the exponential decay of Green functions for space like distances in field theories involving no zero mass particles (see Eqn. (1.2.38')).

2.6.2 The chromoelectric Meissner effect

If the Wilson criterion as in Eqns. (2.6.21) and (2.6.22) were valid, it would mean that pure gluon dynamics describes a potential between a static quark–antiquark pair, which increases linearly with increasing distance. This type of potential is very different from the known Coulomb potential in QED which also results in QCD from a 1-gluon exchange in perturbation theory. Thus, there exists a dynamic process which cannot be simply described by perturbation theory and which may cause the confinement potential in QCD. The physical concepts of this mechanism stem from the theory of superconductivity and are known under the general term 'chromoelectric Meissner effect'.

(a) First of all, let the potential which exists between two magnetic monopoles with opposite magnetic charge in a superconductor of the second kind, be considered as a field theoretical model for a linear potential: '*Nielsen–Olesen string*' [Ni 73, Gi 50]. The relativistic Ginzburg–Landau field equations for a superconductor of the second kind provide the starting point:

$$\begin{aligned}(\partial_\mu - \mathrm{i}eA_\mu)^2 &= \frac{\mu^2}{2}\phi - \frac{\lambda}{2}|\phi|^2\phi, \\ \partial^\nu F_{\nu\mu} = -ej_\mu &= -\mathrm{i}e(\phi^* \partial_\mu\phi - \phi\,\partial_\mu\phi^*) - 2e^2A_\mu\phi^*\phi.\end{aligned} \tag{2.6.23}$$

These equations describe the gauge-invariant coupling of the em field $F_{\mu\nu} = \partial_\mu A_\nu - \partial_\nu A_\mu$ with a self-interacting, scalar, complex field ϕ. In superconduction theory, $|\phi|^2$ describes an order parameter which is proportional to the number of superconducting electrons, i.e. Cooper pairs [Go 58, Ri 65, Pa 69]. The Ginzburg–Landau equations give a macroscopic approximation of the atomistic theories proposed by J. Bardeen, L. N. Cooper, and J. R.

Schrieffer [Ba 57]. It applies for variations in ϕ which are small in comparison with the coherence length ξ, $|\boldsymbol{\partial}\phi|/|\phi|<\xi^{-1}$, i.e. predominantly for temperatures in the critical range.

The field equations (2.6.23) can be derived from the Lagrangian density:

$$\mathscr{L}=[(\partial^\mu+\mathrm{i}eA^\mu)\phi^*(x)](\partial_\mu-\mathrm{i}eA_\mu)\phi(x)-\tfrac{1}{4}F_{\mu\nu}(x)F^{\mu\nu}(x)-\frac{\lambda}{4}\left(|\phi|^2-\frac{\mu^2}{\lambda}\right)^2 \tag{2.6.24}$$

Non-Abelian generalizations of such gauge theories play an important rôle in weak interactions. Details of this will be given in Section 3.2.

In this section, time-independent, cylindrically symmetric solutions of Eqns. (2.6.23) are considered [Pa 69]. The cylindrical coordinates are as follows:

$$x^1=\rho\cos\theta,\quad x^2=\rho\sin\theta,\quad x^3=z,\quad x^0=t, \tag{2.6.25}$$

$$A^1=A_\rho\cos\theta-A_\theta\sin\theta,\quad A^2=A_\rho\sin\theta+A_\theta\cos\theta,\quad A^3=A_z,\quad A^0=A_t$$

The following *ansatz* is made

$$\phi(x)=\mathrm{e}^{in\theta}G(\rho),\quad n=0,\pm1,\pm2,\dots \tag{2.6.26}$$

in view of the uniqueness of function $\phi(x)$ to give

$$A_\theta(x)=A(\rho),\quad A_t=A_\rho=A_z\equiv0. \tag{2.6.26'}$$

This implies that of the em field tensor, only the magnetic induction in the z-direction $B(\rho)$ differs from zero:

$$-F_{12}(x)=B(\rho)=\frac{1}{\rho}\frac{\mathrm{d}}{\mathrm{d}\rho}\rho A(\rho) \tag{2.6.27}$$

and that the field equations are reduced to:

$$\left[-\frac{1}{\rho}\frac{\mathrm{d}}{\mathrm{d}\rho}\rho\frac{\mathrm{d}}{\mathrm{d}\rho}+\left(eA(\rho)-\frac{n}{\rho}\right)^2-\frac{\mu^2}{2}+\frac{\lambda}{2}G^2\right]G(\rho)=0, \tag{2.6.28}$$

$$\frac{\mathrm{d}}{\mathrm{d}\rho}\frac{1}{\rho}\frac{\mathrm{d}}{\mathrm{d}\rho}\rho A(\rho)=ej_\theta(\rho)=2e\left(eA-\frac{n}{\rho}\right)G^2(\rho) \tag{2.6.28'}$$

The energy of such a field configuration is obtained from the Lagrangian density (Eqn. (2.6.24)). In temporal gauge, $A_t=0$, it has the form:

$$\begin{aligned}H&=\int\mathrm{d}^3x\left[\tfrac{1}{2}(\mathbf{B}^2(\mathbf{x})+\mathbf{E}^2(\mathbf{x}))+|\partial_t\phi|^2+|(\boldsymbol{\partial}-\mathrm{i}e\mathbf{A})\phi|^2+\frac{\lambda}{4}\left(|\phi|^2-\frac{\mu^2}{\lambda}\right)\right]\\&=\int\mathrm{d}z\left\{2\pi\int_0^\infty\rho\,\mathrm{d}\rho\left[\frac{1}{2\rho^2}\left(\frac{\mathrm{d}}{\mathrm{d}\rho}\rho A(\rho)\right)^2+\left(\frac{\mathrm{d}}{\mathrm{d}\rho}G(\rho)\right)^2\right.\right.\\&\quad\left.\left.+\left(eA(\rho)-\frac{n}{\rho}\right)^2+\frac{\lambda}{4}\left(G^2(\rho)-\frac{\mu^2}{\lambda}\right)^2\right]\right\}\\&=\int\mathrm{d}zK\{A,G\}.\end{aligned} \tag{2.6.29}$$

K represents the energy per cylinder length. The simplest solution of the field equations (2.6.28) is

$$A(\rho)=\frac{n}{e\rho}, \quad G(\rho)=\frac{\mu}{\sqrt{\lambda}} \, (=\text{constant}). \tag{2.6.30}$$

in which the energy K has the lowest value $K=0$. This is called the classical vacuum solution. Transformation of this solution into Cartesian coordinates (Eqn. (2.6.25)) shows that the associated em potential $A_\mu(x)$ is a pure gauge:

$$A_\mu(x)=\frac{n}{\mathrm{e}}\partial_\mu\theta(x)=\frac{n}{\mathrm{e}}\partial_\mu \arccos\left(\frac{x^1}{\sqrt{(x_1^2+x_2^2)}}\right). \tag{2.6.31}$$

$\phi(x)$ is obtained by the corresponding gauge transformation of the constant $\mu/\sqrt{\lambda}$:

$$\phi(x)=\mathrm{e}^{in\theta(x)}\frac{\mu}{\sqrt{\lambda}}. \tag{2.6.32}$$

Other solutions with finite energy K can now be considered. As K is made up of positive expressions, these must disappear individually for $\rho\to\infty$. Eqn. (2.6.29) shows that $A(\rho)$ and $G(\rho)$ thus approach, for large values of ρ, the vacuum solution (2.6.30). For this reason the solutions can be written as:

$$A(\rho)=\frac{n}{e\rho}+\bar{A}(\rho), \quad G(\rho)=\frac{\mu}{\sqrt{\lambda}}+\bar{G}(\rho). \tag{2.6.33}$$

If this is used in Eqn. (2.6.28) and considered for large values of ρ, then the asymptotic behaviour for $\rho\to\infty$ becomes evident

$$\bar{A}(\rho)\sim\rho^{-1/2}\exp\left[-\sqrt{\left(\frac{2}{\lambda}\right)}e\mu\rho\right], \quad \bar{G}(\rho)\sim\rho^{-1/2}\mathrm{e}^{-\mu\rho}. \tag{2.6.34}$$

According to Stokes' theorem, the following applies for the magnetic flux $\phi_B(\kappa)$ through a large circular surface $\mathscr{K}$ with boundary $\mathscr{K}$:

$$\phi_B(\mathscr{K})=\int_{\mathscr{K}} B\,df=\oint_{\mathscr{K}} A_\mu\,ds^\mu=\lim_{\rho\to\infty}\rho\int_0^{2\pi}A_\theta\,d\theta=\frac{2\pi}{e}n \tag{2.6.35}$$

(based on Eqn. (2.6.33)). Because of the strong decrease of $\bar{A}(\rho)$, only the asymptotic vacuum solution contributes to the boundary integral. This means that the magnetic flux $\phi_B(\mathscr{K})$ is quantized in multiples of $2\pi/e$ [Lo 50, Ab 57]. n is called a 'topological quantum number' because the fact that n is an integer depends on the uniqueness of $\phi(x)$ (Eqn. (2.6.26)), (see Section 2.6.4). So $\int_{\mathscr{K}} B\,df=-2\pi(\rho\bar{A})|_{\rho=0}$, follows from Eqn. (2.6.27) and the boundary condition

$$\bar{A}(\rho)\sim-\frac{n}{e\rho} \tag{2.6.36}$$

is obtained for $\rho\to 0$. This condition establishes the solution's structure by a power series around $\rho=0$. As $A(\rho)\sim$ constant according to Eqns. (2.6.33),

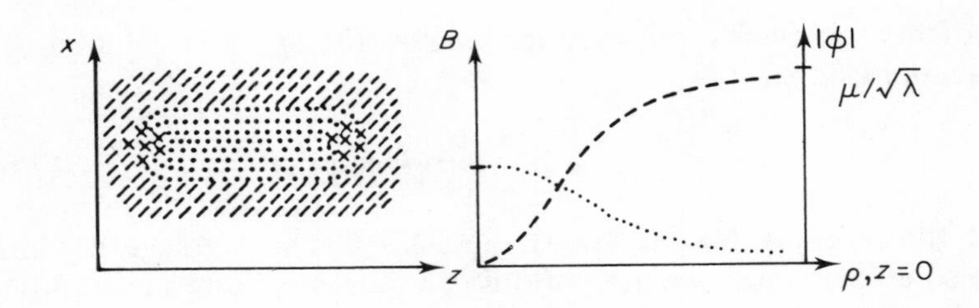

Figure 2.16 Field configuration in the Nielsen–Olesen model. ---, ϕ field; . . . , B field; × × ×, monopole density

(2.6.36), and since the 'repulsive potential' $\sim n^2/\rho^2$ in Eqn. (2.6.28) acts like the centrifugal potential in the Schrödinger equation, the form of $G(\rho)$ at small ρ is

$$G(\rho) = C_n\rho^n + \ldots, \tag{2.6.37}$$

and from this, the magnetic field strength according to Eqns. (2.6.28) and (2.6.29), becomes:

$$B(\rho) = \int_0^\rho \frac{d}{d\rho'}\frac{1}{\rho'}\frac{d}{d\rho'}\rho' A(\rho') + B(0) = B(0) - eC_n^2\rho^{2n} + \ldots .$$

In general, the complete solution of the non-linear equations (2.6.28) cannot be given analytically and so numerical calculations must be brought in. However, our description clarifies the most important points involved. Fig. 2.16 shows the field configuration which emerges for two very distant magnetic monopoles.

Points worthy of note are

(1) The undisturbed superconducting medium is characterized by the value of the order parameter $|\phi| = \mu/\sqrt{\lambda}$, the classical vacuum value of $\phi(x)$.
(2) In this medium, a magnetic flux tube—or 'magnetic string'—is formed between very distant magnetic poles. The property of a superconductor of expelling a magnetic field (Meissner effect) leads in this case to a magnetic induction B being concentrated in a string. As a magnetic field cannot penetrate a superconductor, the latter is therefore regarded as a perfect diamagnet with magnetic permeability $\mu_{\text{magn}} = 0$.
(3) The constant, finite energy per string length (K in Eqn. (2.6.29)) gives a linear potential when the monopoles are very distant.
(4) The magnetic flux is quantized (Eqn. (2.6.35)) and so are the magnetic charges of the monopoles. Quantization of the classical magnetic flux in units of $2\pi/e$ provides clear proof of the non-perturbative character of the mechanism described here.
(5) The continuously distributed magnetic flux is stabilized by a circular current $j_\theta = dB(p)/dp$ (Eqns. (2.6.27), (2.6.28′)).
(6) The medium resumes its normal conducting nature $|\phi| \sim 0$ at the centre

of the magnetic string. The value of the magnetic flux corresponds to the critical field strength.

(7) The magnetic field penetrates the superconductor to a depth of $[\sqrt{(2/\lambda)}e\mu]^{-1}$ (Eqn. (2.6.34)). In dynamic solutions, this corresponds to the Compton wavelength of a 'massive' photon (see the Higgs–Kibble mechanism in Section 3.2.2).

(8) The 'displaced' scalar field is restored to the vacuum value with a restoration length μ^{-1} (Eqn. (2.6.34)). In the Higgs model (Section 3.2.2), μ represents the Higgs meson mass.

(b) If the above classical model of a linear potential is applied to the problem of quark confinement, then the picture of a *gluon string* emerges [tH 74, Ko 75]: the gluon field between the colour charges of a separated, static quark–antiquark pair is compressed to a string in the QCD vacuum, which acts as a confinement medium through a chromoelectric Meissner effect. The constant field energy of the string per unit length gives the linear potential required by the Wilson confinement criterion.

In Table 2.3 the analogy is carried further. As the quark colour charges couple with the electric components of the gluon field, magnetic properties of the Nielsen–Olesen string correspond to the chromoelectric properties of the gluon string.

The assumption that QCD contains a confinement mechanism which provides an adequate description of a classical gluon string is only a fruitful speculation at the present time. This picture is partly realized in the strong coupling limit of QCD on a lattice (Section 2.6.3). On the other hand, it is known that the quantum mechanical description of a string modifies the picture somewhat, as string thickness increases with increasing string length

Table 2.3

	Magnetic Nielsen–Olesen string	Chromoelectric gluon string
Sources	Magnetic monopoles with opposite magnetic charge	Non-commutative colour charges of quarks and anti-quarks
String made of	Magnetic flux between magnetic charges	Chromoelectric flux between colour charges
Embedded in	Perfect diagmagnets $\mu_{magn}=0$ (superconductor)	Confinement vacuum $\mu_{el}=\infty$ ($\varepsilon = 1/\mu_{el}$ = dielectric constant)
Produced by	(Magnetic) Meissner effect	Chromoelectric Meissner effect
Internal phase	Normal conducting	Asymptotically free

[Lü 80]. When the quark–antiquark pair is separated by a moderate distance, the internal (asymptotically free) phase is in the shape of a bubble. This is called a 'bag' [Ch 74] and is the basis for these phenomenological models [Ha 78].

(c) With the em fields in superconductors of the second kind and gluon fields forming strings gauge fields configurations were considered which differ qualitatively from the photons in QED or from the gluons in the asymptotically free region of QCD. These are known as the various *phases of gauge fields* and can be roughly categorized as follows:

(1) Higgs phase: massive physical gauge particles (e.g. $W^{\pm}$, Z^0 bosons, etc.), linear, magnetic potentiai;
(2) Confinement phase: massive physical gauge particles (e.g. 'glue balls'), linear, electrical potential;
(3) Coulomb phase: zero mass gauge bosons (e.g. photons), Coulomb potential.

The phase in which a gauge field occurs depends on the details of the theory, such as local symmetry group structure, possible coupling with dynamic particle fields and their charge structure, space–time dimension and the strength of the various couplings. Systematic investigation of the various possible phases of gauge fields received attention in elementary particle physics because of the problems associated with confinement in strong interactions and the short range of forces involved in weak interactions (Section 3.2). The investigations carried out so far imply that the roughly defined phases above are the most important phases of gauge fields [tH 78]. Certain mixed forms of Higgs phases in which not all gauge bosons are massive occur in Chapter 3. Literature gives exact results based on lattice gauge theories [Fr 79, Ko 79]. Just a few points are singled out below to characterize the confinement phase. (Recent results [Ca 82N, Dr 82N])

A phase is often characterized by an order parameter. This is a field quantity which indicates a type of long-range order, i.e. whose vacuum expectation value characterizes the physical behaviour over large distances. Thus, in a quantized form of the relativistic Ginzburg–Landau theory (Eqn. (2.6.24)), the quantum mechanical expectation value $\langle\phi(x)\rangle = \mu/\sqrt{\lambda}$, is equal to the classical vacuum value (2.6.30) (see Section 3.2.2). The 2-point function is $\langle\phi^*(x)\phi(0)\rangle \sim \mu^2/\lambda + \text{constant}.\ |x|^{-2}e^{-\mu|x|}$. The expectation value of the order parameter $\phi(x)$, which does not disappear at long range characterizes the superconducting phase. This was explained in (a) above.

One order parameter which partly characterizes a pure gauge theory is the Wilson loop (2.6.1). Its potential interpretation means that in the expansion of $W(\mathscr{C})$ with respect to enclosed surface $Ar[\mathscr{C}]$ and length $|\mathscr{C}|$ etc.

$$-\log W(\mathscr{C}) = \kappa\, Ar[\mathscr{C}] + \Delta\, |\mathscr{C}| + \ldots, \tag{2.6.38}$$

a non-vanishing string constant $\kappa \neq 0$ represents a characteristic for the confinement phase as described in Eqn. (2.6.21). In the Coulomb phase, $\kappa = 0$ (Eqn. (2.6.12)). In the confinement phase, $\kappa \neq 0$, the strong decrease of the gauge-invariant field: $\mathrm{Tr}\, P \exp \int \mathrm{d}s_\mu \mathcal{A}_\mu(s)$ describes a short-range correlation similar to the exponential decrease of 2-point functions in field theories containing massive particles only. Thus, in the confinement phase, there seems to be a gap in the energy spectrum between the vacuum and the lowest, gauge-invariant excitation of the gluon field. These excitations which are presumably like particles, are known as 'glue balls' and are, therefore, massive (see Eqn. (2.6.79)). Compared with these short-range vacuum fluctuations of chromoelectric gluon strings, it takes a great deal of energy to build up a coherent, chromoelectric field between external sources at large distances, consistent with the linearly increasing potential in the confinement phase. Another interpretation of the Wilson loop as an order parameter is derived by defining a type of chromomagnetic flux $\phi_B(\mathscr{C})$ by

$$\mathrm{Tr}\, P \exp\left(\mathrm{i} \int_{\mathscr{C}} \mathrm{d}s_\mu \mathcal{A}_\mu(s)\right) = \mathrm{Tr}\, W(\mathscr{C}) = \exp(\mathrm{i}\phi_B(\mathscr{C})) \qquad (2.6.39)$$

[tH 79a], in a similar way to Eqn. (2.6.35). Although this flux is not a strictly additive quantity, it allows the confinement phase to be characterized by a type of constant magnetic flux density in the vacuum according to the area Law of Wilson's confinement criterion. The chromomagnetic flux is said to have been condensed in the confinement phase.

If the superconduction model is taken as an example of a (complete) Higgs phase, then in the comparison between the confinement and Higgs phase in Table 2.3, there appears to be a certain duality between the electric and magnetic components of the gauge field. Thus G. 't Hooft [tH 78] introduced a 'magnetic loop' $M(\mathscr{C})$ as a dual order parameter for the characterization of the Higgs phase. $M(\mathscr{C})$ can be defined formally for an $SU(N)$ gauge theory, using the Wilson loop and commutation relations:

$$W(\mathscr{C})M(\mathscr{C}') = M(\mathscr{C}')W(\mathscr{C}) \exp(\mathrm{i}n/N) \qquad (2.6.40)$$

in which $\mathscr{C}$ and $\mathscr{C}'$ are closed paths in a 3-dimensional space (constant time). n is the number of windings for $\mathscr{C}'$ and $\mathscr{C}$. The factor $\exp(\mathrm{i}n/N)$ is a central element of the gauge group $SU(N)$ (Eqn. (1.2.17)). $M(\mathscr{C})$ can be constructed explicitly in a lattice gauge theory [Ma 80]. The following expansion is used again:

$$-\log\langle \mathrm{Tr}\, M(\mathscr{C})\rangle = \kappa_M Ar[\mathscr{C}] + \Delta_M |\mathscr{C}| \qquad (2.6.41)$$

$\kappa_M \neq 0$ then characterizes the complete Higgs phase, i.e. the condensation of a type of non-Abelian electric flux density and the associated linear potential between an external magnetic monopole–antimonopole pair. $\kappa_M = 0$ in the confinement and Coulomb phase.

The phase of a gauge theory is a function of the theoretical parameters for a given symmetry group and charge structure of particle fields [Fr 79, Ko 79]. The points in parameter space at which the phases change are called critical points.* At these critical points, the free energy or order parameter does not depend analytically on the parameters. An example of this is the string constant $\kappa(g_0)$ as an order parameter of a pure gauge theory, and its dependence on the coupling constant g_0. At a critical point g_0^c, which separates the Coulomb phase $\kappa(g_0) \equiv 0$ in $g_0 < g_0^c$ from the confinement phase $\kappa(g_0) > 0$ in $g_0 > g_0^c$, $\kappa(g_0)$ is obviously not an analytical function of g_0. This point will be described in more detail later on.

No particles which have colour charges and therefore no free gluons or no free quarks are expected in the confinement phase. But the reasons for this are different for quarks and gluons. The physical excitations of the gauge field in the confinement phase, the glue balls, are gauge-invariant and massive. The gluon correlations have a short range in this case, therefore concentrated, gauge-invariant states carry no colour charge [Ma 78a]. Physically the gluons seem to agglomerate into glue ball states and thus show a mutual gluon colour screening.

The colour charges of static quarks have a triality which differs from zero (Eqn. (1.2.17)). In a pure gauge theory, these cannot be screened by colour charges of gluons with zero triality. A quark charge is always a source of a chromoelectric flux which can end only in a colour charge of opposite triality, e.g. an antiquark. As mentioned above, the chromoelectric Meissner effect concentrates this flux into a string and thus produces a confinement potential. Hence, the different representation of the centre of the colour group characterized by triality, is an important factor in the dynamics of quark confinement [Ma 78a, tH 78].

Colour charges of quarks and antiquarks can be screened in complete QCD incorporating dynamical quarks. The states formed in this way should then be the conventional, colourless hadrons which can be considered as bound states of $q\bar{q}$ and qqq in the simple quark model. In the case of a hypothetical QCD theory, in which the quarks have a zero triality colour charge, e.g. an octet charge, the quark colour charge could be screened by the gluon colour charge. The existence of free, colourless quarks which would continue to carry the broken flavour charges would then be expected.

The above considerations merely form a qualitative description of a possible confinement phase. Whether such a phase is actually realized in QCD and which dynamical mechanisms are responsible for it, these questions form the essence of the still unsolved confinement problem. Bringing in the superconduction analogy once again, the question as to the existence and formation of a dual equivalent to electrically charged Cooper pairs, is still open. This has all to do with the condensation of magnetic 'structures' (Eqn. (2.6.39)). Chromomagnetic monopoles [Ma 79], vortices [Ma 78b, tH 79, tH 79a], and similar configurations [Sa 77, Ni 78] are discussed in the literature. Section 2.6.3 investigates the extent to which the above ideas can actually be realized in lattice gauge theories.

* More accurately speaking, only continuous phase transitions are regarded as critical points [St 71].

(d) A dynamic description of the gluon string can also be given by means of a *string equation* [Na 79, Po 79]. To this end, the vacuum expectation value of the Wilson loop $W(\mathscr{C})$ (Eqn. (2.6.1)) is regarded as a functional of the closed path $\mathscr{C}$, i.e. a functional of the coordinate functions of $\mathscr{C}$: $x_\mu(\sigma)$, $0 \leqslant \sigma \leqslant 2\pi$, which is invariant against reparametrization $\sigma \to \sigma'$. The variation of the functional $W(\mathscr{C})$ for an infinitesimal deformation of $\mathscr{C}$ into a neighbouring curve $\bar{\mathscr{C}}$: $x_\mu(\sigma)+\varepsilon v_\mu(\sigma)$ can then be written as a functional differentiation:

$$W(\bar{\mathscr{C}}) = W(\mathscr{C}) + \varepsilon \int_0^{2\pi} \mathrm{d}\sigma \frac{\delta W(\mathscr{C})}{\delta x^\mu} v_\mu(\sigma) + O(\varepsilon^2). \tag{2.6.42}$$

$\bar{x}_\mu = x_\mu + \varepsilon f(\sigma)\dot{x}_\mu$, $\dot{x}_\mu = \mathrm{d}x_\mu/\mathrm{d}\sigma$ corresponds to an infinitesimal coordinate change $\sigma' = \sigma + \varepsilon f(\sigma)$. The following equation is a consequence of the general reparametrization invariance condition and Eqn. (2.6.42):

$$\dot{x}_\mu(\sigma) \frac{\delta W}{\delta x_\mu(\sigma)} = 0. \tag{2.6.43}$$

A two-fold variation of Eqn. (2.6.1) gives [Ma 79a]

$$\frac{\delta^2 W(\mathscr{C})}{\delta x^\mu(\sigma)\, \delta x_\mu(\sigma)} = \langle 0 |\, \mathrm{Tr}[\dot{x}^\rho \mathscr{F}_{\mu\rho}(x(\sigma))]^2 U(\mathscr{C}_{x(\sigma)})\, | 0 \rangle. \tag{2.6.44}$$

$U(\mathscr{C}_{x(\sigma)})$ is the parallel displacement of $x(\sigma)$ to $x(\sigma)$ along $\mathscr{C}$. It is assumed that the generalized path $\mathscr{C}$ is not intersecting. In a simplified picture $(\dot{x}^\rho \mathscr{F}_{\mu\rho})(\sigma)$ has the physical significance of a chromoelectric flux in a string along $\mathscr{C}$. If this flux is put constant, then

$$\left(-\frac{\delta^2}{\delta x(\sigma)^2} + \kappa^2 \dot{x}^2(\sigma) \right) W(\mathscr{C}) = 0. \tag{2.6.45}$$

is derived from Eqn. (2.6.44). Based on our concepts of the gluon string in Fig. 2.16 and Table 2.3, this approximation is best suited to long, smooth paths $\mathscr{C}$. Eqns. (2.6.43) and (2.6.45) apply to a quantum mechanical string [Sc 75]. Confirmation that these equations do actually apply for the Wilson loop functional in QCD is given by calculations using the $1/N$ approximation in addition to the above heuristic explanations. In this approximation, QCD shows signs of similarity with the dual resonance model [Re 74, Sc 75].

For large strings $\lambda\mathscr{C}$, $\lambda \to \infty$ an asymptotic form of the solutions of the string equation is found, with certain restrictions on $\mathscr{C}$ [Lü 80a]:

$$\log W(\lambda \cdot \mathscr{C}) = -\lambda^2 \kappa\, Ar[\mathscr{C}], \quad \lambda \to \infty.$$

This means that the string equation reproduces the Wilson confinement criterion with κ as a string constant. The formal Eqn. (2.6.45) must be renormalized for higher approximations [Lü 80b]. The next approximation modifies the confinement potential for large distances r, for paths $\mathscr{C}$ of the shape as shown in Fig. 2.13:

$$V(r) = \kappa r - \frac{\pi}{12}\frac{1}{r} + \text{constant}. \tag{2.6.46}$$

2.6.3 Lattice gauge theory

In the limit of strong coupling, the Wilson criterion for quark confinement is satisfied for a lattice approximation of QCD. The gluon strings, described in the previous section under the heading of 'chromoelectric Meissner effect' as a model for a linear potential are found in this approximation. Replacement of the 3-dimensional space or the 4-dimensional Euclidean space–time continuum by a space lattice is obviously only a rough approximation. However, it can be regarded as a regularization procedure, as part of the renormalization scheme. The transition to the continuum, i.e. for lattice constants $\varepsilon \to 0$, then requires subtraction and renormalization. The decisive questions are, of course, whether confinement is conserved during this limiting process and also whether this limit of QCD on the lattice is identical to the QCD which is asymptotically free at small distances in the continuum. Tests carried out so far are very promising. Initial results have given a relation between the physical string constant κ (Eqn. (1.1.15)) and the asymptotic scale parameter Λ (Eqn. (2.5.24)).

(a) First of all, the lattice approximation of formula (2.6.1) must be formulated for the vacuum expectation value of the Wilson loop. The geometric aspect of gauge theories can be used to find an expression for the *action in lattice approximation.*

The lattice Γ^d can be regarded as the usual mathematical division into cells of the d-dimensional, physical space or of the $(d+1)$-dimensional, Euclidean space–time continuum. For the sake of simplicity, a cubic lattice with lattice constant ε and N^d lattice points where N is finite or infinite, will be used. The various lattice elements are:

points x: coordinates $x = \varepsilon(n^1, n^2, \ldots, n^d)$, $n^i = 1, 2, \ldots, N$
links b: coordinates $b = (x_2, x_1) = (x_1 + e_\mu, x_1) = [x_1, \mu]$;
these are oriented in μ direction from x_1 to x_2.
plaquettes $\mathscr{P}$: coordinates

$$\begin{aligned} \mathscr{P} &= (x_4, x_3, x_2, x_1) = [x_1; \mu, \nu] \\ &= (x_1 + e_\nu, x_1 + e_\mu + e_\nu, x_1 + e_\mu, x_1); \end{aligned} \tag{2.6.47}$$

these are oriented by this sense of rotation.
cubes $\mathscr{K}$: with corresponding orientation
super-cubes: with corresponding orientation

A path $\mathscr{C}$ in the lattice is the sum of links $\mathscr{C} = \sum b_i$, a surface $\mathscr{B}$ is the sum of plaquettes $\mathscr{B} = \sum \mathscr{P}_i$. The boundary of a surface is denoted by $\dot{\mathscr{B}}$ and that of a

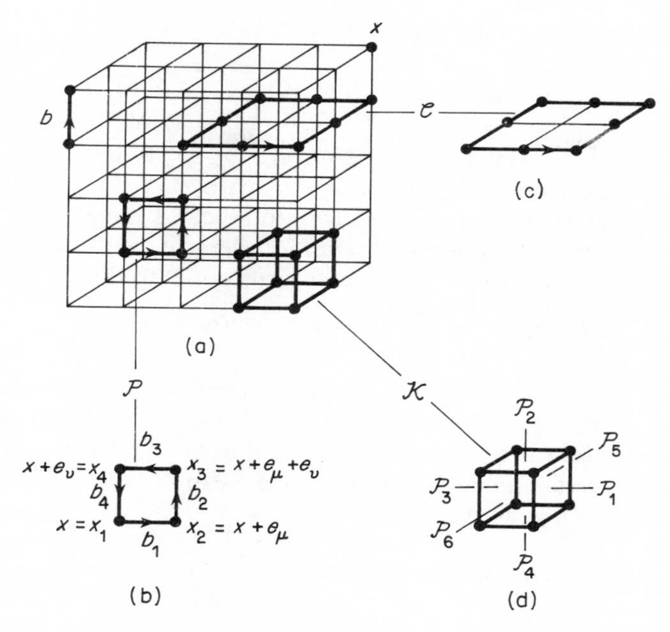

Figure 2.17 Notations of lattice geometry

path by $\mathscr{C}$. Fig. 2.17 illustrates this and anyone requiring further details should refer to a textbook of algebraic topology [Al 56, Si 76a].

The geometry of the local gauge symmetry described in Section 2.2.1 can be transferred point by point on to the lattice. This then provides the basis for the definitions of lattice gauge field theory:

(1) Particle fields $\psi_c(x)$ are scalars or spinors bearing charges Q_a of the symmetry group G, which are defined on the lattice points, e.g. quark fields:

$$\psi_{\alpha,f,c}(x) \rightsquigarrow \psi_{\alpha,f,c}(x)|_{x=\varepsilon(n^1,\ldots,n^4)}. \tag{2.6.48}$$

For historical reasons $(\psi_c(n^i))$ are often called spin variables.

(2) Gauge fields describe the infinitesimal parallel displacement of charged fields (2.2.3). Hence, parallel displacement along lines $b \in \Gamma$ (Eqn. (2.2.20)) correspond to the gauge fields on the lattice:

$$P\exp\left(+\mathrm{i}\int_b \mathrm{d}t_\mu \mathscr{A}_\mu(t)\right) \rightsquigarrow U(b) \equiv \mathrm{e}^{+\mathrm{i}\mathscr{A}(b)}. \tag{2.6.49}$$

Lattice gauge fields $U(b)$ are, therefore, functions which are defined on the links of the lattice and their values are group elements of G or their representations. They possess the following property:

$$U(-b) = U^{-1}(b) \quad \text{with } -(x_1, x_2) = (x_2, x_1). \tag{2.6.50}$$

(3) Parallel displacement along a curve $\mathscr{C}$ is therefore

$$P\exp\left(+\mathrm{i}\int_{\mathscr{C}} \mathscr{A}_\mu(t)\,\mathrm{d}t^\mu\right) \rightsquigarrow U(\mathscr{C}_L) = P\prod_{b_i\in\mathscr{C}_L} U(b_i) \qquad (2.6.51)$$

in lattice approximation (2.2.20), $\mathscr{C}_L = \sum_i b_i \subset \Gamma$ is a path which is approximately equal to the continuous path $\mathscr{C}$.

(4) In agreement with Eqn. (2.2.10), the field strength describes the parallel displacement around a surface element. Parallel displacement around the boundary of a plaquette corresponds to this on the lattice:

$$U(\dot{\mathscr{P}}_x) = P\prod_{b_i\in\dot{\mathscr{P}}} U(b_i) = U(b_4)U(b_3)U(b_2)U(b_1). \qquad (2.6.52)$$

As a result of path ordering, the initial point x is distinguished on the boundary $\dot{\mathscr{P}}$; the notions used in Fig. 2.17(b) are also used here. If the Baker–Hausdorff formula for the multiplication of exponential functions of non-commuting factors, i.e. $\exp A\ \exp B = \exp(A + B + \frac{1}{2}[A, B] + \ldots)$, is used, then the result is

$$U(\dot{\mathscr{P}}_x) = \exp\left[+\mathrm{i}\left(\sum_i \mathscr{A}(b_i) - \frac{i}{2}\sum_{i>k}[\mathscr{A}(b_i), \mathscr{A}(b_k)] + \ldots\right)\right] =: \mathrm{e}^{+\mathrm{i}\mathscr{F}(\mathscr{P})}$$

hence

$$\mathscr{F}_{\mu\nu}(x) \rightsquigarrow \mathscr{F}(\mathscr{P}), \quad \text{if } \mathscr{P} = [x; \mu, \nu]. \qquad (2.6.53)$$

(5) Local gauge transformations as in Eqn. (2.2.2) are defined at lattice points: $\mathrm{g}(x)\in G$, $x\in\Gamma$. The various geometric quantities then transform in a similar way to the continuum:

$$\psi(x) \to \mathrm{g}(x)\psi(x) \qquad (2.2.2)$$

$$U(b) \to \mathrm{g}(x_1)U(b)\mathrm{g}^{-1}(x_2) \quad \text{if } b = (x_1, x_2), \qquad (2.2.19)$$

then (2.6.54)

$$U(\dot{\mathscr{P}}_x) \to \mathrm{g}(x)U(\dot{\mathscr{P}}_x)\mathrm{g}^{-1}(x) \qquad (2.2.14)$$

$$U(\mathscr{C}) \to \mathrm{g}(x)U(\mathscr{C})\mathrm{g}^{-1}(x'), \quad \text{if } \dot{\mathscr{C}} = x - x' \qquad (2.2.19)$$

$$(\psi^\dagger(x)U(\mathscr{C})\psi(x')) \to \psi^\dagger(x)U(\mathscr{C})\psi(x').$$

The dynamics of a lattice gauge theory is determined by the definition of an action. According to K. Wilson [Wi 74] the following expression is considered for a pure Euclidean gauge theory with gauge group $SU(n)$ [Ba 74]:

$$S_E = \frac{\beta}{2n}\sum_{\mathscr{P}} \mathrm{Tr}\{U(\dot{\mathscr{P}}) + U^{-1}(\dot{\mathscr{P}}) - 2\}. \qquad (2.6.55)$$

The sum extends over all non-oriented plaquettes of the Euclidean space time lattice. Trace $U(\dot{\mathscr{P}}_x)$ permits a cyclic permutation of the factors in Eqn. (2.6.52) and is therefore independent of x. The gauge invariance of the action S_E follows from Eqn. (2.6.54), and the same trace property. The correspondence with the continuum case (Eqns. (2.6.49)–(2.6.53)) show formally that S_E is an approximation of the action $S_E = \int d^4x \mathscr{L}_{\text{gauge}}$ (Eqn. (2.2.30)), for small lattice constants ε. From this correspondence follows:

$$\mathscr{A}(b) \sim -g_0 \varepsilon A^a_\mu \frac{\lambda_a}{2}, \quad \mathscr{F}(\mathscr{P}[x;\mu,\nu] \sim -\varepsilon^2 g_0 F^a_{\mu\nu} \frac{\lambda_a}{2}$$

and then with Eqns. (1.2.5) and (2.6.53):

$$\frac{\beta}{2n} \sum_{\mathscr{P}} \text{Tr}\{U(\dot{\mathscr{P}}) + U^{-1}(\dot{\mathscr{P}}) - 2\} = -\sum_x \varepsilon^4 \frac{g_0^2}{4} \frac{\beta}{2n} (F^a_{\mu\nu}(x) F^a_{\mu\nu}(x))$$

$$\sim -\frac{1}{4} \int d^4x F^a_{\mu\nu} F^a_{\mu\nu}, \quad \text{if } \beta = \frac{2n}{g_0^2} \tag{2.6.56}$$

(QCD sign convention for g_0 see p. 74).

Although the gauge invariance of expressions $(\psi^\dagger U(\mathscr{C})\psi)$ would allow the definition of the dynamics of gauge theories with particle fields based on Section 2.2.2, this aspect will not be discussed here. On the other hand, lattice gauge theories are only applied here to the discussion of the Wilson criterion, which describes a pure gauge theory property and on the other, the formulation of Dirac fields on the lattice is tied up with many unsolved problems [Su 77, Wi 77a], so that no satisfactory lattice approximation for complete QCD can yet be given. For new results see Appendix of 2. Ed.

The path integral in a quantized lattice gauge theory can be formulated with help of the action defined above. If $\Omega(U)$ is an observable which is given as a gauge-invariant functional of the lattice gauge fields, then its vacuum expectation value can be calculated from the general path integral formula given in Section 2.3.2:

$$\langle \Omega[U] \rangle = \frac{1}{Z} \int \mathscr{D}[U] \Omega[U] e^{S_E[U]},$$

$$Z = \int \mathscr{D}[U] e^{S_E[U]}. \tag{2.6.57}$$

The Euclidean formulation (Section 2.3.2(c) and Eqn. (2.6.1)) is used here and the measure $\mathscr{D}(U)$ must be defined in the space of gauge fields in order to complete it. This can be given as

$$\mathscr{D}[U] = \prod_{b \in \Gamma} d\mu(U(b)), \tag{2.6.58}$$

in which $d\mu(g)$ is Haar's measure (Eqns. (1.2.12)–(1.2.14)), for the symmetry group $G = SU(n)$. The invariance of the Haar measure implies that a gauge-invariant quantum theory is described by the basic equation (2.6.57). The gauge invariance of the expectation values follows from the gauge invariance of $\Omega(U)$. Gauge fixing terms as in Eqn. (1.3.129′) are not necessary for integration with the finite Haar measure. The basic equation (2.6.57) together with Eqn. (2.6.54) for the parallel displacement, gives the desired lattice approximation for the Wilson loop (Eqn. (2.6.1)):

$$W(\mathscr{C}) = \frac{1}{Z} \int \mathscr{D}[U] \operatorname{Tr} U(\mathscr{C}) e^{S_E[U]}. \qquad (2.6.59)$$

This forms an essential starting point for further discussion of the confinement problem in the framework of lattice gauge theory. However, before any systematic evaluation of this formula is made, another aspect of lattice gauge theory which gives a better understanding of the significance of gluon strings for quark confinement shall be considered.

(b) In *the Hamiltonian formulation for pure QCD* on the lattice [Ko 75, Ko 77], the excited states are described as gluon strings in the strong coupling limit. This method is not explicitly covariant either with respect to general gauge transformations or to Lorentz transformations in the continuum. Therefore, serious problems are anticipated when attempting to formulate a consistent, renormalized, relativistic field theory on this basis. However, the Hamiltonian formulation of the gauge theory is described briefly because it does provide intuitive ideas and stimulates phenomenological calculations [Ba 77a, Ko 76].

The temporal gauge $A_a^0 = 0$ (Eqn. (2.2.23)) is chosen in the continuum and the electric component of the gluon field strength (2.2.12) $F_a^{i0}(x) = E_a^i(x) = -\mathrm{d}A_a^i(x)/\mathrm{d}t$ is regarded as the (negative) canonically conjugate quantity of the gauge potential $A_a^i(x)$:

$$A_a^0(x) = 0, \quad E_a^i(x) = -\frac{\mathrm{d}}{\mathrm{d}t} A_a^i(x), \quad B_a^i = \varepsilon^{ikl}\left(\nabla^k A_a^l - \frac{g_0}{2} f_{abc} A_b^k A_c^l\right),$$
$$[A_a^i(t,\mathbf{x}), A_b^k(t,\mathbf{x}')] = 0, \quad [E_a^i(t,\mathbf{x}), E_b^k(t,\mathbf{x}')] = 0, \qquad (2.6.60)$$
$$[E_a^i(t,\mathbf{x}), A_b^k(t,\mathbf{x}')] = \mathrm{i}\, \delta_{ab}\, \delta^{ik}\, \delta(\mathbf{x}-\mathbf{x}').$$

The formal Hamiltonian operator H is expressed in $\mathbf{E}_a(x)$ and the magnetic component of the gluon field strength $\mathbf{B}_a(x)$, as follows:

$$H = \frac{1}{2} \int \mathrm{d}^3x \sum_a (\mathbf{E}_a^2 + \mathbf{B}_a^2). \qquad (2.6.61)$$

The field equations (2.2.29), $D^\mu \mathscr{F}_{\mu\nu}(x) = 0$, follow only partly from Heisenberg's equations of motion $\mathrm{d}F/\mathrm{d}t = \mathrm{i}\,[H, F]$ as postulated by canonical quantum mechanics. That part of the field equations which contains no time derivatives,

$$\partial_i E_a^i + g_0 E_b^i f_{abc} A_c^i = 0, \qquad (2.6.62)$$

does not result from the Heisenberg equations. However, as the operator

$$Q[\theta] = \int \mathrm{d}^3x \left(\theta_a(x) \frac{1}{g_0} \partial_i E_a^i(x) + E_a^i(x) f_{abc} A_b^i(x) \theta_c(x)\right) \qquad (2.6.63)$$

generates infinitesimal gauge transformations corresponding to Eqn. (2.2.7):

$$\delta A_a^i = \mathrm{i}[Q[\theta], A_a^i] = -\left(\frac{1}{g_0}\,\partial^i\theta_a - f_{abc}A_b^i\theta_c\right), \tag{2.6.64}$$

then

$$Q[\theta]\,|\mathrm{Phys}\rangle = 0 \tag{2.6.65}$$

applies for physical and gauge-invariant states and the field equations (2.6.62) are also satisfied. Moreover, as $Q[\theta]$ commutes with H as a result of the gauge invariance of H, the operator H transforms physical states to physical states. The most important result of this short description is that a Hamiltonian operator for pure QCD is given for gauge-invariant states by H according to Eqn. (2.6.61).

The general rules for the lattice approximation of a gauge theory can be used to transfer the Hamiltonian formulation direct to the lattice [Ko 75]. Time remains a continuous coordinate and only 3-dimensional space is approximated by a lattice Γ^3. The following quantities correspond to each other:

(1) gauge fields in temporal gauge $\rightsquigarrow$ lattice gauge fields $U(b)$ (Eqn. (2.6.49)), defined on spatial links;
(2) magnetic field strength, i.e. $\mathscr{F}_{\mu\nu}(x)$ with spatial indices $\mu, \nu = 1, 2, 3$, $\rightsquigarrow$ plaquette terms $U(\dot{\mathscr{P}}_x)$ defined by Eqn. (2.6.52) on plaquettes of the space lattice;
(3) electric field strength, i.e. $\mathscr{F}^{\mu 0}(x)$, $\mu = 1, 2, 3 \rightsquigarrow$ fields $E_a(b)$ defined on oriented links. As time is continuous in the Hamiltonian formalism, of the plaquette terms in Eqn. (2.6.53) for a space–time plaquette $\mathscr{P} = [x, \mu, 0]$ in temporal gauge ($U(b) = 1$ for $b = (x, 0)$), only the temporal differential quotient of a link function remains. The relation

$$E_a(b) = -\sum_{a'} \tilde{U}_{aa'}(b)E_{a'}(-b). \tag{2.6.66}$$

applies between $E_a(b)$ and $E_a(-b)$, in which $\tilde{U}_{aa}$ represents the lattice gauge field in the octet (= adjoint) representation (see Eqn. (2.2.15)).
(4) Canonical commutation relations as in Eqn. (2.6.60) $\rightsquigarrow$

$$\begin{aligned}[][E_a(b_j), E_{a'}(b_k)] &= \mathrm{i}f_{aa'a''}E_{a''}(b_k)\,\delta_{b_j,b_k},\\ [U_{\alpha\beta}(b_j), U_{\gamma\delta}(b_k)] &= 0,\end{aligned} \tag{2.6.67'}$$

$$[E_a(b_j), U(b_k)] = -\frac{\lambda_a}{2}\,U(b_k)\delta_{b_j,b_k} + \delta_{-b_j,b_k}U(b_k)\,\frac{\lambda_a}{2}. \tag{2.6.67''}$$

Similar to the explanations of Eqns. (2.6.56), this is essentially a transfer of the canonical commutation relations on to the lattice, so that the limit for small lattice constants ε gives the continuum commutation relation formally. ($U(b) \approx \exp(-\mathrm{i}g_0\varepsilon A_a^l(x)\lambda_a/2)$, $E_a(b) \approx -\varepsilon^2 E_a^l(x)/g_0$, $\varepsilon^3\,\delta(\mathbf{x}-\mathbf{x}')\,\delta_{lk} \approx \delta_{b_l,b_k'}$ for $b_l = [x, l]$, $b_k' = [x', k]$.) There is a certain difference in that the $E_a(b)$ do not commute but for fixed b they do satisfy the commutation relations for local gauge symmetry. However, the commutator is of the order of ε for $\varepsilon \to 0$. The matrix expression $\sum_\beta (\lambda^a)_{\alpha\beta} U_{\beta\gamma}(b) \equiv (\lambda^a U(b))_{\alpha\gamma}$ is used in Eqn. (2.6.67″) and no confusion between the products and commutators of quantum mechanical operators $E_a(b)$, $U_{\alpha\beta}(b)$ and these matrix products is expected. Formally, Eqn. (2.6.67″) means that the $E_a(b)$ or $E_a(-b)$ act on the group-valued gauge field $U(b)$ as infinitesimal right or left transformations. The relation (2.6.66) [Ch 46, Ha 62a] follows from this in a known way. All this means that:

(5) gauge transformations can be represented by a generator similar to Eqn. (2.6.63):

$$Q[\theta] = \sum_{x \in \Gamma^3} (\nabla E_a)(x) \theta^a(x) \tag{2.6.68}$$

with $(\nabla E_a)(x) = \sum_\mu E_a(b = [x, \mu])$, in which summation takes place over all links which have an end point x. $(\nabla E_a)(x)$ is the divergence of a vector $\boldsymbol{\partial}\mathbf{E}(x)$ on the lattice. In fact, as a result of commutation relations (2.6.67) $Q[\theta]$ does produce the gauge transformations:

$$e^{+iQ[\theta]} U(b) e^{-iQ[\theta]} = \exp\left(-i\frac{\lambda_a}{2}\theta^a(x)\right) U(b) \exp\left(+i\frac{\lambda_a}{2}\theta^a(x')\right), \quad b = (x, x'), \tag{2.6.69}$$

$$e^{+iQ[\theta]} (\lambda_a E_a(b)) e^{-iQ[\theta]} = \exp\left(-i\frac{\lambda_a}{2}\theta^a(x')\right) (\lambda_b E_b) \exp\left(i\frac{\lambda_a}{2}\theta_a(x)\right).$$

(6) The physical states are defined as gauge-invariant states similar to Eqn (2.6.65):

$$Q[\theta]\,|\text{Phys}\rangle = 0 \tag{2.6.70}$$

(7) The Hamiltonian operator is

$$H = \frac{1}{2\varepsilon}\left\{g_0^2 \sum_{b \in \Gamma^3} \sum_{a=1}^{8} E_a^2(b) - \frac{2}{g_0^2} \sum_{\mathcal{P} \in \Gamma^3} \text{Tr}(U(\dot{\mathcal{P}}) + U^{-1}(\dot{\mathcal{P}}) - 2)\right.$$

$$\equiv \frac{g_0^2}{2\varepsilon}(H_0 + \bar{x} H_I), \quad \bar{x} = 2/g_0^4; \tag{2.6.71}$$

in agreement with Eqn. (2.6.61), in which the magnetic term H_I similar to Eqn. (2.6.56) emerges as the sum of all non-oriented $\mathcal{P}$. As $\sum_a E_a^2$ is invariant under the transformation described by Eqn. (2.6.66), summation of the electric field strengths is over the non-oriented links, too.

Eqns. (2.6.66)–(2.6.71) represent the basic equations of lattice gauge theory in the Hamiltonian formulation and form the basis on which a perturbation theory for strong coupling $g_0^2 \to \infty$ can be worked out. For this end the eigenstates of the 'unperturbed' Hamiltonian operator H_0 must be considered. According to Eqn. (2.6.70), the gauge-invariant groundstate $|0\rangle_s$ c

H_0 has the property:

$$E_a(b_i)\,|0\rangle_s = 0 \quad \text{for all } b_i \in \Gamma^3. \tag{2.6.72}$$

In agreement with the commutation relations (2.6.67″), the $U_{\alpha\beta}(b_j)$ produce gluon strings from the vacuum, i.e. the state $|b_j;\alpha\beta\rangle \equiv U_{\alpha\beta}(j)\,|0\rangle$ is an eigenstate of the chromoelectric field strength $E_a(b_k)$:

$$\begin{aligned} E_a(b_k)\,|b_j;\alpha\beta\rangle &= E_a(b_k)U_{\alpha\beta}(b_j)\,|0\rangle \quad \text{see (2.6.67″), (2.6.72)} \\ &= -\tfrac{1}{2}(\lambda_a)_{\alpha\gamma}\,|b_j;\gamma\beta\rangle\,\delta_{b_j,b_k} + \tfrac{1}{2}(\lambda_a)_{\gamma\beta}\,|b_j;\alpha\gamma\rangle\,\delta_{-b_k,b_j}. \end{aligned} \tag{2.6.73}$$

Thus, the state $|b_j;\alpha\beta\rangle$ describes a chromoelectric flux which is concentrated on the link b_j. In a strong coupling limit, the energy of this state is:

$$H\,|b_j;\alpha\beta\rangle \approx \left(\frac{g_0^2}{2\varepsilon}\sum_{a,b_k} E_a^2(b_k)\right)|b_j;\alpha\beta\rangle = \frac{g_0^2}{2\varepsilon}\left(\tfrac{4}{3}\right)|b_j;\alpha\beta\rangle. \tag{2.6.74}$$

based on Eqns. (2.6.71) and (2.6.73). In this formula, $\frac{4}{3}$ is the value of the 'local' Casimir operator $\sum_a E_a^2(b_j) \approx \frac{1}{4}\sum_a \lambda_a^2 = \frac{4}{3}$; it determines the energy of an elementary flux quantum on a link. Link b_j is further excited by multiple application of $U(b)_j$. Here the energies of the higher excited states are proportional to the values of the Casimir operator for higher-dimensional representations. The excitation energies are, therefore, not proportional to the powers of the creation operator as is the case for the commutation relations of free particles (Eqn. (1.2.19)), but follow from the Clebsch–Gordon rules for products of representations of the gauge group $S \cong SU(3)$ [Li 78]. The various lattice links can be independently excited since operators $U_{\alpha\beta}(b)_j$, (Eqn. (2.6.67′)) of different links commute. So, the state space of lattice gauge theory can be fully constructed by multiple application of $U_{\alpha\beta}$, $b_i \in \Gamma^3$ on $|0\rangle_s$. This complex structure is not described further in itself for the present, but is used to discuss the problems of physical, gauge-invariant states. Two physically relevant examples are used to illustrate the arguments.

The state $|\mathbf{y},\alpha;\mathbf{z},\beta\rangle$ of a static quark–antiquark pair situated at lattice points $\mathbf{z}$ and $\mathbf{y}$ is transformed under gauge transformations of the quarks, according to

$$\begin{aligned} \exp(+\mathrm{i}Q^a[\theta])\,|\mathbf{y},\alpha;\mathbf{z},\beta\rangle &= \sum\left[\exp\left(-\mathrm{i}\frac{\lambda_a}{2}\,\theta^a(\mathbf{y})\right)\right]_{\alpha\alpha'} \\ &\quad\times\left[\exp\left(\mathrm{i}\frac{\lambda_a}{2}\,\theta^a(\mathbf{z})\right)\right]_{\beta'\beta}|\mathbf{y},\alpha';\mathbf{z},\beta'\rangle \end{aligned} \tag{2.6.75}$$

(see Eqn. (2.2.2)). Thus, the state in which the quarks are connected by a gluon string:

$$|\mathbf{y},\mathbf{z};\mathscr{C}\rangle = \sum U_{\beta\alpha}(\mathscr{C})\,|\mathbf{y},\alpha;\mathbf{z},\beta\rangle, \quad \begin{array}{l}\mathscr{C} \text{ from } \mathbf{y} \text{ to } \mathbf{z}, \text{ i.e.} \\ \dot{\mathscr{C}} = \mathbf{z}-\mathbf{y},\end{array} \tag{2.6.76}$$

is gauge-invariant: $\exp(+\mathrm{i}(Q[\theta]+Q^q[\theta]))\,|\mathbf{y},\mathbf{z};\mathscr{C}\rangle=|\mathbf{y},\mathbf{z};\mathscr{C}\rangle$ according to Eqns. (2.6.54) and (2.6.69). $U(\mathscr{C})$ is the parallel displacement operator as in Eqn. (2.6.51). If M_q is the quark mass operator, then a strong coupling approximation would give the energy ε of this state as

$$\left(M_q+\frac{g_0^2}{2\varepsilon}H_0\right)|\mathbf{y},\mathbf{z};\mathscr{C}\rangle=\left(2m_q+\frac{4}{3}\frac{g_0^2}{2\varepsilon^2}L\right)|\mathbf{y},\mathbf{z};\mathscr{C}\rangle. \tag{2.6.77}$$

based on the calculation of Eqn. (2.6.74). L is the physical length, i.e. ε times the number of links of $\mathscr{C}$. For a given quark position $\mathbf{y}$, $\mathbf{z}$ the state with the lowest energy is that with the shortest $\mathscr{C}$.

Thus, the production of a linear potential by a chromoelectric potential has actually been achieved in a primitive fashion, as a strong coupling approximation. The string constant κ can be read off direct from Eqn. (2.6.77): $\varepsilon^2\kappa=\frac{4}{3}g_0^2$. This result can be improved by a systematic perturbation calculation in $\bar{x}\cdot H_\mathrm{I}$ (Eqn. (2.6.71)). We give the result worked out by J. B. Kogut *et al.* [Ko 79a, Ko 80], $\bar{x}=2/g_0^4$:

$$\kappa=\frac{g_0^2}{2\varepsilon^2}\left(\tfrac{4}{3}-\tfrac{11}{153}\bar{x}^2-\tfrac{61}{1632}\bar{x}^3-0.041\,378\bar{x}^4-0.034\,436\bar{x}^5+\ldots\right). \tag{2.6.78}$$

Section (e) below deals with the question of what significance this calculation has in continuum QCD.

A gauge-invariant state without quarks is produced from a closed curve $\mathscr{C}$ based on Eqn. (2.6.69):

$$|\mathscr{C}\rangle=\mathrm{Tr}\,U(\mathscr{C})\,|0\rangle,\quad H\,|\mathscr{C}\rangle=\left(\tfrac{4}{3}\right)\frac{g_0^2}{\varepsilon^2}L(\mathscr{C})\,|\mathscr{C}\rangle+\ldots. \tag{2.6.79}$$

The lowest physical state is obtained when $\mathscr{C}$ encloses one plaquette only: $\mathscr{C}=\dot{\mathscr{P}}$. Its energy is $\frac{16}{3}g_0^2/2\varepsilon+\ldots$ and this must be interpreted in this primitive model as the glue ball mass [Ko 76].

Finally, one more observation on the Wilson criterion in the Hamiltonian formulation of $SU(3)$ gauge theory must be made. With temporal gauge $\mathscr{A}_0=0$ and the notations used in Fig. 2.13, the following formula applies

$$\left\langle 0\left|\mathrm{Tr}\,P\exp\left(-\mathrm{i}\int_{\mathscr{C}}\mathrm{d}t^\mu\mathscr{A}_\mu(t)\right)\right|0\right\rangle\sim\langle 0|\,U^\dagger_{\beta\alpha}(L)\mathrm{e}^{\mathrm{i}HT}U_{\alpha\beta}(L)\,|0\rangle \tag{2.6.80}$$
$$\sim\mathrm{e}^{\mathrm{i}V(r)T}\sim\mathrm{e}^{\mathrm{i}\kappa LT},$$

a result which uses only the leading order in Eqn. (2.6.77) for the estimate given in the second line; this confirms the general relation of Eqn. (2.6.3) in this order.

(c) Evaluation of the Wilson loop in Eqn. (2.6.59) in a lattice approximation of Euclidean field theory has formal similarities with the calculation of expectation values in statistical mechanics. Section 2.3.2(g) has already referred to this aspect. Now, if the action (2.6.55): $S_\mathrm{E}=-\beta\mathscr{H}_\mathrm{E}$, $\beta=2n/g_0^2$, i considered proportional to a 4-dimensional, analogous classical energy function $\mathscr{H}_E$—which is not to be confused with H in Eqn. (2.6.71)—then an inverse temperature $\beta \approx 1/kT$ corresponds to the inverse coupling constant $\beta = 2n/g_0^2$ and the high temperature expansion in statistical mechanics corresponds to a *strong coupling approximation* of field theory [Ba 75a, Dr 78, Dr 78a, Ko 79a]. This method, taken from statistical mechanics, is described briefly below. It is demonstrated first for a simple model in which all stages can be carried out explicitly. By this an impression of the structure of all kind of similar models, the methodical investigation of which plays such an important part in this context, is also given.

The model adopted is the Z_2 gauge theory on a 2-dimensional lattice with N^2 points. The gauge group G is in this case the finite, discrete group with two elements $Z_2 = \{\sigma \mid \sigma = \pm 1\}$. Lattice gauge fields $U(b) \equiv \sigma(b)$ are functions on links $b \in \Gamma^2$, which have values ± 1. In view of the commutativity of group multiplication, the 'field strength' $\sigma(\dot{\mathscr{P}}) = \sigma(b_4)\sigma(b_3)\sigma(b_2)\sigma(b_1)$, (Eqn. (2.6.52)), are pure plaquette functions. The following definitions correspond to Eqns. (2.6.55), (2.6.59), and (2.6.57):

Action $$S_E[\sigma] = \beta \sum (\sigma(\dot{\mathscr{P}}) - 1) = -\beta \mathscr{H}_E[\sigma], \tag{2.6.81}$$

Wilson-Loop: $$W(\mathscr{C}) = \frac{1}{Z} \int \mathscr{D}[\sigma] \sigma(\mathscr{C}) \exp(-\beta \mathscr{H}_E[\sigma]), \tag{2.6.82}$$

Partition sum $$Z = \int \mathscr{D}[\sigma] \exp(S_E[\sigma]), \tag{2.6.83}$$

Haar's measure $$\int d\mu(\sigma) f(\sigma) = \frac{1}{2} \sum_{\sigma = \pm 1} f(\sigma). \tag{2.6.84}$$

The integral for the partition sum and the Wilson loop for small β, i.e. the high temperature expansion, is best evaluated in the following steps. The Boltzmann factor for an individual plaquette in a 'Z_2 Fourier transformation' is worked out first

$$e^{\beta(\sigma-1)} = e^{-\beta}(\cosh \beta + \sigma \sinh \beta) = C(1 + v\sigma)$$

with

$$C = \int d\mu(\sigma) e^{\beta(\sigma-1)}, \quad Cv = \int d\mu(\sigma) e^{\beta(\sigma-1)}. \tag{2.6.85}$$

As $v = \tanh \beta \approx \beta \ldots$ applies, a power series in v can be considered instead of one in β. When the total Boltzmann factor is expanded:

$$\exp\left(\beta \sum_{\mathscr{P} \in \Gamma^2} (\sigma(\dot{\mathscr{P}}) - 1)\right) = C^{N^2} \prod_{\mathscr{P} \in \Gamma^2} (1 + v\sigma(\dot{\mathscr{P}}))$$

$$= C^{N^2} \sum_{n=0}^{N^2} v^n \left(\sum_{G^{(n)}} \prod_{\mathscr{P} \in G^{(n)}} \sigma(\dot{\mathscr{P}}) \right) \tag{2.6.86}$$

after evaluation of the product $\prod_{\mathscr{P} \in \Gamma^2}$, the coefficient of v^n-emerges as a sum of the contributions of all possible lattice graphs $G^{(n)}$. A lattice graph $G^{(n)}$ is a set of n non-oriented plaquettes and its contribution to the Boltzmann factor is the product of all $\sigma(\dot{\mathscr{P}})$ of these plaquettes. High temperature expansion for this partial sum is obtained by substituting from Eqn. (2.6.86) in Eqn. (2.6.83):

$$Z = C^{N^2} \sum_n v^n \Big(\sum_{G^{(n)}} Z(G^{(n)}) \Big), \tag{2.6.87}$$

in which $Z(G^{(n)})$ is the contribution of a lattice graph to the partition sum

$$Z(G^{(n)}) = \int \mathscr{D}[\sigma] \prod_{\mathscr{P} \in G^{(n)}} \sigma(\dot{\mathscr{P}}), \quad \mathscr{D}[\sigma] = \prod_{b \in \Gamma^2} \mathrm{d}\mu(b). \tag{2.6.88}$$

Now, if

$$\int \mathrm{d}\mu(b) = 1, \quad \int \mathrm{d}\mu(b)\sigma(b) = 0, \quad \sigma^2(b) = 1, \tag{2.6.85'}$$

is taken into account, it is clear that $Z(G^{(n)})$ only differs from zero when $\sigma(b_i)$ in $\prod_{\mathscr{P} \in G^{(n)}} \sigma(\dot{\mathscr{P}})$ is present in even powers. In view of $\sigma(\dot{\mathscr{P}}) = \prod_{b_i \in \dot{\mathscr{P}}} \sigma(b_i)$, this means that each link must be in the boundary of at least two plaquettes of $G^{(n)}$, i.e. $Z(G^{(n)}) \neq 0$ only when the boundary $\dot{G}^{(n)}$ of $G^{(n)}$ vanishes. In a 2-dimensional, finite lattice, this only applies for $G^{(0)}$, i.e. the empty lattice graph and hence, in agreement with Eqn. (2.6.87): $Z = C^{N^2}$. Taking a periodic lattice in which opposite links of Γ^2 are identified, then $\dot{\Gamma}^2$ is also equal to 0 and the following result is obtained:

$$Z = C^{N^2}(1 + v^{N^2}) \approx C^{N^2} \quad \text{for } N^2 \gg 1,\ v < 1. \tag{2.6.89}$$

The influence of the boundary disappears for large N! The corresponding strong coupling expansion for the Wilson integral is obtained by substituting the Boltzmann factor of Eqn. (2.6.86) into Eqn. (2.6.82):

$$W(\mathscr{C}) = \sum_n v^n \Big(\sum_{G^{(n)}} W(\mathscr{C}; G^{(n)}) \Big), \quad N \gg 1,$$

with

$$W(\mathscr{C}; G^{(n)}) = \int \mathscr{D}[\sigma]\sigma(\mathscr{C}) \prod_{\mathscr{P} \in G^{(n)}} \sigma(\dot{\mathscr{P}}). \tag{2.6.90}$$

The conclusion is similar to that from the discussion of the contributions to the partition sum, i.e. that only those $G^{(n)}$ make a contribution in which $\dot{G}^{(n)} = \mathscr{C}$ applies. In two dimensions and with no periodic boundary conditions, this is a single graph consisting of $Ar[\mathscr{C}]$ plaquettes. The result obtained is:

$$W(\mathscr{C}) = v^{Ar[\mathscr{C}]} = e^{\log v \, Ar[\mathscr{C}]}. \tag{2.6.91}$$

Thus, the Wilson confinement criterion is satisfied with a string constant $\kappa = -\log v$, in lattice constant units, for a 2-dimensional Z_2 lattice gauge theory.

The individual steps towards a strong coupling approximation, viz:

(1) group theoretical expansion of the Boltzmann factor for the individual plaquette;

(2) expansion of the total Boltzmann factor into lattice graph contributions;
(3) determination and calculation of graphs which make a non-vanishing contribution; and
(4) graph summation;

are the essential components of the method also when applied to QCD. The complicated structure of the $SU(3)$ group and the extension to four Euclidean dimensions complicate the individual steps so that frequent references will be made to the literature for further details on the following outline of the strong coupling approximation for $SU(3)$ gauge theory.

The starting points for a group theoretical Fourier transformation are the orthogonality and completeness relations of characters $\chi^{(k)}(g) = \operatorname{Tr} D^{(k)}(g)$ of the irreducible, unitary representations (k) [Ch 46] of the symmetry group $G = SU(3)$:

$$\int_G d\mu(f)\chi^{(k)}(g' \cdot f^{-1})\chi^{(k')}(f \cdot g^{-1}) = \frac{\delta_{(k),(k')}}{\dim(k)}\chi^{(k)}(g'g^{-1}) \qquad (2.6.92)$$

or, more especially:

$$\int_G d\mu(f)\chi^{(k)}(f)^*\chi^{(k')}(f) = \delta_{(k),(k')}. \qquad (2.6.93)$$

In a similar way to Eqn. (2.6.85), this leads to the expansion of the Boltzmann factor for a plaquette $\mathcal{P}$:

$$\begin{aligned}\exp\left(\frac{\beta}{6}(\chi^{(3)}(U)+\chi^{(3)}(U)^*)\right) &= \sum_k C_k(\beta)\chi^{(k)}(U)\\ &= C_0(1+3v_{(3)}\chi^{(3)}(U)+3v_{(\bar{3})}\chi^{(\bar{3})}(U)+\ldots), \quad U \equiv U(\dot{\mathcal{P}}) \in SU(3), \quad (2.6.94)\end{aligned}$$

with

$$C_0 \dim(k) v_{(k)} = C_k = \int d\mu(U)\chi^{(k)}(U)^* \exp\left(\frac{\beta}{3}\operatorname{Re}\chi^{(3)}(U)\right). \qquad (2.6.95)$$

In view of the reality of the action $\frac{1}{6}\beta(\chi^{(3)}+\chi^{(3)*})$, the $v_{(k)}$ of complex-conjugated representations, e.g. for quark (3) and antiquark $(\bar{3})$ representation are equal: $v_{(k)} = v_{(\bar{k})}$. Group theoretical integrals are not easy to calculate in closed form [Cr 78, Ba 79b], but Clebsch–Gordon series for products of characters can be used for series expansion, e.g.

$$\begin{aligned}v_{(3)} \approx \tfrac{1}{3}C_3 &= \frac{1}{3}\sum_{n=0}^{\infty}\int d\mu(U)\chi^{(3)}(U)^*\frac{1}{n!}\left(\frac{\beta}{6}\right)^n(\chi^{(3)}(U)+\chi^{(3)}(U)^*)^n \qquad (2.6.96)\\ &= \tfrac{1}{3}(g_0^{-2}+\tfrac{1}{2}g_0^{-4}+g_0^{-6}+\tfrac{5}{24}g_0^{-8}+\ldots); \quad \beta = 6/g_0^2.\end{aligned}$$

As in Eqn. (2.6.86) above, expansion in lattice graphs is carried out with the

Boltzmann factor $\exp S_E$, written as the product*

$$\exp\left(\frac{\beta}{3}\sum_{\mathscr{P}} (\mathrm{Re}\,\chi^{(3)}(U(\dot{\mathscr{P}})))\right) = C_0^{(\frac{d}{2})N^d} \prod_{\mathscr{P}\in\Gamma} (1+v F(\mathscr{P})) = C_0^{N^d} \sum_n v^n \sum_{G^{(n)}} \prod_{\mathscr{P}\in G^{(n)}} F(\mathscr{P}) \tag{2.6.97}$$

with

$$v F(\mathscr{P}) = \sum_{(k)\neq(1)} \dim(k) v_{(k)} \chi^{(k)}(U(\dot{\mathscr{P}})). \tag{2.6.98}$$

The effective coupling constant $v_{(3)}$ associated with the fundamental representation is often used to advantage as general expansion parameter $v \sim \beta$ for the power series. Using the form in Eqn. (2.6.97), each plaquette of a graph is associated with a factor $F(\mathscr{P})$ in the contribution $\mathscr{B}(G^{(n)})$ to the Boltzmann factor. If $F(\mathscr{P})$ is then expanded as in Eqn. (2.6.98), the result is

$$\mathscr{B}(G^{(n)}) := \prod_{\mathscr{P}\in G^{(n)}} F(\mathscr{P}) = \sum_{k(\mathscr{P}_i)} \mathscr{B}(G^{(n)}; k(\mathscr{P}_1), \ldots, k(\mathscr{P}_n)) \tag{2.6.99}$$

with

$$\begin{aligned}\mathscr{B}(G^{(n)}; k(\mathscr{P}_1), \ldots, k(\mathscr{P}_n)) = \bar{v}_{(k(\mathscr{P}_1))} \cdot \ldots \cdot \bar{v}_{(k(\mathscr{P}_n))} \chi^{(k(\mathscr{P}_1))}(U(\dot{\mathscr{P}}_1)) \cdot \ldots \\ \ldots \cdot \chi^{(k(\mathscr{P}_n))}(U(\dot{\mathscr{P}}_n))\end{aligned} \tag{2.6.100}$$

and $\bar{v}_{(k)} = \dim(k) v_{(k)}/v$. $\mathscr{B}(G^{(n)}; k(\mathscr{P}_1), \ldots, k(\mathscr{P}_n))$ is the contribution to the Boltzmann factor of a group theoretical graph in which a non-trivial, irreducible representation is associated with each plaquette. The Boltzmann factor is obtained as the sum of all group theoretical graphs by substituting Eqn. (2.6.99) in Eqn. (2.6.97)).

The fact that the strong coupling approximation of $SU(3)$ gauge theory on a 4-dimensional lattice is more complex than the above example is due to the greater number of $SU(3)$ group representations and to the geometric diversity of the simple lattice graphs when four dimensions are involved. Some of these points are elaborated below, using the example of contributions of individual group theoretical graphs to the partition sum

$$Z\{G^{(n)}; k(\mathscr{P}_i)\} = \int \mathscr{D}[U] \mathscr{B}(G^{(n)}; k(\mathscr{P}_1), \ldots, k(\mathscr{P}_n)). \tag{2.6.101}$$

By way of illustration:

(1) The boundary $\dot{G}^{(n)}$ must be equal to 0 so that $Z\{G^{(n)}; \kappa(\mathscr{P}_i)\} \neq 0$; this results from $\int d\mu(U) \chi^{(k)}(U) = 0$, $((k) \neq (0), (k') = (1)$ in Eqn. (2.6.93)) by an argument similar to that used in the example above, Eqn. (2.6.85′)ff.

(2) Two plaquettes of $G^{(n)}$ are called connected when they have a common link. A general lattice graph can be decomposed into connected components: $G^{(n)} = \bigcup_i H^i$. The direct result of the product representation in Eqns. (2.6.99) and (2.6.101) is:

$$Z\{G^{(n)}\} = \prod_i Z(H^i). \tag{2.6.102}$$

* In the rest of the text, a physically unimportant factor $\exp(-\beta \sum_{\mathscr{P}} 1) C_0^{(\frac{d}{2})N^d}$ is frequently ommitted.

(3) Connected graphs $G^{(n)}$ without boundary form closed surfaces which may branch out. Links b_i of $G^{(n)}$, which connect exactly two plaquettes are called internal links and those which are common to several plaquettes are called singular. The integral in Eqn. (2.6.101) over the gauge field $U(b)$ with a specific link b which connects two specific plaquettes $\mathscr{P}_1$ and $\mathscr{P}_2$ with representations (k_1) and (k_2) can be written as follows:

$$\int d\mu(U(b))\chi^{(k_1)}(U(\dot{\mathscr{P}}_1))\chi^{(k_2)}(U(\dot{\mathscr{P}}_2))$$

$$= \int d\mu(U(b))\chi^{(k_1)}(U(b_1^1)\ldots U^{-1}(b))\chi^{(k_2)}(U(b))\ldots U(b_3^2))$$

$$= \chi^{(k_1)}(U(b_1^1)U(b_2^1)U(b_3^1)U(b_1^2)U(b_2^2)U(b_3^2))\ \delta_{(k_1),(k_2)} \qquad (2.6.103)$$

(see Fig. 2.17(d)) and can be carried out with help of Eqn. (2.6.92). It differs from zero only when $(k_1)=(k_2)\equiv(k)$. The result is $\chi^{(k)}(U(\mathscr{C}))$ with $\mathscr{C}=\dot{\mathscr{P}}_1+\dot{\mathscr{P}}_2$. Thus, the additional integrals in Eqn. (2.6.101) for the gauge fields of other links b_i have the same form and can be evaluated with Eqn. (2.6.92).

In the case of the integral for $U(b)$ with a singular link b which is common to plaquettes $\mathscr{P}_1, \ldots, \mathscr{P}_r$, the following equation:

$$\int d\mu(U(b))\chi^{(k_1)}(\dot{\mathscr{P}}_1)\cdot\ldots\cdot\chi^{(k_r)}(\dot{\mathscr{P}}_r)\neq 0 \qquad (2.6.104)$$

puts restrictions on the representations associated with the graph. The conditions are satisfied when the trivial representation is contained in the product representation $(k_i)\otimes\ldots\otimes(k_r)$ (assuming that b is identically oriented in all $\dot{\mathscr{P}}_i$). The result is that connected, group theoretical graphs with no singular links make a contribution to the partion sum which differs from zero only when the plaquette representations are equal. Branching and orientability impose further conditions on the group theoretical graphs.

(4) The most simple example is a graph in the form of an elementary cube in a 3- or more dimensional lattice. It has internal links only. Its contribution can be evaluated according to the method described for Eqn. (2.6.103). The corresponding notations are given in Fig. 2.17(d). The result of multiple application of Eqn. (2.6.92) is

$$Z\{\mathscr{K}^{(6)};k\} = \bar{v}_{(k)}^{6}\int\ldots\int d\mu(b_1)\ldots\mu(b_{12})\chi^{(k)}(\dot{\mathscr{P}}_1)\cdot\ldots\cdot\chi^{(k)}(\dot{\mathscr{P}}_6)$$

$$= [\dim(k)]^2 v_{(k)}^6/v^6. \qquad (2.6.105)$$

(5) If the contribution of all graphs of a certain order v^n is considered in a finite, 4-dimensional lattice Γ_N^4 with periodic boundary conditions: $v^n: \sum_{G^{(n)}} Z\{G^{(n)};k\}$, then summation must be taken with respect to graphs in the various positions which follow from each other through N^4 translations. For example, Z is given in lowest order by the cubic graph described by

Eqn. (2.6.105); multiplicity by translation and 4-fold orientation of the 3-dimensional cube in 4-dimensional space results in:

$$Z = 1 + 4N^4 \sum_{(k)} v^6_{(k)} [\dim(k)]^2 + \dots \qquad (2.6.106)$$

Even the N^4 dependence of the lowest-order term shows that the convergence of series (2.6.97), i.e. of Eqn. (2.6.106), and the transition to the infinite lattice, $N \to \infty$ ('thermodynamic limit') are a cause of non-trivial problems. Two non-connected cubic graphs $\mathcal{K}^{(6)}$ and $\mathcal{K}^{(6)'}$ contribute in the order v^{12}. Their number is:

$$A[\mathcal{K}^{(6)}, \mathcal{K}^{(6)'}] = 16N^8 - b[\mathcal{K}^{(6)}, \mathcal{K}^{(6)'}], \qquad (2.6.107)$$

in which $b[\mathcal{K}^{(6)}, \mathcal{K}^{(6)'}]$ represents the number of configurations in which $\mathcal{K}^{(6)} \cup \mathcal{K}^{(6)'}$ form a connected graph.* In any case, $A[\mathcal{K}^{(6)}, \mathcal{K}^{(6)'}] \sim N^8$ makes the convergence problem more serious. This can be solved by considering the strong coupling expansion for the free energy instead of that for Z:

$$F = -\frac{1}{\beta N^4} \log Z \qquad (2.6.108)$$

In this case, the thermodynamic limit $N \to \infty$ exists in any order and the resultant power series in $\beta = 6/g_0^2$ converges for sufficiently small β [Os 78]. The transition from Z to F graph expansion leads to a representation of the strong coupling approximation for F by means of 'graph clusters' [Dr 78, Mü 80]. In the Feynman graphs, factorization of the Green functions in Eqn. (2.3.38) gave a representation of $\phi\{j\}$ by connected graphs, as result of a corresponding logarithmic relationship between the functionals $\phi\{j\}$ and $T\{j\}$ (Eqn. (2.3.68)). The problem of overlapping in the counting of non-connected graphs has the consequence that despite factorization (Eqn. (2.6.102)), 'graph clusters' are more complex than connected graphs for higher orders. However, their multiplicity rises not stronger than N^4.

A corresponding rule [Os 78] which guarantees the existence of the thermodynamic limit and convergence of the strong coupling approximation, applies for the expectation values of local observables Ω:

$$\langle \Omega[U] \rangle = \frac{1}{Z} \int \mathcal{D}[U] \Omega[U] \exp(S_E[U]). \qquad (2.6.57)$$

$\Omega[U]$ should depend only on gauge fields $U(b)$ on a finite number of links. Roughly speaking, the N-dependence in the numerator and denominator in Eqn. (2.6.57) is compensated in the thermodynamic limit.

The most important example for describing the expectation value of a local observable is the Wilson loop $W(\mathscr{C})$. There is a convergent, strong coupling expansion of $W(\mathscr{C})$ on the infinite lattice, based on the above

* It is when $\kappa^{(6)}$, $\kappa^{(6)'}$ are overlapping.

theorem due to K. Osterwalder and E. Seiler [Os 78]. If the expansion of the Boltzmann factor (2.6.97) in terms of group theoretical graphs (2.6.99), (2.6.100) is substituted in Eqn. (2.6.59), then the contribution of a graph to the Wilson loop is:

$$W(\mathscr{C}; G^{(n)}; k(\mathscr{P}_i)) = \frac{1}{Z}\left(\prod_i v_{(k(\mathscr{P}_i))}\right)\int \mathscr{D}[U]\chi^{(q)}(\mathscr{C})$$
$$\times \chi^{(k(\mathscr{P}_1))}(U(\dot{\mathscr{P}}_1))\cdot \ldots \cdot \chi^{(k(\mathscr{P}_n))}(U(\dot{\mathscr{P}}_n)). \qquad (2.6.109)$$

Integration can be carried out using the convolution formula (2.6.92) following the method described in connection with Eqn. (2.6.103).

The orthogonality relations impose conditions which must be satisfied if a group theoretical graph is to make a non-vanishing contribution to the Wilson loop. Thus, in agreement with Eqn. (2.6.104), $(k(\mathscr{P}_1))\otimes \ldots \otimes (k(\mathscr{P}_i))\otimes(q)$ must contain the trivial representation.

This is responsible for the basic difference between the confinement of quarks with normal (3)-colour charge and the screening of quarks with (8)-colour charge, which was discussed in Section 2.6.2. For $(q)=(8)$, a graph as in Fig. 2.18 with $(k(\mathscr{P}_i))\equiv(8)$ will give a non-vanishing contribution of the lowest order:

$$W_8(\mathscr{C}) \approx C v_{(8)}^{4|\mathscr{C}|} + \ldots \approx C \exp(4|\mathscr{C}| \log v_{(8)}) + \ldots, \qquad (2.6.110)$$

in which $|\mathscr{C}|$ represents the length of $\mathscr{C}$ in lattice constants. The Wilson loop for octet quarks satisfies a length rule; there is screening. [Ma 80, Gl 80].

For $(q)=3$, triality flux conservation [Os 78, Ma 78a] can be used to show that the lowest order, non-vanishing contribution to the Wilson loop $W(\mathscr{C})$ comes from the graph $G^{(Ar[\mathscr{C}])}$, which has $\mathscr{C}$ as its boundary: $G^{(m)}=\mathscr{C}$ and which consists of the lowest number m of plaquettes (see Fig. 2.17(c)); in this case $(k(\mathscr{P}_i))=(3)$. Its contribution is:

$$W(\mathscr{C}) = \text{constant.}\ v^m = \exp(\varepsilon^{-2} Ar[\mathscr{C}] \log v). \qquad (2.6.111)$$

and from this, the leading approximation to the string constant can be read off: $\kappa = -\varepsilon^{-2}\log v \approx \varepsilon^{-2}\log 3g_0$ (see Eqn. (2.6.96)). Higher approximations,

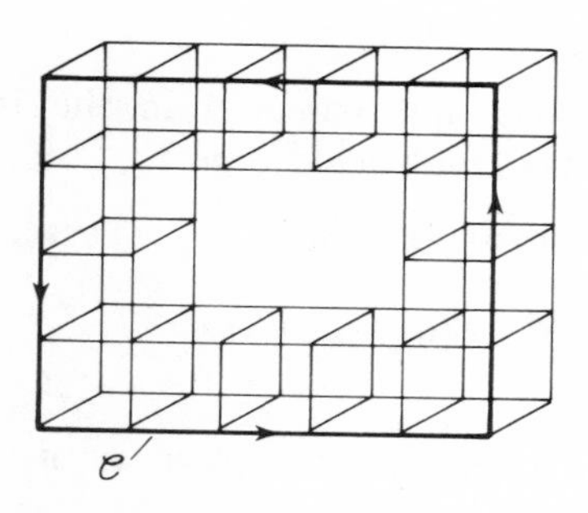

Figure 2.18 Lattice graph describing octet charge screening

which are systematically based on the method used for strong coupling approximation described above, give the following result:

$$\varepsilon^2\kappa = -\left(\log v + 4v^4 + 12v^5 - 10v^6 - 36v^7 + \frac{391}{2}v^8 + \frac{1131}{10}v^9 + \frac{2550837}{5120}v^{10} + \cdots\right) \quad (2.6.112)$$

Details of this and the relevant graphs can be found in the literature. As the coupling constant g belonging to the Hamiltonian scheme is not directly related to that of the path integral formulation, this result can be compared with Eqn. (2.6.78) in an indirect way only.

In conclusion to this section on strong coupling expansion, the outcome which has the most relevance to the quark confinement programme must be emphasised once again. When coupling is sufficiently strong, this method gives a convergent series expansion for the Wilson loop on the lattice. This proves quark confinement under the given conditions. There is a method available for calculating the string constant, the accuracy of which is restricted only by limitations of effort invested in the calculations.

(d) Approximation of QCD by means of a *lattice gauge theory corresponds to a regularization* by 'momentum cut-off', as described in the discussions on divergent Feynman graphs in Eqn. (2.4.4). Before discussing the continuum limit within the renormalization programme, this aspect of lattice approximation will be illustrated in Hamiltonian formulation, using a free, scalar field as an example. In the canonical quantum theory of a scalar field $A(x)$, the Hamilton operator is

$$H = \frac{1}{2}\int d^3x\{\pi^2(t, \mathbf{x}) + (\boldsymbol{\nabla}A(t, \mathbf{x}))^2 + M^2A^2(t, \mathbf{x})\} \quad (2.6.113)$$

and the commutation relations for the canonically conjugate quantities $A(t, \mathbf{x})$ and $\pi(t, \mathbf{x}) = \partial A(t, \mathbf{x})/\partial t$ are

$$[\pi(t, \mathbf{x}), A(t, \mathbf{x}')] = -\mathrm{i}\,\delta(\mathbf{x}-\mathbf{x}'), \quad [A(t, \mathbf{x}), A(t, \mathbf{x}')] = [\pi(t, \mathbf{x}), \pi(t, \mathbf{x}')] = 0. \quad (2.6.114)$$

These basic equations are approximated in the following way on the 3-dimensional space lattice, with fixed time t:

$$A(\mathbf{n}) = \varepsilon A(t, \mathbf{x})|_{\mathbf{x}=\varepsilon\mathbf{n}}, \quad \pi(\mathbf{n}) = \varepsilon^2\pi(t, \mathbf{x})|_{\mathbf{x}=\varepsilon\mathbf{n}}, \quad [\pi(\mathbf{n}), A(\mathbf{n}')] = -\mathrm{i}\,\delta_{\mathbf{n},\mathbf{n}'} \text{ etc.},$$

$$H = \frac{1}{2\varepsilon}\sum_{\mathbf{n}}\{\pi^2(\mathbf{n}) + (\boldsymbol{\nabla}A)^2(\mathbf{n}) + \mu^2A^2(\mathbf{n})\} \quad (2.6.115)$$

with

$$(\nabla^1 A)(\mathbf{n}) = A(\mathbf{n}^1+1, \mathbf{n}^2, \mathbf{n}^3) - A(\mathbf{n}^1, \mathbf{n}^2, \mathbf{n}^3), \quad \mu = \varepsilon M.$$

As in the continuum the free Hamiltonian operator on the lattice is diagonalized, using the Fourier transformation:

$$A(\mathbf{n}) = \frac{1}{(2\pi)^{3/2}} \iiint_{-\pi}^{+\pi} \tilde{A}(\boldsymbol{\beta}) e^{-i\mathbf{n}\boldsymbol{\beta}} \, d^3\beta, \quad \pi(\mathbf{n}) = \frac{1}{(2\pi)^{3/2}} \iiint \tilde{\pi}(\boldsymbol{\beta}) e^{i\mathbf{n}\boldsymbol{\beta}} \, d^3\beta,$$

$$A^{\dagger}(\boldsymbol{\beta}) = A(-\boldsymbol{\beta}) \quad \text{etc.,} \tag{2.6.116}$$

and introducing creation and annihilation operators:

$$a(\boldsymbol{\beta}) = \sqrt{\left(\frac{\omega(\boldsymbol{\beta})}{2}\right)} \tilde{A}(\boldsymbol{\beta}) + \frac{i}{\sqrt{(2\omega(\boldsymbol{\beta}))}} \tilde{\pi}(-\boldsymbol{\beta}),$$

$$a^{\dagger}(\boldsymbol{\beta}) = \sqrt{\left(\frac{\omega(\boldsymbol{\beta})}{2}\right)} \tilde{A}(-\boldsymbol{\beta}) - \frac{i}{\sqrt{(2\omega(\boldsymbol{\beta}))}} \tilde{\pi}(\boldsymbol{\beta}) \tag{2.6.117}$$

A direct calculation, in which the essential steps are based on the orthogonality and completeness relations of the lattice Fourier transformation, in accordance with the following formula:

$$\frac{1}{2\pi} \int_{-\pi}^{+\pi} e^{i(n-m)\beta} \, d\beta = \delta_{n,m}, \quad n, m = 0, \pm 1, \pm 2, \ldots,$$

$$\frac{1}{2\pi} \sum_{n=-\infty}^{+\infty} e^{in(\beta-\beta')} = \delta(\beta - \beta'), \quad \text{with } -\pi < \beta \leqslant +\pi \text{ as } e^{in(\beta+2\pi)} = e^{in\beta}. \tag{2.6.118}$$

shows that the canonical commutation relations (2.6.115) are equivalent to the commutation relations for the creation and annihilation operators

$$[a(\boldsymbol{\beta}), a^{\dagger}(\boldsymbol{\beta}')] = \delta(\boldsymbol{\beta} - \boldsymbol{\beta}'), \quad [a(\boldsymbol{\beta}), a(\boldsymbol{\beta}')] = [a^{\dagger}(\boldsymbol{\beta}), a^{\dagger}(\boldsymbol{\beta}')] = 0$$

(see Eqn. (1.2.19)). Moreover, if $\nabla^i e^{-i\boldsymbol{n}\boldsymbol{b}} = -2i \sin(\beta_i/2) e^{i\beta_i/2} e^{-i\mathbf{n}\boldsymbol{\beta}}$ is used, the form:

$$H = \frac{1}{\varepsilon} \iiint_{-\pi}^{+\pi} a^{\dagger}(\boldsymbol{\beta}) \omega(\boldsymbol{\beta}) a(\boldsymbol{\beta}) \, d^3\beta \quad \text{with } \omega(\boldsymbol{\beta}) = \left[\mu^2 + 4 \sum_i \sin^2(\beta_i/2)\right]^{1/2}. \tag{2.6.119}$$

follows for the Hamiltonian operator. If $\exp(-i\boldsymbol{n}\boldsymbol{\beta}) = \exp(-i\boldsymbol{p}\boldsymbol{x})$ is written in Eqn. (2.6.116) in physical quantities $\boldsymbol{x} = \varepsilon\boldsymbol{n}$, $\boldsymbol{p} = \boldsymbol{\beta}/\varepsilon$, (Eqn. (2.6.47)), then the restraint on $\boldsymbol{\beta}$ in the completeness relation (2.6.118) corresponds to the maximal physical momentum on the lattice: $|p_i| \leqslant \pi/\varepsilon$. Using the physical quantities which approximate the continuum quantities on the lattice $a(\boldsymbol{p}) = \varepsilon^{3/2} a(\boldsymbol{\beta})|_{\boldsymbol{\beta}=\boldsymbol{p}/\varepsilon}$, $\omega(\boldsymbol{p}) = \omega(\boldsymbol{\beta})/\varepsilon$ the Hamiltonian operator then becomes:

$$H = \iiint_{-\pi/\varepsilon}^{+\pi/\varepsilon} d^3p \, a^{\dagger}(\boldsymbol{p}) \sqrt{\left[\frac{\mu^2}{\varepsilon^2} + \frac{4}{\varepsilon^2} \sum_i \sin^2\left(\frac{p_i\varepsilon}{2}\right)\right]} a(\boldsymbol{p}). \tag{2.6.120}$$

It is seen that also a maximal energy follows from maximal momentum. This energy and momentum cut-off applies to all quantities derived from creation and annihilation operators and the Hamiltonian operator, e.g. to the Feynman propagator as described in Section 2.3.1(c). Finally, Eqn. (2.6.120) shows direct how the Hamiltonian operator is approximated for ever-decreasing lattice constant $\varepsilon \to 0$ and variable $\mu^2 = \varepsilon^2 M^2$.

(e) The Wilson criterion for quark confinement is satisfied for lattice gauge theories with strong coupling. The question is, does this $SU(3)$ gauge theory property remain when the renormalization process approaches the *continuum limit*? The following section attempts to answer this in the light of current knowledge.

Comparison of numerical calculations [Cr 80, Wi 80a] of the Wilson integral (Eqn. (2.6.59)) on a finite lattice with results from calculations of strong and weak coupling approximation [Mü 80] has given considerable insight into the problem. Eqn. (2.6.112) gives the result of calculations for the string constant κ in strong coupling approximation; Fig. 2.19 shows the function $\varepsilon^2\kappa = F_s(g_0)$. Assuming that confinement is conserved when going to the continuum, i.e. for lattice constant $\varepsilon \to 0$, then the string constant κ must differ from zero in this limit. Under these assumptions, the physical quantity κ can be kept constant in this limiting process as part of the renormalization programme ('renormalization condition'), whereas the unrenormalized coupling constant $g_0(\varepsilon)$ gets adapted at the same time to this condition: $\varepsilon^2\kappa = F_s(g_0(\varepsilon))$ ('subtraction'). The form of $F_s(g_0)$ shows that

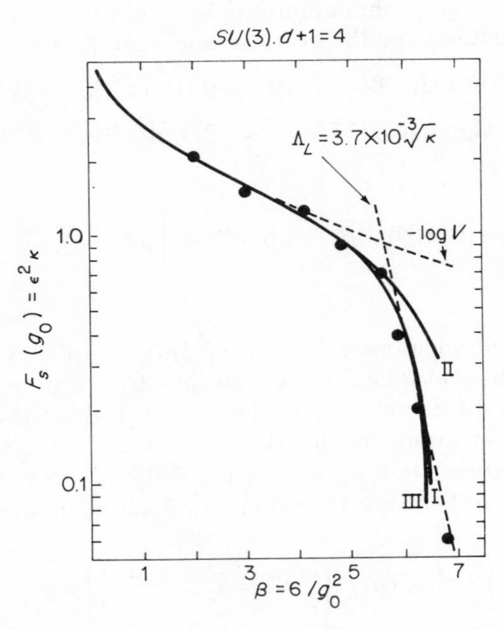

Figure 2.19 String constant as a function of the unrenormalized coupling constant based on [Mii 80]. I, II, III represent respectively the 10th, 11th, and 12th orders of the strong coupling approximation

$g_0(\varepsilon)$ becomes smaller with decreasing ε, so that eventually the range of application for the strong coupling approximation is left. The problem is then whether a method for the calculation of $\varepsilon^2\kappa \equiv F_s(g_0)$ with small g_0 can be found. As pointed out in the introduction to the quark confinement problem, this cannot be achieved directly by a perturbative calculation. However, the renormalization group describes the general features of the form of $F_s(g_0)$ in asymptotically free QCD. Section 2.5.2(e) showed that $g_0(\varepsilon)$ satisfies differential equations (2.5.8) and (2.539) for the running couplng constant and can therefore be expressed as:

$$\frac{g_0^2(\varepsilon)}{4\pi} = \frac{4\pi}{\beta_0 \log(1/\varepsilon^2\Lambda_L)} - \frac{4\pi\beta_1}{\beta_0^3} \frac{\log\log(1/\varepsilon^2\Lambda_L^2)}{\log^2(1/\varepsilon^2\Lambda_L^2)} + \ldots . \tag{2.6.121}$$

Λ_L is the asymptotic scale parameter referring to the lattice approximation as renormalization method (see p. 121). On the other hand, the physical quantity κ is a renormalization group invariant and is expressed as:

$$\kappa = C\varepsilon^{-2}\left(\frac{\beta_0}{16\pi^2} g_0^2(\varepsilon)\right)^{\beta_1/\beta_0^2} \exp\left(-\frac{16\pi^2}{\beta_0 g_0^2(\varepsilon)}\right). \tag{2.6.122}$$

based on Eqn. (2.5.14) and the approximation in Eqn. (2.5.29). $F_s(g_0)$ must converge smoothly to this behaviour of $\varepsilon^2\kappa$ if the asymptotically free $SU(3)$ gauge theory has the property of quark confinement and this can be described by a transition to the continuum from the confinement phase of lattice gauge theory for strong coupling. Numerical calculations have confirmed this state of affairs. In Fig. 2.19 a dashed line shows a fit of Eqn. (2.6.122) to $F_s(g_0)$. The points reproduce the result of the numerical calculation of the Wilson integral (2.6.59) with subsequent string constant extraction. Integrations over gauge fields on links was carried out for lattices with 6^4 lattice points using the Monte Carlo method. These calculations indicate strongly that the above assumption of a smooth transfer from the strong to the weak coupling region with fixed string constant $\kappa \neq 0$, is correct and thus pure $SU(3)$ gauge theory in the continuum is in fact in a confinement phase. In this way, the function $\varepsilon^2\kappa = F_s(g_0)$ in Fig. 2.19 determines formally an extrapolation of the β function of the unrenormalized coupling constant, in agreement with Eqn. (2.5.39)

$$-\varepsilon \frac{d}{d\varepsilon} g_0(\varepsilon) = \tilde{\beta}(g_0) = -\frac{2F_s(g_0)}{\partial F_s/\partial g_0} \tag{2.6.123}$$

from the asymptotically free region to the strong coupling region without a zero indicating a fixed point (see Section 2.5.1(c)).

This indication of quark confinement is based essentially on the result of numerical calculations [Cr 79] and its conclusive force is, therefore, somewhat restricted. Thus, it is all the more gratifying that the numerical calculations for cases in which a different phase in the regime of strong and weak coupling is expected, show a phase transition. This applies in particular to calculations for the theory with the $U(1)$ gauge group which should give the free QED in the continuum. In the weak coupling limit, free QED is in the Coulomb phase (Eqns. (2.6.38)ff. and (2.6.12)). On the other hand, a convergent high temperature expansion similar to Eqn. (2.6.111) demonstrates the confinement phase for strong coupling. Indeed, the numerical calculations show a hysteresis effect for $\beta = 0.99$ in the transition region between strong and weak coupling, which in itself is an indication of a phase transition [Cr 79, Mo 80]. An intuitive picture of the dynamical approximation to the continuum by a sequence of refined space lattices can be drawn from the Hamiltonian formulation of lattice gauge theories (see point (b) above). This describes, in simplest approximation a concentration of the chromo-electric flux between quarks on the shortest bond along links (see Eqn. (2.6.77) and Fig. 2.20(a)). States with excitations of a larger number of links have much greater energy and are therefore suppressed in strong coupling perturbation theory. However, if the coupling constant becomes smaller as the lattice becomes finer—$\varepsilon \to 0$, $g_0 \to 0$ with fixed physical string constant—then more and more link excitations contribute to the perturbation series with respect to $\bar{x} \cdot H_1$ (Eqn. (2.6.71)), until the wavefunction of the state is represented as a coherent superposition of link excitations in the region of the chromoelectric string in the continuum (Fig. 2.20(c)). Formal treatment of such a limit process with lattice refinement can be based on the Wilson blockspin method [Wi 75, Ko 74a, Ka 77]. This is mentioned also here as a method for calculating phase transitions.

This picture is probably an oversimplification. There are indications that as the coupling constant decreases, the chromoelectric string wavefunction becomes less compact and expands somewhat with increasing length [Lü 80]. This expansion of the boundary layer between various phases is known in statistical mechanics as a 'roughening transition' and is being thoroughly investigated as part of lattice gauge theory at the present time [It 80, Ha 80, Mü 80c, Lü 80b]. This phenomena could possibly be linked with a weak non-analyticity of the function $F_s(g_0)$ and thus it limits the validity of the

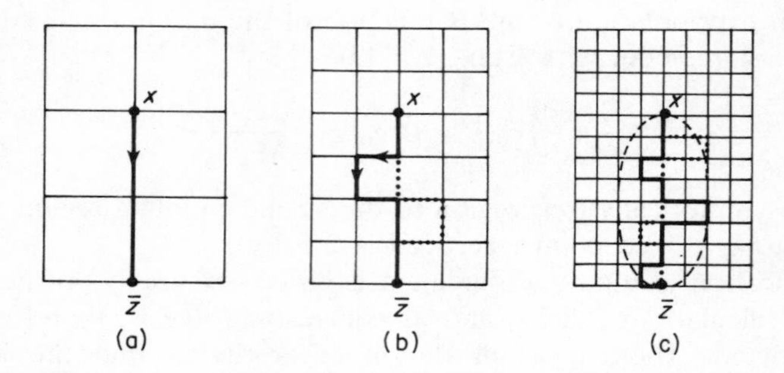

Figure 2.20 Symbolic representation of the structure of a string wavefunction in a sequence of refined lattices

strong coupling approximation somewhere in the transition region between strong and weak coupling (Fig. 2.19).

Finally, some attention must be given to the practical significance of calculations based on Fig. 2.19. The constant C in Eqn. (2.6.122) is determined by fitting the string constant κ as a renormalization group invariant to a high order strong coupling approximation. If the comparable representation (2.5.28) for the asymptotic scale parameter Λ_L is compared with (2.6.122) then $\Lambda_L = C^{-1/2}\sqrt{\kappa}$ can be deduced. In order to compare this Λ_L, which is based on the lattice as a renormalization scheme, with Λ_{MOM} obtained by dimensional regularization with coupling constant renormalization (2.5.15″), both forms must be transformed into each other (see p. 156). This gives a relatively large factor [Ha 80a]. When this is included, then the result is:

$$\Lambda_{\text{MOM}} \approx (0.35 \pm 0.1)\sqrt{\kappa} \qquad (2.6.124)$$

[Mü 80b, Ko 80, Cr 80]. The theoretical errors have been estimaed by the authors. Considering that quark couplings. etc. have been ignored a string constant of $\sqrt{\kappa} \approx 450$ MeV gives a reasonable figure for Λ. So, this method for the treatment of confinement problems consisting of:

(1) a high order of strong coupling approximation on the lattice;
(2) extrapolation to the continuum using the renormalization group; and
(3) control by Monte Carlo calculations;

has produced physically interesting results. The method has already been used for calculating the glue ball mass $M_G = (2.9 \pm 1.2)\sqrt{\kappa}$ [Mü 80d] (Eqn. (2.6.79)). Only time will tell whether further results will emerge, from an application to a complete QCD including quarks in particular (see Literature of Appendix).

2.6.4 Semiclassical approximation

Another aspect of non-Abelian gauge theories based on non-perturbative techniques follows from the consideration of the classical field equations. Solutions characterized by topological quantum numbers have been discovered for the field equations in the Euclidean space–time [Be 75a]. In a semiclassical approximation of the path integral representation in quantum gauge field theory, these 'instanton solutions' make contributions which modify the structure of the vacuum state in a way different from perturbation theory [Ja 76]. A brief description of this aspect of gauge theory will be given only, as although this research helps to clarify further the concept of quark confinement, the current view is that this method does not make the most significant contribution to the whole [Co 77, Ol 79].

(a) For the sake of simplicity, *instanton solutions* are considered in a pure $SU(2)$ gauge theory in Euclidean space–time. These can be represented by a 3-component gauge field $A_\mu^a(x)$ for which the matrix notation of Eqn. (2.2.4), $\mathcal{A}_\mu = \bar{g}A_\mu^a\tau_a/2$, is used together with the Pauli matrices (1.2.3).

Definition of field strength, action, and field equations follows from discussions in Sections 2.2.1 and 2.2.2:

field strength $\quad \mathscr{F}_{\mu\nu}(x) = \partial_\mu \mathscr{A}_\nu(x) - \partial_\nu \mathscr{A}_\mu(x) + \mathrm{i}[\mathscr{A}_\mu(x), \mathscr{A}_\nu(x)],$ (2.6.125)

action
$$S_E[\mathscr{A}] = -\frac{1}{4}\int \mathrm{d}^4x F^a_{\mu\nu}(x) F^a_{\mu\nu}(x)$$
$$= -\frac{1}{2\bar{g}^2}\int \mathrm{d}^4x \, \mathrm{Tr}\, \mathscr{F}_{\mu\nu}(x)\mathscr{F}_{\mu\nu}(x) \equiv -\frac{1}{4\bar{g}^2}(\mathscr{F}, \mathscr{F}), \tag{2.6.126}$$

field equations $\quad \partial_\mu \mathscr{F}_{\mu\nu} + \mathrm{i}[\mathscr{A}_\mu, \mathscr{F}_{\mu\nu}] = 0.$ (2.6.127)

The space–time indices refer to the Euclidean metric: $\mu = 1, 2, 3, 4$, $x_\mu y_\mu = x_1y_1 + x_2y_2 + x_3y_3 + x_4y_4$, etc. Our interest centres on solutions of the classical field equations with finite action. These have a finite norm $\|\mathscr{F}\|^2 = (\mathscr{F}, \mathscr{F}) < \infty$, $(\mathscr{F}, \mathscr{F})$ defined in Eqn. (2.6.126). This means that for $x^2 \to \infty$; $\mathscr{F}_{\mu\nu}(x) \to 0$, or that $\mathscr{A}_\mu(x)$ approaches a pure gauge of the form (2.2.25):

$$\mathscr{A}_\mu(x) \to -\mathrm{i}g(x)\,\partial_\mu g^{-1}(x) \quad \text{for } x^2 \to \infty. \tag{2.6.128}$$

The so-called Pontryagin density $\mathscr{P}_E$ plays an important part in classifying the solutions of Eqn. (2.6.127) which satisfy these boundary conditions:

$$\mathscr{P}_E = \mathrm{Tr}\, \mathscr{F}_{\mu\nu} \breve{\mathscr{F}}_{\mu\nu} \tag{2.6.129}$$

with $\breve{\mathscr{F}}_{\mu\nu} = \frac{1}{2}\varepsilon_{\mu\nu\rho\sigma}\mathscr{F}_{\rho\sigma}$, $\varepsilon_{1234} = 1$, $\varepsilon_{\mu\nu\rho\sigma}$ antisymmetrical. $\breve{\mathscr{F}}_{\mu\nu}$ is the dual field strength to $\mathscr{F}_{\mu\nu}$. As $\mathscr{P}_E$ can be written as a divergence:

$$\mathscr{P}_E(x) = \partial_\mu G_\mu(x), \quad G_\mu = 2\mathrm{i}\varepsilon_{\mu\nu\rho\sigma}\, \mathrm{Tr}(\mathscr{A}_\nu(\partial_\rho \mathscr{A}_\sigma + \tfrac{2}{3}\mathscr{A}_\rho \mathscr{A}_\sigma)), \tag{2.6.130}$$

according to Gaussian law, the integral $(\mathscr{F}, \breve{\mathscr{F}})$ is determined solely by the pure gauge on the boundary (2.6.128):

$$\frac{1}{32\pi^2}(\mathscr{F}, \breve{\mathscr{F}}) = \frac{1}{16\pi^2}\int \mathscr{P}_E\, \mathrm{d}^4x = r = \frac{1}{16\pi^2}\int_{|x|\to\infty} \mathrm{d}\sigma_\mu G_\mu$$
$$\approx -\frac{1}{12\pi^2}\int_{|x|\to\infty} \mathrm{d}\sigma_\mu \varepsilon_{\mu\nu\rho\sigma}\, \mathrm{Tr}\{g(x)\,\partial_\nu g^{-1}(x) \cdot g(x)\,\partial_\rho g^{-1}(x) \cdot g(x)\,\partial_\sigma g^{-1}(x)\}. \tag{2.6.131}$$

Evaluation of this integral gives the topological quantum number $r = 0, \pm 1, \pm 2, \ldots$. Quantization of the magnetic flux of the solutions of the field equations (2.6.23) or (2.6.28), which describe a 'Nielsen–Olesen string', follows from a completely analogous argument. In complete agreement with Eqn. (2.6.128), these solutions for large distances ρ are characterized by a pure gauge, (Eqns. (2.6.33), (2.6.30), and (2.6.31)). If the magnetic flux is regarded as a two-dimensional Pontryagin density [Dr 77]: $B \sim \mathscr{P}_E^{(2)} = \partial_\mu \varepsilon_{\mu\nu} A_\nu = \partial_\mu G_\mu^{(2)}$, $\mu = 1, 2$, then Eqn. (2.6.35) which determines flux quantization corresponds exactly to Eqn. (2.6.131). The abstract, topological significance of the flux quantum number n is the fact that a continuous mapping of the infinite circle in the plane $|x| = \rho$, $\rho = \infty$ on the group $U(1) = \{e^{i\alpha} \mid 0 \leqslant \alpha < 2\pi\}$ is defined by the asymptotic gauge transformation in Eqn. (2.6.32): $e^{i\alpha} = e^{in\theta(x)}$, $\theta(x) \equiv$ planar polar angle (Eqn. (2.6.25)), which covers the $U(1)$ group n times. The Pontryagin number r introduced by Eqn. (2.6.131) also has the meaning of a winding number. If the asymptotic gauge $g(x)$ in Eqn. (2.6.128) is regarded as a continuous mapping of the infinitely distant sphere S_3 in 4-dimensional Euclidean space on the group $SU(2)$, then $|r|$ describes how often this mapping covers the group. The following is an example of such a mapping for $S_3 \to SU(2)$:

$$g(x) = \frac{1}{|x|}(x_4 + i\boldsymbol{\tau}\mathbf{x}) \tag{2.6.132}$$

In agreement with the general representation of a 2-dimensional unitary matrix with det $g = 1$ (Eqn. (1.2.11)), $g(x)$ maps the sphere S_3 uniquely onto $SU(2)$. An exact solution of the non-linear field equations (2.6.127) which approaches asymptotically this pure gauge and thus belongs to the topological quantum number $r = 1$ will be discussed below. Powers of $g(x)$: $g^r(x)$ are examples of mappings with an arbitrary winding number r. The formula of Haar's measure (Eqn. (1.2.14)) for $SU(2)$:

$$d\mu(g) = \frac{1}{12\pi^2} \operatorname{Tr}(g\, \partial_i g^{-1} \cdot g\, \partial_j g^{-1} \cdot g\, \partial_k g^{-1})\varepsilon^{ijk}\, d\theta_1\, d\theta_2\, d\theta_3$$

with (2.6.133)

$$\theta_1 = \theta, \quad \theta_2 = \phi, \quad \theta_3 = \vartheta$$

is used to confirm this result on r by evaluation of the integral (2.6.131).

For self-dual fields

$$\mathscr{F}_{\mu\nu}(x) = \breve{\mathscr{F}}_{\mu\nu}(x) = \tfrac{1}{2}\varepsilon_{\mu\nu\rho\sigma}\mathscr{F}_{\rho\sigma}(x) \tag{2.6.134}$$

the action agrees with the Pontryagin number up to a factor (Eqn. (2.3.131)):

$$S_E = -\frac{1}{4\bar{g}^2}(\mathscr{F}, \mathscr{F}) = -\frac{1}{4\bar{g}^2}(\mathscr{F}, \breve{\mathscr{F}}) = -\frac{8\pi^2}{\bar{g}^2}\, r. \tag{2.6.135}$$

Moreover, as the relationship

$$(\mathscr{F}, \mathscr{F}) \geqslant (\mathscr{F}, \breve{\mathscr{F}}) \tag{2.6.136}$$

follows from Schwarz's inequality $(\mathscr{F}, \breve{\mathscr{F}}) \leqslant \sqrt{[(\mathscr{F}, \mathscr{F})(\breve{\mathscr{F}}, \breve{\mathscr{F}})]}$ for arbitrary fields and $(\mathscr{F}, \mathscr{F}) = (\breve{\mathscr{F}}, \breve{\mathscr{F}})$, it is clear that self-dual fields within the class of fields with Pontryagin number r ('Chern class' [Bo 65]) give the minimal value of $|S_E|$. Thus, in agreement with the principle of 'minimal' action (Eqns. (1.2.31)–(1.2.34)), self-dual fields are in fact solutions to the field equations (2.6.127). This result follows also directly from the Bianchi identity (2.2.17) and the self-duality equation (2.6.134).

In order to solve the self-duality equation (2.6.134), the *ansatz*:

$$\mathscr{A}_\mu(x) = -\mathrm{i}f(x^2)g(x)\,\partial_\mu g^{-1}(x) = -2f(x^2)\sigma_{\mu\nu}\frac{x_\nu}{x^2} \tag{2.6.137}$$

is made with $g(x)$ from Eqn. (2.6.132) as asymptotic gauge. The matrix tensor $\sigma_{\mu\nu}$ is self-dual:

$$\sigma_{\mu\nu} = \frac{1}{4\mathrm{i}}(s_\mu\bar{s}_\nu - s_\nu\bar{s}_\mu), \quad (s_\mu) = (\mathrm{i}\boldsymbol{\tau}, 1), \quad (\bar{s}_\mu) = (-\mathrm{i}\boldsymbol{\tau}, 1), \tag{2.6.138}$$

$$\sigma_{\mu\nu} = \breve{\sigma}_{\mu\nu}.$$

Then, calculation of the field strength belonging to these $\mathscr{A}_\mu(x)$ (Eqn. (2.6.125)):

$$\mathscr{F}_{\mu\nu}(x) = \partial_\mu\mathscr{A}_\nu(x) - \partial_\nu\mathscr{A}_\mu(x) + \mathrm{i}[\mathscr{A}_\mu(x), \mathscr{A}_\nu(x)]$$
$$= -4\left[f(1-f)\frac{\sigma_{\mu\nu}}{x^2} + [f'x^2 - f(1-f)]\left(\sigma_{\nu\rho}\frac{x_\mu x_\rho}{x^4} - \sigma_{\mu\rho}\frac{x_\nu x_\rho}{x^4}\right)\right],$$

shows that because of the self-duality of $\sigma_{\mu\nu}$, $\mathscr{F}_{\mu\nu}$ is self-dual when the second term disappears, i.e.

$$f'x^2 + f(1-f) = 0 \tag{2.6.139}$$

The solution of this equation is $f(x^2) = x^2(x^2+\lambda^2)^{-1}$, which gives the field strength:

$${}^{(1)}\mathscr{F}_{\mu\nu}(x) = -\frac{4\lambda^2}{(x^2+\lambda^2)^2}\sigma_{\mu\nu}, \quad {}^{(1)}\mathscr{A}_\mu(x) = \frac{-2x^2}{\lambda^2+x^2}\sigma_{\mu\nu}\frac{x_\nu}{x^2}. \tag{2.6.140}$$

(Solution $\lambda^2/(x^2+\lambda^2)$ gives a singular gauge field for $x=0$.) These $\mathscr{F}_{\mu\nu}(x)$, $\mathscr{A}_\mu(x)$ of Eqn. (2.6.140) give an exact solution to the non-linear field equations [Be 75a]. In view of the translation invariance of field equations, ${}^{(1)}\mathscr{F}_{\mu\nu}(x-a)$, ${}^{(1)}\mathscr{A}_\mu(x-a)$ are also solutions. It is easy to see that in this 5-parameter family of solutions, all belong to the Pontryagin number $r=1$. Intuitively, these solutions represent field concentrations in 4-dimensional Euclidean space–time, which are localized around the space–time point a and have a mean width λ. These objects in Euclidean space–time, which are not actually physical, are known as instantons and are often described in quasiparticle terminology.

Other similar types of solution, the so-called anti-instanton solutions, are obtained from calculation of antidual fields $\breve{\mathscr{F}}_{\mu\nu} = -\mathscr{F}_{\mu\nu}$. Antidual fields are solutions of the field equations by the same argument as described above and the simple anti-instanton solution can be written as

$${}^{(\bar{1})}\mathscr{F}_{\mu\nu}(x) = -\frac{4\lambda^2}{(x^2+\lambda^2)^2}\bar{\sigma}_{\mu\nu} \quad {}^{(\bar{1})}\mathscr{A}_\mu(x) = \frac{-2x^2}{\lambda^2+x^2}\bar{\sigma}_{\mu\nu}\frac{x_\nu}{x^2} \tag{2.6.141}$$

with

$$\bar{\sigma}_{\mu\nu} = \frac{1}{4\mathrm{i}}(\bar{s}_\mu s_\nu - \bar{s}_\nu s_\mu), \quad \bar{\sigma}_{\mu\nu} = -\breve{\bar{\sigma}}_{\mu\nu}.$$

The Pontryagin number is -1.

Self-(anti)dual solutions for higher Pontryagin numbers r can be obtained from the *ansatz*: [Ja 77] ${}^{(r)}\mathscr{A}_\mu(x)=\sigma_{\mu\nu}\,\partial_\nu \log\rho(x)$ and ${}^{(\bar{r})}\mathscr{A}(x)=\bar{\sigma}_{\mu\nu}\,\partial_\nu \log\rho(x)$ as a solution for the self-(anti)duality equation $(1/\rho)\,\partial_\mu\,\partial_\mu\rho=0$ in the form:

$$ {}^{(r)}\mathscr{A}_\mu=\begin{Bmatrix}\sigma_{\mu\nu}\\ \bar{\sigma}_{\mu\nu}\end{Bmatrix}\partial_\nu \log\left(\sum_{i=1}^{r+1}\frac{\lambda_i^2}{(x-y_i)^2}\right). \tag{2.6.142} $$

If we consider that some of these solutions can be transformed into each other by gauge transformations, then there are five independent paameters in these solutions for $r=1$, 13 for $r=2$ and $5r+4$ for $r\geqslant 3$. They can be interpreted as r instantons which are located at positions y_i and have a mean width λ_i. The most general self-dual solution with topological charge r is characterized by $8r-3$ parameters α_i^r [Sc 77]. No simple explicit representation has been found for these. However, there are algebraic methods which allow the construction of all solutions of the self-duality equation. This also applies for self-dual $SU(n)$ gauge fields [Ch 78, At 78, Co 78].

(b) Following this short account of the remarkable recently acquired insight into a large class of solutions of a complicated, non-linear field equation, their importance to quantized gauge theory can now be assessed. This leads to an evaluation of the path integral formula

$$ \begin{aligned} \langle\Omega[\mathscr{A}]\rangle&=\frac{1}{Z}\int\mathscr{D}[\mathscr{A}]\Omega[\mathscr{A}]\exp(S_E[\mathscr{A}]),\\ Z&=\int\mathscr{D}[\mathscr{A}]\exp(S_E[\mathscr{A}]), \end{aligned} \tag{2.6.143} $$

in the Euclidean region, using the saddle-point method. This is called a *semiclassical approximation* for reasons which will soon become apparent.

The saddle-point method [De 61] describes the approximation of integrals of the form:

$$ I(g)=\int_{-\infty}^{+\infty}\mathrm{d}x\phi(x)\mathrm{e}^{-f(x)/g} $$

by Gaussian integrals, for small g. The most simple example of an ordinary integral will be considered first. In general, the most important contribution to such an integral is expected to come from the region around x_0, in which $f(x)$ assumes its minimum value (a more precise statement must be based on specific assumptions): so, if $f(x)$ is expanded in this region

$$ \begin{aligned} f(x)&=f(x_0)+f''(x_0)(x-x_0)^2/2+\dots,\\ f'(x_0)&=0,\quad f''(x_0)>0, \end{aligned} \tag{2.6.144} $$

this leads to:

$$ \begin{aligned} I(g)&\approx\phi(x_0)\int_{-\infty}^{+\infty}\mathrm{d}x\exp\left(-\frac{1}{g}[f(x_0)+f''(x_0)(x-x_0)^2/2+\dots]\right)\\ &=(2\pi/f''(x_0))^{1/2}\phi(x_0)\exp(-f(x_0)/g). \end{aligned} \tag{2.6.145} $$

If the function takes on a minimum at several places, summation with respect to the various contributions of all these minima must take place.

This method can be generalized for evaluation of the path integral formula. For this one has to find first of all the maxima of the Euclidean action ($S_E < 0$). Field configurations which minimize $|S_E|$ are, however, solutions of the Euclidean field equations. This is especially true for self-dual or antidual solutions, which assume the absolute minimum of $|S_E|$ for field configurations with a specific topological quantum number r, in agreement with Eqns. (2.6.135) and (2.6.136). If field configurations similar to Eqn. (2.6.144) are expanded around classical solutions (and for the given case this expansion is performed around self-dual solutions):

$$\mathcal{A}_\mu = \mathcal{A}_\mu^{\text{cl}} + \eta_\mu$$
$$S_E[\mathcal{A}] = S_E[\mathcal{A}^{\text{cl}}] - (\eta\, \Delta\eta), \quad S_E[\mathcal{A}^{\text{cl}}] = \frac{8\pi^2}{\bar{g}^2}\, r. \tag{2.6.146}$$

then the partition sum Z can be evaluated as a sum of Gaussian integrals (Eqn. (2.3.58)):

$$Z = \sum_r \exp\left(-\frac{8\pi^2}{\bar{g}^2}\, r\right) \int_{M_r} d\mu(\alpha^r)(\det \Delta)^{-1/2} \tag{2.6.147}$$

and similarly the expectation values are:

$$\langle \Omega[\mathcal{A}] \rangle = \frac{1}{Z} \sum_r \exp\left(-\frac{8\pi^2}{\bar{g}^2}\, r\right) \int_{M_r} d\mu(\alpha^r)(\det \Delta)^{-1/2}\Omega[\mathcal{A}^{\text{cl}}(x, \alpha^r)]. \tag{2.6.147'}$$

M_r denotes the $(8r-3)$-dimensional space of self-dual solutions over which integration must be performed as all these solutions belong to the same minimal action. In general, the fluctuation matrix Δ is a function of the $(8r-3)$ instanton parameters. An eigenvalue zero of the matrix Δ is linked with each instanton parameter which must be split off when $\det \Delta$ is evaluated (see Eqn. (2.3.121)). This can be done by introducing collective instanton coordinates [Ge 76a].

The fluctuation determinant for the 1-instanton sector $r = 1$ was calculated initially by G. 't Hooft [tH 76a]. After renormalization, splitting off the zero modes etc. the partition sum in the 1-instanton sector emerges as an integral over position a and range λ:

$$Z^{(1)} = C \int \cdots \int \frac{d^4a\, d\lambda}{\lambda^5}\, \bar{g}^{-8}(\lambda) \exp\left(-\frac{8\pi^2}{\bar{g}^2(\lambda)}\right) \equiv \int d^4a \frac{d\lambda}{\lambda}\, n_0(\lambda). \tag{2.6.148}$$

C is a constant dependent on the renormalization scheme [Ca 79]. Essentially, renormalization means that $\bar{g}^2$ in Eqn. (2.6.147) is replaced by the running coupling constant $\bar{g}^2(\lambda)$. $n_0(\lambda)$ is regarded as the phase space density of an instanton of size λ.

The general evaluation of the partition sum in Eqn. (2.6.147) and the corresponding Gaussian approximation for the physical expectation values in Eqn. (2.6.147′) has not yet been fully investigated. There are representations for fluctuation determinants in higher instanton sectors [Be 79a] but it is not clear whether the restriction on fluctuations around exact, self(anti)dual field configurations is sufficient for a physical approximation [Lü 80c]. One way of looking at the problem is based on the analogy with 4-dimensional statistical mechanics and makes Z in Eqn. (2.6.147) the grand, canonical distribution function for a pure interacting 'instanton gas'. It appears now that the description of a partition sum is more complete when a distribution function for an instanton–anti-instanton gas is used [Ca 78, Ca 79]. Assuming a dilute instanton–anti-instanton gas, the quantum mechanical effects of classical instanton solutions can be described by averaging over a phase space density

$$\sim \frac{1}{n_+!\,n_-!} \prod_{j=1}^{n_+ + n_-} \left(\mathrm{d}^4 a_j \frac{\mathrm{d}\lambda_j}{\lambda_j} n_0(\lambda_i) \right)$$

for all instanton and anti-instanton degrees of freedom a_j, λ_j and for all instanton numbers n_+ and anti-instanton numbers n_-. $n_0(\lambda)$ is the individual instanton density of Eqn. (2.6.148). As concrete calculations are in fact burdened with many details such as problems involving thermodynamic transition to infinite volume, infrared singularities, validity of ideal gas approximation etc. the rough description given above seems satisfactory for the present.

Quantum mechanical instanton effects were studied in order to clarify several physical problems. They did not lead to confinement in the ideal gas approximation when calculating the Wilson loop [Ca 78]. However, they did cause a strong, albeit not perfect, 'chromo-paramagnetism' $\mu_{\mathrm{el}} \gg 1$ [Ca 79, Il 79] as expected in connection with the chromoelectric Meissner effect (see p. 169). Instantons cause confinement in $(2+1)$ dimensions [Po 77a). According to G. 't Hooft, [tH 76], instanton contributions can be discussed in connection with a Bell–Jackiw anomaly

$$\partial^\mu j^5_\mu(x) = -\mathrm{i}\left(\frac{Ng^2}{16\pi^2} F^a_{\mu\nu} \breve{F}^a_{\mu\nu}\right)$$

(N = number of flavour degrees of freedom, see Section 2.4.3(f)). The relatively large mass of flavour-neutral, pseudoscalar η particles seems to be related to this problem ('$U(1)$ problem' [We 75], see also [Ol 79, Cr 78a]).

(c) A note on the non-perturbative character of these contributions will serve to conclude this short survey of the problems associated with semiclassical approximations of quantum mechanical instanton effects. Semiclassical approximations such as the saddle-point method are based on the assumption that g^2 is small, but contributions are not analytical in the coupling constant. The deeper reason for this is as follows: a Gaussian approximation around the classical field configuration $\mathscr{A}_\mu \equiv 0$ was used in the derivation of the Feynman rules for perturbation theory calculations (Section 2.3.2(e)). However, in a non-Abelian field theory there are classical field configurations $\mathscr{A}_\mu(x) = g(x)\,\partial_\mu g^{-1}(x)$, which, as pure gauges such as $\mathscr{A}_\mu \equiv 0$, belong to the Euclidean action $S_E[\mathscr{A}^{cl}] = 0$, but which cannot be regarded as fluctuations of $g(x) \equiv 1$ because they are topologically different from this gauge. Degeneracy of the classical ground state configurations is counteracted by a tunnel effect in quantum mechanics and hence the instanton contributions represent the tunneling amplitudes [Po 77a, Ja 76].

The lattice approximation of QCD, the semiclassical approximation, and the method of string equations all result in mutually complementary non-perturbative procedures for dealing with the confinement problem. The most far reaching results to date have emerged from the lattice approximation. However, there is as yet no way of describing the coupling of gluon strings and dynamic quarks and as this coupling is responsible for string decay in hadrons, the problem must be solved before any complete concept of confinement in QCD can be formed.

2.7 Phenomenological application of quantum chromodynamics [PQ]

The previous sections have established QCD as a strong interaction theory and discussed its most important structural properties such as asymptotic freedom and quark confinement. Now the question arises as to what extent these QCD concepts can be confirmed by experiment and whether phenomenological analysis of experiments can be put on a firmer footing by QCD.

The most important elements of QCD are the quanta of the field mediating the strong interaction, i.e. the gluons. However, 'tracks' of gluons were not actually confirmed until 1979 in electron–positron annihilation at PETRA [Ta 79, Pl 79, Ja 80, Ba 79c]. In the same way as quarks appear in the form of jets rather than as free particles, so the gluons are seen as gluon jets and provide direct confirmation that hard gluon bremsstrahlung of quarks does exist. Its frequency acts as a measure for the running, strong, fine structure constant $\bar{\alpha}_s(Q^2)$. Section 2.7.1 describes the jets in e^+e^- annihilation and discusses the experimental results.

As mentioned several times already, the asymptotic freedom of QCD provides a basis for the quark-parton model. However, its results are modified by field theoretical radiative corrections found in logarithmic deviations from simple scaling. In Section 2.7.2 the field theoretical treatment of deep inelastic lepton scattering is presented. A semiquantitative application of perturbation theory on hard scattering processes is permitted at high momentum transfers because of the decreasing running coupling constant. No precise formulation or strictly field theoretical derivation has yet been drawn up for this method. Its advantages are that it is based on intuitive picture and has a wide range of application. The perturbative methods in particular are used for the calculation of quark and gluon jets (Section 2.7.3).

Results of the discussion of quark confinement and asymptotic freedom form the starting point for a phenomenological treatment of heavy quark bound systems. In addition to giving information on quarkonia spectra (Section 1.1), QCD provides a dynamical foundation and quantification of the Zweig rule which explains the decay properties of the quarkonia systems (Section 2.7.4).

2.7.1 Gluons and gluon couplings

The fundamental field quanta in QCD, quarks and gluons, have not as yet been observed as free particles and may never be in the future either if QCD does actually have confinement properties. Despite this, quarks and gluons leave clear experimental 'tracks': in the form of pointlike scattering centres (Section 1.4) in deep inelastic lepton scattering, and as jets with corresponding quantum numbers in e^+e^- annihilation. Evidence which comes from e^+e^- jets, for quarks, gluons, and their coupling is now given.

(a) The first step in e^+e^- annihilation in hadrons proceeds via a virtual photon creating a *quark–antiquark* pair. These primary quarks radiate mostly soft, almost colinear gluons, which together with the primary quarks, hadronize in two *jets*

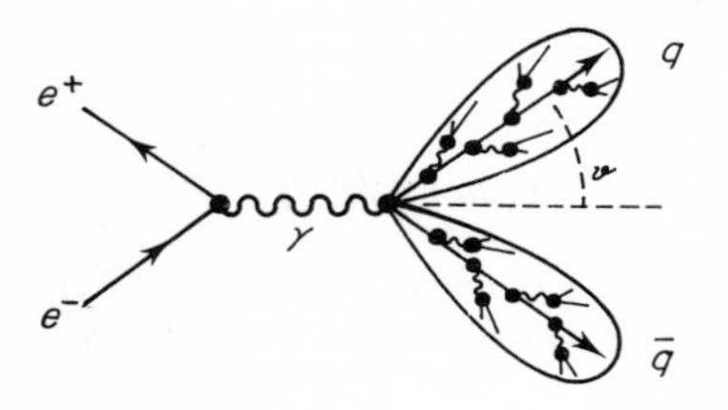

Jet axes are, therefore, determined by the direction of the primary quark pair. They have the characteristic angular distribution of spin 1/2 particles (see Eqn. (1.2.52)):

$$\frac{\mathrm{d}\sigma}{\mathrm{d}\Omega}(e^+e^- \to q\bar{q}) = \frac{3\alpha^2}{4s}\sum_i Q_i^2(1+\cos^2\vartheta).$$

2-jet events with this angular distribution have been observed by Spear [Ha 75] and PETRA [Ha 82N, Cr 82N]. Fig. 2.21(a)–(c) shows a 2-jet event at 30 GeV, by way of example [Br 79a].

(b) For the *phenomenological analysis of jet* events a sphericity tensor can be constructed from the momenta $p_{i\alpha}$ of the r particles produced in the event, $\alpha = 1, \ldots, r$:

$$M_{ij} = \frac{\sum_\alpha p_{i\alpha}p_{j\alpha}}{\sum_\alpha (\mathbf{p}_\alpha)^2}.$$

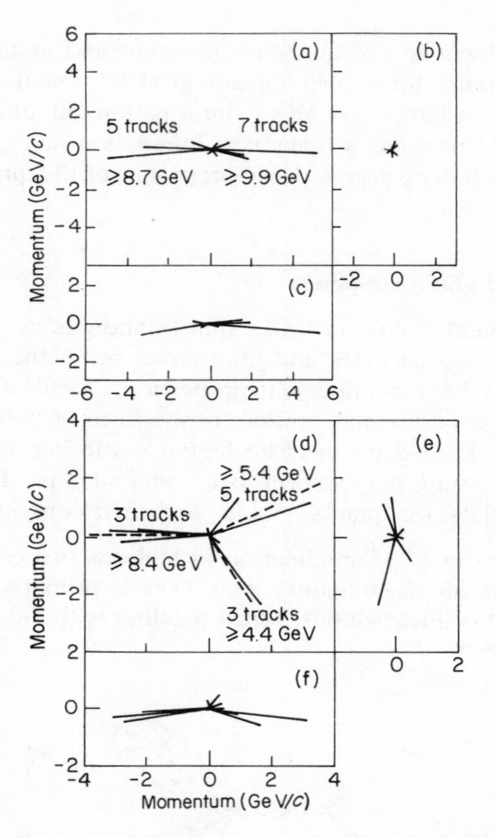

Figure 2.21 Momenta of particles in a 2-jet ((a)–(c)) and a 3-jet ((d)–(f)) event projected onto the $(\hat{n}_2, \hat{n}_3)$ plane ((a), (d)), $(\hat{n}_1, \hat{n}_2)$ plane ((b), (e)), and $(\hat{n}_1, \hat{n}_3)$ plane ((c), (f)) [Wo 80]

Its eigenvalues $\lambda_1, \lambda_2, \lambda_3$ with $\lambda_1+\lambda_2+\lambda_3=1$, $0\leqslant\lambda_1\leqslant\lambda_2\leqslant\lambda_3$, define the sphericity [Bj 70]:

$$S' = \tfrac{3}{2}(\lambda_1+\lambda_2) = \tfrac{3}{2}\,\underset{\{e\}}{\mathrm{Min}}\,\frac{\sum_\alpha (\mathbf{p}_\alpha - \hat{e}(\hat{e}\mathbf{p}_\alpha))^2}{\sum_\alpha (\mathbf{p}_\alpha)^2}, \tag{2.7.1}$$

where $\hat{e}$ is an arbitrary unit vector, and the acoplanarity

$$A = \tfrac{3}{2}\lambda_1. \tag{2.7.2}$$

These parameters S', A characterize the shape of an event. $S'=1$, $A=1/2$ for isotropically distributed momenta and $S'=0$, $A=0$ for an ideal jet. The principal axes of the sphericity tensor are $\hat{n}_1$, $\hat{n}_2$, $\hat{n}_3$; $\hat{n}_3$ is called the sphericity axis and the event plane is spanned by $\hat{n}_2$, $\hat{n}_3$.

Apart from sphericity, in view of the infrared problem in QCD which will be discussed in more detail in Section 2.7.3, various quantities which are linear in the momenta are used [Ge 77, Fo 78]. The most common of these shape parameters is thrust [Fa 77]

$$T = 2 \operatorname*{Max}_{\{e\}} \frac{\sum_\alpha |\hat{e}\mathbf{p}_\alpha|}{\sum_\alpha |\mathbf{p}_\alpha|} \tag{2.7.3}$$

with $T = 1/2$ for isotropic and $T = 1$ for perfect jet events. In practice, there is a slight difference between jet axes, defined by different prescriptions.

(c) With increasing centre of mass energy, the jet structure of the events becomes more pronounced, i.e. the angular apertures of the cone in which the particles are emitted in the jet decrease. Reduction in sphericity as a function of energy is shown in Fig. 2.23. This fact means that several jets can be isolated in one event and thus *3- and 4-jet events* can be sought. These are expected in QCD, as one of the quarks generated in the primary process can also emit a highly energetic gluon over a large angle, i.e. with large momentum transfer Q^2, with a probability $\bar{\alpha}_s(Q^2)$. This hard gluon then radiates soft, almost colinear gluons or quark–antiquark pairs which form a gluon jet. Hence n-jet events ($q\bar{q}G$, $q\bar{q}GG$, $q\bar{q}q\bar{q}$, ...) with frequencies:

$$\sigma(n \text{ jets})/\sigma(2 \text{ jets}) = O((\bar{\alpha}_s(Q^2))^{n-2}) \tag{2.7.4}$$

should exist. Figs. 2.21(d)–(f) and 2.22 show two 3-jet events at 30 GeV.

A detailed analysis of the frequency of 3-jet events using quark and gluon fragmentation models then allows the determination of the strong coupling constant. As Eqn. (2.5.31) showed, comparison with a calculation of up to the second-order perturbation theory at the very least is required. About 100 Feynman graphs have to be calculated for this analysis [El 80, Fa 80].

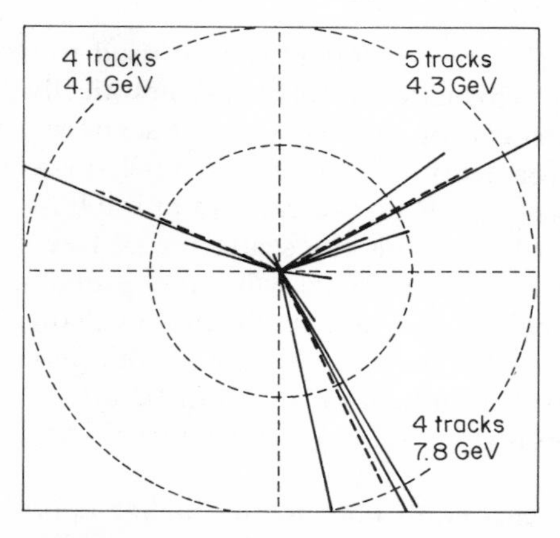

Figure 2.22 Projection of a 3-jet event onto the event plane [TASSO, Wo 80]

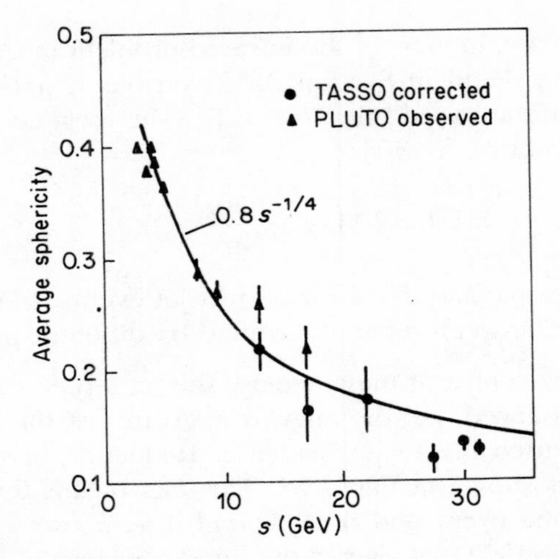

Figure 2.23 Average sphericity as a function of energy [Wo 80] (PLUTO [Be 79] and TASSO [Br 80] measurements)

The soft and collinear partons complicate the association of gluons and quarks to jets of a given extension. This implies difficulties for the comparison of theoretical calculations with experiments. A new discussion of these problems [Kr 82N] in the analysis of the data Ba 82bN results in the value

$$\bar{\alpha}_s = 0.16 \pm 0.015 \text{ (stat)} \pm 0.03 \text{ (syst.)} \tag{2.7.5}$$

for the running coupling constant at $Q^2 = (30 \text{ GeV})^2$. Combined with the value of $\bar{\alpha}_s$ from $\delta_{tot}(e^+e^- \to \text{hadrons})$ this gives a clear hint for the non-Abelian character of the QCD gauge theory.

2.7.2 Parton model and violation of scaling in deep inelastic lepton–nucleon scattering

Being a renormalizable theory, QCD can be used to work out a consistent perturbation theory. This latter becomes asymptotically free at high momentum transfers and so can be applied to hard scattering processes. This will now be described for deep inelastic lepton–nucleon scattering [PQ].

Fig. 2.24(a) shows once again the concept at the basis of the parton model (Section 1.4). According to this the scattering of free, zero mass quarks off leptons is calculated and averaged with the probability of finding these quarks with momentum fraction x in the nucleon, thus obtaining the inclusive cross-section as a result. In QCD, this model is modified in g^2 order by gluon bremsstrahlung (Fig. 2.24(b)), quark pair creation by the gluons in the nucleon (Fig. 2.24(c)), and internal radiative corrections (Fig. 2.24(d)).

Very many diagrams contribute to the higher approximations. The ones which make the most significant contributions at high momentum transfers

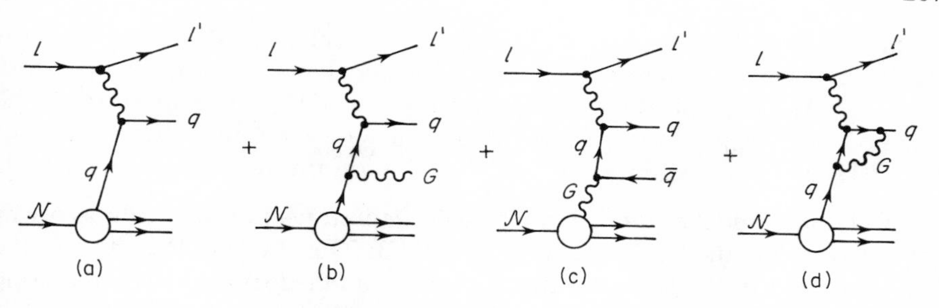

Figure 2.24 Deep inelastic lepton–nucleon scattering in the parton model (a) and QCD corrections in g^2 order ((b), (c), (d))

will be discussed in Section 2.7.3. Instead the RG equation method described in Section 2.5 is used now for an effective summation of the perturbation series. One point in favour of this method is that it allows the problem to be suitably divided up into the hard quark–lepton or gluon–lepton scattering process which can be dealt with perturbatively and the process of quark–binding in the nucleon which is not covered by perturbation theory.

As in Section 1.4.1 the starting point is the hadronic tensor $W_{\mu\nu}(\nu, q^2)$ (Eqn. (1.4.3) and Fig. 1.9). Instead of the product $j_\mu(y)j_\nu(0)$, the commutator can be used, as the term $j_\nu(0)j_\mu(y)$ for spacelike q^2 and $q^0 = E - E' \geqslant 0$ makes no contribution to $W_{\mu\nu}$:

$$\begin{aligned} W^{abr}_{\mu\nu}(\nu, q^2) &= \int \frac{d^4y}{(2\pi)^4} e^{iqy} \frac{1}{2} \sum_{\text{spin}} \langle \mathcal{N}, P| [j^{ar}_\mu(y), j^{br}_\nu(0)] |\mathcal{N}, P\rangle \\ &= \left(-g_{\mu\nu} + \frac{q_\mu q_\nu}{q^2}\right) W^{abr}_1(\nu, q^2) \\ &\quad + \left(P_\mu - q_\mu \frac{qP}{q^2}\right)\left(P_\nu - q_\nu \frac{qP}{q^2}\right) \frac{W^{abr}_2(\nu, q^2)}{M^2_{\mathcal{N}}} \\ &\quad - \frac{i}{2} r\varepsilon_{\mu\nu\alpha\beta} P^\alpha q^\beta W^{abr}_3(\nu, q^2)/\mathcal{M}^2_{\mathcal{N}}. \end{aligned} \tag{2.7.6}$$

The following, slightly modified structure functions are used in addition to the structure functions W_i: $F^{abr}_1(\nu, q^2) = W^{abr}_1(\nu, q^2)$, $F^{abr}_{2,3}(\nu, q^2) = \nu W^{abr}_{2,3}(\nu, q^2)/M^2_{\mathcal{N}}$, since these have a more simple scaling behaviour (see Eqn. (1.4.5)). The inverse of the decomposition into covariants (2.7.6) reads for Q^2, $\nu \gg M^2_{\mathcal{N}}$ $(Q^2 = -q^2)$:

$$P^\mu P^\nu W^{abr}_{\mu\nu} = \frac{Q^2}{8x^3}(F^{abr}_2 - 2xF^{abr}_1) =: \frac{Q^2}{8x^3} F^{abr}_L,$$

$$\left(-g^{\mu\nu} + 12x^2 \frac{P^\mu P^\nu}{Q^2}\right) W^{abr}_{\mu\nu} = \frac{1}{x} F^{abr}_2, \quad i\varepsilon^{\mu\nu\alpha\beta} P_\alpha q_\beta W^{abr}_{\mu\nu} = -r\nu F^{abr}_3. \tag{2.7.7}$$

If the Bjorken limit $\nu \to \infty$, $x = -q^2/2\nu$ = constant in Eqn. (2.7.6) is considered in the rest system of the target nucleon $P^\mu = (M_{\mathcal{N}}, \mathbf{0})$, $q^\mu = [\nu, 0, 0, \nu\sqrt{(1+2xM_{\mathcal{N}}^2/\nu)}]/M_{\mathcal{N}}$, then: $\varepsilon^{iqy} \sim \exp[i\nu(y^0 - y^3)/M_{\mathcal{N}}]$. It follows from the method of stationary phase that for $\nu \to \infty$, the behaviour of $\langle \mathcal{N}, P | [j_\mu(y), j_\nu(0)] | \mathcal{N}, P \rangle$ for $y^0 - y^3 \to 0$, i.e. on the light cone is essential.

(a) *Parton model results* are once again *derived field theoretically* as a way of introducing the techniques involved [Br 71a, Fr 71]. Results on the Bjorken limit of the structure functions are obtained from the commutation relations for free quark fields (1.2.22), for calculating the light cone behaviour ($y^2 \approx 0$) of the commutator of quark currents:

$$j_\mu^{ar}(y) = \bar{\psi}(y)\gamma_\mu \frac{1 - r\gamma_5}{2} \frac{\lambda^a}{2} \cdot \mathbf{1} \cdot \psi(y),$$

where

$$a = 1, 2, \ldots, N_F^2 - 1,$$
$$\mathbf{1} = 1\text{-matrix in colour space}, \tag{2.7.8}$$
$$r = \pm 1.$$

The result is as follows [Gr 71]:

$$\begin{aligned}
[j_\mu^{ar}(y), j_\nu^{br'}(0)] \underset{y^2 \approx 0}{\approx} & \left(\partial^\rho \frac{\varepsilon(y^0)}{2\pi} \delta(y^2)\right) \delta_{rr'} \\
& \times \left[\tfrac{1}{2} s_{\mu\rho\nu\alpha} \frac{1}{2} \left(\bar{\psi}(y)\gamma^\alpha \frac{1 - r\gamma_5}{2} \frac{\lambda^a}{2} \frac{\lambda^b}{2} \psi(0)\right.\right. \\
& \left. - \bar{\psi}(0)\gamma^\alpha \frac{1 - r\gamma_5}{2} \frac{\lambda^b}{2} \frac{\lambda^a}{2} \psi(y)\right) \\
& + \frac{i}{2} r\varepsilon_{\mu\rho\nu\alpha} \frac{1}{2} \left(\bar{\psi}(y)\gamma^\alpha \frac{1 - r\gamma_5}{2} \frac{\lambda^a}{2} \frac{\lambda^b}{2} \psi(0)\right. \\
& \left.\left. + \bar{\psi}(0)\gamma^\alpha \frac{1 - r\gamma_5}{2} \frac{\lambda^b}{2} \frac{\lambda^a}{2} \psi(y)\right)\right] \\
= & \left(\partial^\rho \frac{\varepsilon(y^0)}{2\pi} \delta(y^2)\right) \delta_{rr'} \cdot \tfrac{1}{2} [s_{\mu\rho\nu}{}^\alpha j_\alpha^{abr,A}(y, 0) \\
& + ir\varepsilon_{\mu\rho\nu}{}^\alpha j_\alpha^{abr,S}(y, 0)], \qquad (2.7.9)
\end{aligned}$$

with

$$s_{\mu\rho\nu\alpha} = g_{\mu\rho}g_{\nu\sigma} + g_{\mu\sigma}g_{\nu\rho} - g_{\mu\nu}g_{\rho\sigma},$$
$$\varepsilon_{\mu\rho\nu\alpha} = \text{total antisymmetric tensor}, \quad \varepsilon_{0123} = +1.$$

This expression is composed of the light cone singularity of the free quark commutator (see Eqns. (1.2.24), (1.2.28)) and the bi-local operators

$$j_\mu^{abr,\hat{s}}(x, y) = \tfrac{1}{2}\bar{\psi}(x)\gamma_\mu \frac{1 - r\gamma_5}{2} \frac{\lambda^a}{2} \frac{\lambda^b}{2} \psi(y) \mp \tfrac{1}{2}\bar{\psi}(y)\gamma_\mu \frac{1 - r\gamma_5}{2} \frac{\lambda^b}{2} \frac{\lambda^a}{2} \psi(x) \tag{2.7.10}$$

For the sake of simplicity, the abbreviation $f=(a, b, r)$ is now introduced; colour indices are not written down.

The Bjorken limit of the structure functions can now be determined. To this end, Eqn. (2.7.9) is substituted in Eqn. (2.7.5):

$$W^f_{\mu\nu}(\nu, q^2)=\int\frac{d^4y}{(2\pi)^4}\,e^{iqy}\left(\partial^\rho\,\frac{\varepsilon(y^0)}{2\pi}\,\delta(y^2)\right)$$

$$\times\frac{1}{2}\sum_{\text{spin}}\tfrac{1}{2}\{s_{\mu\rho\nu}{}^{\alpha}\langle\mathcal{N}, P|\,j^{f,A}_\alpha(y,0)\,|\mathcal{N}, P\rangle$$

$$+ir\varepsilon_{\mu\rho\nu}{}^{\alpha}\langle\mathcal{N}, P|\,j^{f,S}_\alpha(y,0)\,|\mathcal{N}, P\rangle$$

and the spin-averaged matrix element of the bi-local operator $j^f_\mu(y, 0)$ is decomposed into Lorentz covariants:

$$(2\pi)^{-3}\frac{1}{2}\sum_{\text{spin}}\langle\mathcal{N}, P|\,j^f_\alpha(y,0)\,|\mathcal{N}, P\rangle=2P_\alpha\tilde{g}^f(Py, y^2)+y_\alpha\tilde{h}^f(Py, y^2). \tag{2.7.11}$$

As only the first term is important for small distances $y^2=0$, its Fourier transform is introduced:

$$\tilde{g}^f(Py, 0)=\int_{-\infty}^{+\infty}d\xi g^f(\xi)e^{i\xi(Py)}. \tag{2.7.12}$$

By using the formula:

$$\int\frac{d^4y}{(2\pi)^4}\,e^{iqy}\varepsilon(y^0)\,\delta(y^2)=\frac{+i}{(2\pi)^2}\,\varepsilon(q^0)\,\delta(q^2)$$

the following result emerges for the hadronic tensor:

$$W^f_{\mu\nu}(\nu, q^2)=\int\frac{d^4y}{(2\pi)^2}\int_{-\infty}^{+\infty}d\xi e^{iy(q+\xi P)}(\partial^\rho\varepsilon(y^0)\,\delta(y^2))$$

$$\times P^\alpha[s_{\mu\rho\nu\alpha}g^{f,A}(\xi)+ir\varepsilon_{\mu\rho\nu\alpha}g^{f,S}(\xi)]$$

$$=-i\int_{-\infty}^{+\infty}d\xi i(q+\xi P)^\rho\varepsilon(q^0+\xi P^0)\,\delta((q+\xi P)^2)P^\alpha$$

$$\times[s_{\mu\rho\nu\alpha}g^{f,A}(\xi)+ir\varepsilon_{\mu\rho\nu\alpha}g^{f,S}(\xi)].$$

Inserting intermediate states to provide a more thorough investigation of the spin-averaged matrix element of the bi-local operators shows that $g^f(\xi)$ differs from zero only for $0\leqslant\xi\leqslant 1$. Thus, integration only has to take place between $\xi=0$ and $\xi=+1$. Moreover, as a result of $q^0+\xi P^0\geqslant 0$:

$$\varepsilon(q^0+\xi P^0)\,\delta((q+\xi P)^2)=\delta(q^2+2\xi\nu+\xi^2M^2_{\mathcal{N}})\approx\delta(q^2+2\xi\nu)$$

$$=\frac{1}{2\nu}\,\delta(\xi-x)\quad\text{with } x=-q^2/2\nu. \tag{2.7.13}$$

Thus, the Bjorken limit for the hadronic tensor can be calculated as

$$W^f_{\mu\nu}(\nu, q^2) = \frac{1}{2\nu}(q+xP)^\rho P^\alpha [s_{\mu\rho\nu\alpha} g^{f,A}(x) + \mathrm{i} r \varepsilon_{\mu\rho\nu\alpha} g^{f,S}(x)]$$

$$= \left(-g_{\mu\nu} + \frac{q_\mu q_\nu}{q^2}\right)\tfrac{1}{2}g^{f,A}(x)$$

$$+ \left(P_\mu - q_\mu \frac{qP}{q^2}\right)\left(P_\nu - q_\nu \frac{qP}{q^2}\right)\frac{x}{\nu}\, g^{f,A}(x) \tag{2.7.14}$$

$$+ \frac{\mathrm{i}}{2} r \varepsilon_{\mu\nu\rho\sigma} q^\rho P^\sigma \frac{1}{\nu}\, g^{f,S}(x).$$

A comparison with Eqn. (2.7.6) gives the following results for the structure functions:

$$\begin{aligned} \text{Bj-lim}\, F^f_1(\nu, q^2) &= F^f_1(x) = \tfrac{1}{2} g^{f,A}(x), \\ \text{Bj-lim}\, F^f_2(\nu, q^2) &= F^f_2(x) = x g^{f,A}(x), \\ \text{Bj-lim}\, F^f_3(\nu, q^2) &= F^f_3(x) = g^{f,S}(x). \end{aligned} \tag{2.7.15}$$

This shows that in the Bjorken limit a free quark field type of light cone singularity gives Bjorken scaling, i.e. that F_1, F_2, and F_3 are scale-invariant functions. The Callan–Gross relation $F_2(x) = 2xF_1(x)$ emerges once again. In addition, a field-theoretical definition of the parton distribution functions is obtained: these are the Fourier transforms $g^f(x)$ of the reduced matrix elements of the bi-local operators on the light cone.

(b) If the bi-local operator $j^f_\mu(y, 0)$ is expanded in a power series in y:

$$\begin{aligned} j^{abr,{A \atop S}}_\mu(y, 0) &= \sum_{n=0}^{\infty} \frac{1}{n!} y^{\mu_1} \ldots y^{\mu_n} \left\{ \bar\psi(0) \overleftarrow{\partial}_{\mu_1} \ldots \overleftarrow{\partial}_{\mu_n} \gamma_\mu \frac{1 - r\gamma_5}{2} \frac{\lambda^a}{2} \frac{\lambda^b}{2} \psi(0) \right. \\ &\quad \left. \mp \bar\psi(0) \overrightarrow{\partial}_{\mu_1} \ldots \overrightarrow{\partial}_{\mu_n} \gamma_\mu \frac{1 - r\gamma_5}{2} \frac{\lambda^b}{2} \frac{\lambda^a}{2} \psi(0) \right\} \\ &= \sum_{n=0}^{\infty} \frac{1}{n!} y^{\mu_1} \ldots y^{\mu_n} \mathcal{O}^{abr,{A \atop S}}_{\mu_1 \ldots \mu_n \mu}(0) \end{aligned} \tag{2.7.16}$$

and this is then inserted in Eqn. (2.7.9), the *Wilson expansion for the free current commutator*, i.e. its representation by singular functions, Lorentz tensors and local operators ('Wilson operators') is obtained [Wi 69]:

$$\begin{aligned} [j_\mu(y), j_\nu(0)]^f \underset{y^2 \approx 0}{\approx} &\sum_{n=0}^{\infty} \frac{1}{2} \left(\partial^\rho \frac{\varepsilon(y^0)}{2\pi} \delta(y^2) \right) s_{\mu\rho\nu}{}^{\alpha} \frac{y^{\mu_1} \ldots y^{\mu_n}}{n!} \mathcal{O}^{f,A}_{\mu_1 \ldots \mu_n \alpha}(0) \\ &+ \sum_{n=0}^{\infty} \frac{\mathrm{i}}{2} r \left(\partial^\rho \frac{\varepsilon(y^0)}{2\pi} \delta(y^2) \right) \varepsilon_{\mu\rho\nu}{}^{\alpha} \frac{y^{\mu_1} \ldots y^{\mu_n}}{n!} \mathcal{O}^{f,S}_{\mu_1 \ldots \mu_n \alpha}(0). \end{aligned} \tag{2.7.17}$$

The mass dimension of the local operators $\mathcal{O}^{f,K}_{\mu_1\ldots\mu_n\mu}(0)$, $K=A, S$, can be calculated as $\mathcal{O}_n = n+3$ from the above equation. On the other hand, they are tensors of rank $(n+1)$ and thus contain the maximum angular momentum $l_{\max}=n+1$. The difference between dimension and angular momentum of an operator is called twist

$$t = \dim \mathcal{O}_n - l. \tag{2.7.18}$$

Hence, the operators $\mathcal{O}_n$ have minimal twist 2. When the spin-averaged nucleon matrix element of the Wilson operators is decomposed into covariants and reduced matrix elements:

$$\begin{aligned}(2\pi)^{-3}\frac{1}{2}\sum_{\text{spin}} \langle \mathcal{N}, P|\, \mathcal{O}^{f,K}_{\mu_1\ldots\mu_n\mu}(0)\,|\mathcal{N}, P\rangle \\ = \mathrm{i}^n P_{\mu_1}\ldots P_{\mu_n}P_\mu B^{f,K}_n + g_{\mu_1\mu_2}M^2_{\mathcal{N}}P_{\mu_3}\ldots P_{\mu_n}P_\mu C^{f,K}_{n,1} + \ldots\end{aligned} \tag{2.7.19}$$

then the terms $g_{\mu_1\mu_2}M^2_{\mathcal{N}}$ have a greater twist $t \geqslant 4$. In the Bjorken limit, they contribute $(M^2_{\mathcal{N}}/\nu)C^{f,K}_{n,i}$, to the structure functions, their effect is negligible when compared with the leading terms with twist 2, i.e. $B^{f,K}_n$.

A comparison between Eqns. (2.7.11) and (2.7.16), (2.7.19) shows that the coefficients B^f_n are obtained from the invariant functions $\tilde{g}^f(Py, 0)$ by power series expansion:

$$B^{f,K}_n = 2\left(\frac{\mathrm{d}}{\mathrm{d}(\mathrm{i}Py)}\right)^n \tilde{g}^{f,K}(Py, 0)|_{Py=0}$$

If the Fourier representtion of $\tilde{g}$ and the relation of Eqn. (2.7.15) with the structure functions is inserted, then:

$$B^{f,K}_n = 2\left(\frac{\mathrm{d}}{\mathrm{d}(\mathrm{i}Py)}\right)^n \int_0^1 \mathrm{d}x g^{f,K}(x)\mathrm{e}^{\mathrm{i}Py\cdot x} = 2\int_0^1 \mathrm{d}x x^n g^{f,K}(x)$$

or

$$B^{f,A}_n = \int_0^1 \mathrm{d}x x^n F^f_1(x) = \frac{1}{2}\int_0^1 \mathrm{d}x x^{n-1} F^f_2(x),$$

$$B^{f,S}_n = \int_0^1 \mathrm{d}x x^n F^f_3(x). \tag{2.7.20}$$

This is the result: the reduced matrix elements B^f_n of the Wilson operators $\mathcal{O}^f_{\mu i}(0)$ determine the structure function moments in the Bjorken limit. Only operators with twist 2 make a non-vanishing contribution in this limit.

(c) The first step towards obtaining QCD results for structure functions is a generalization of the *Wilson expansion for interacting fields* [Wi 69]. This represents the current commutator by singular functions—Wilson coefficients—and local operators $\mathcal{O}^{\{\mu_1\ldots\mu_n\}}_{f,i}$—Wilson operators. The Wilson coefficients $C_{n,i}^{f,K}(y^2, g, \mathcal{M})$ are now functions of the coupling constant g and the renormalization point $\mathcal{M}$. They are at least as singular as those in a free theory. For the sake of simplicity, the Fourier transformed functions of the current commutator are directly expanded:

$$\int \frac{d^4y}{(2\pi)^4} e^{iqy}[j_\mu(y), j_\nu(0)]^f$$

$$= \frac{(2\pi)^{-2}}{2} \sum_{n=2}^{\infty} \sum_i \Big[(-g_{\mu\nu} q_{\mu_1} q_{\mu_2} + g_{\mu\mu_1} q_\nu q_{\mu_2} + g_{\nu\mu_2} q_\mu q_{\mu_1}$$

$$- g_{\mu\mu_1} g_{\nu\mu_2} q^2) C_{n,i}^{f,A}(Q^2, g, \mathcal{M})$$

$$- \frac{i}{2} \varepsilon_{\mu\nu\alpha\beta} g_{\alpha\mu_1} q_\beta q_{\mu_2} C_{n,i}^{f,S}(Q^2, g, \mathcal{M})$$

$$+ \left(g_{\mu\nu} - \frac{q_\mu q_\nu}{q^2} \right) q_{\mu_1} q_{\mu_2} C_{n,i}^{f,L}(Q^2, g, \mathcal{M}) \Big] \{ q_{\mu_3} \ldots q_{\mu_n} \} \mathcal{O}_{f,i}^{\{\mu_1 \ldots \mu_n\}}(0) \left(\frac{2}{Q^2} \right)^n . \qquad (2.7.21)$$

The Wilson operators $\mathcal{O}_{f,i}^{\{\mu_1 \ldots \mu_n\}}(0)$ are defined so that they become transformed in accordance with the irreducible representation $(n/2, n/2)$ of the homogenous Lorentz group ({ } indicates symmetrization and subtraction of all trace terms [Ca 71]). The index i serves to distinguish operators with equal f, n. The factor $(2/Q^2)^n$ is taken from the coefficient functions $C_{n,i}^{f,K}(Q^2, g, \mathcal{M})$ thus making them dimensionless. Diagrams of order g^2 in Fig. 2.24(b)–(d) and those of higher orders give a non-vanishing longitudinal cross-section, i.e. $F_L = F_2 - 2xF_1 \neq 0$. The final term in Eqn. (2.7.21) takes this fact into account. The spin-averaged matrix element of the Wilson operators can be represented by Lorentz covariants and unknown coefficients determined by the proton structure:

$$(2\pi)^{-3} \frac{1}{2} \sum_{\text{spin}} \langle \mathcal{N}, P | \mathcal{O}_{f,i}^{\{\mu_1 \ldots \mu_n\}}(0) | \mathcal{N}, P \rangle = \{ P_{\mu_1} \ldots P_{\mu_n} \} B_{n,i}^f(g, \mathcal{M}). \qquad (2.7.22)$$

Equations (2.7.16), (2.7.21) and (2.7.22) give

$$W_1^f(Q^2, x) = F_1^f(Q^2, x) = \frac{\pi}{2} \sum_{n,i} x^{-n} (C_{n,i}^{f,A}(Q^2, g, \mathcal{M}) - C_{n,i}^{f,L}(Q^2, g, \mathcal{M}) B_{n,i}^f,$$

$$\frac{\nu}{M_{\mathcal{N}}^2} W_2^f(Q^2, x) = F_2^f(Q^2, x) = \pi \sum_{n,i} x^{-n+1} C_{n,i}^{f,A}(Q^2, g, \mathcal{M}) B_{n,i}^f, \qquad (2.7.23)$$

$$\frac{\nu}{M_{\mathcal{N}}^2} W_3^f(Q^2, x) = F_3^f(Q^2, x) = \pi \sum_{n,i} x^{-n} C_{n,i}^{f,A}(Q^2, g, \mathcal{M}) B_{n,i}^f$$

and conversely [Ch 72]:

$$M^f_{1,n+1}(Q^2)=\int_0^1 dx x^{n-1} F^f_1(Q^2,x)$$
$$=\tfrac{1}{2}\sum_i (C^{f,A}_{n,i}(Q^2,g,\mathcal{M})-C^{f,L}_{n,i}(Q^2,g,\mathcal{M})B^f_{n,i}),$$
$$M^f_{2,n}(Q^2)=\int_0^1 dx x^{n-2} F^f_2(Q^2,x)=\sum_i C^{f,A}_{n,i}(Q^2,g,\mathcal{M})B^f_{n,i}, \tag{2.7.24}$$
$$M^f_{3,n+1}(Q^2)=\int_0^1 dx x^{n-1} F^f_3(Q^2,x)=\sum_i C^{f,S}_{n,i}(Q^2,g,\mathcal{M})B^f_{n,i}, \quad n\geq 2$$
$$M^f_{L,n}(Q^2)=\int_0^1 dx x^{n-2} F_L(Q^2,x)=\sum_i C^{f,L}_{n,i}(Q^2,g,\mathcal{M})B^f_{n,i}.$$

The structure function moments determine the Wilson coefficients for the current product and the nucleon matrix elements for the Wilson operators. In contrast to the factors $B^f_{n,i}$ which cannot be calculated in perturbation theory, the Q^2 dependence of the coefficients can be described in QCD by means of perturbation theory and the Gell–Mann–Low equations [Gr 73].

With regard to flavour symmetry, the current commutator $[j_\mu(y), j_\nu(0)]^f$ can be split up into a non-singlet component which transforms according to the adjoint representation of the flavour group, and a singlet component. Other contributions are unimportant in the Bjorken limit. The non-singlet QCD operators with twist 2 are

$$\mathcal{O}^{\{\mu_1\ldots\mu_n\}}_{f,r,NS}(0)=\left\{\bar{\psi}(0)\gamma^{\mu_1}\frac{1-r\gamma_5}{2}D^{\mu_2}\ldots D^{\mu_n}\frac{\lambda^f}{2}\psi(0)\right\}, \quad f=1,2,\ldots,N_F^2-1. \tag{2.7.25}$$

The colour-covariant derivative D^μ appears here because of gauge invariance. In the singlet case, two sets of Wilson operators with twist 2 exist:

$$\mathcal{O}^{\{\mu_1\ldots\mu_n\}}_{r,\psi}(0)=\left\{\bar{\psi}(0)\gamma^{\mu_1}\frac{1-r\gamma_5}{2}D^{\mu_2}\ldots D^{\mu_n}\cdot 1\cdot\psi(0)\right\}, \tag{2.7.26}$$
$$\mathcal{O}^{\{\mu_1\ldots\mu_n\}}_{G}(0)=\{F^{\mu_1}_{\alpha}(0)D^{\mu_2}\ldots D^{\mu_{n-1}}F^{\alpha\mu_n}(0)\}.$$

$F_{\mu\nu}(0)$ is the gluon field strength tensor. Colour indices are omitted in the above equations.

Quark currents j^f_μ and Wilson operators $\mathcal{O}^{\{\mu_1\ldots\mu_n\}}_{f,i}$ are local operators based on quark and gluon fields to which the multiplicative renormalization procedure of QCD (Section 2.4.3(b)) [Di 74, Jo 76] can be generalized. However, mixing of operators $\mathcal{O}_{n,i}$ occurs when calculating the corresponding loop diagrams, and hence a renormalization constant matrix $(Z_{\mathcal{O}_n})_{ik}$ links the renormalized Wilson operators $\mathcal{O}^R_{n,i}(g,\mathcal{M})$ with the non-renormalized ones:

$$j_\mu^R(g, \mathcal{M}) = \lim_{\varepsilon\to 0} Z_j^{-1}(g, \mathcal{M}, \varepsilon) j_\mu,$$
$$\mathcal{O}_{n,i}^R(g, \mathcal{M}) = \lim_{\varepsilon\to 0} \sum_k [Z_{\mathcal{O}_n}(g, \mathcal{M}, \varepsilon)]_{ik}^{-1} \mathcal{O}_{n,k}. \tag{2.7.27}$$

As quark currents and Wilson operators undergo multiplicative renormalization, the renormalized vertex functions $\Gamma_{\psi\bar{\psi}jj}$, $\Gamma_{\psi\bar{\psi}\mathcal{O}_{n,i}}$ derived from their Green functions, e.g.

$$\tau_{\psi\bar{\psi}jj} = \langle 0|\, T\psi(x_1)[j_\mu(x), j_\nu(0)]^f \bar{\psi}(x_2)\, |0\rangle,$$
$$\tau_{\psi\bar{\psi}\mathcal{O}_{n,i}} = \langle 0|\, T\psi(x_1)\mathcal{O}_{f,i}^{\{\mu_1\dots\mu_n\}}(0)\bar{\psi}(x_2)\, |0\rangle \tag{2.7.28}$$

satisfy the RG equations:

$$\left(\mathcal{M}\frac{\partial}{\partial \mathcal{M}} + \beta(g)\frac{\partial}{\partial g} - 2\gamma_\psi - 2\gamma_j\right)\Gamma_{\psi\bar{\psi}jj} = 0, \tag{2.7.29}$$

$$\sum_k \left[\left(\mathcal{M}\frac{\partial}{\partial \mathcal{M}} + \beta(g)\frac{\partial}{\partial g} - 2\gamma_\psi\right)\delta_{ik} - \gamma_{n,ik}\right]\Gamma_{\psi\bar{\psi}\mathcal{O}_{n,k}} = 0. \tag{2.7.29$'$}$$

This defines the anomalous dimensions in a way similar to Eqn. (2.5.5):

$$\gamma_j(g) = \tfrac{1}{2} \lim_{\varepsilon\to 0} \mathcal{M} Z_j^{-1} \frac{\partial}{\partial \mathcal{M}} Z_j,$$
$$\gamma_{n,ik}(g) = \tfrac{1}{2} \lim_{\varepsilon\to 0} \mathcal{M} \sum_l (Z_{\mathcal{O}_n}^{-1})_{il} \frac{\partial}{\partial \mathcal{M}} (Z_{\mathcal{O}_n})_{lk}. \tag{2.7.30}$$

The anomalous dimensions of the currents $\gamma_j = 0$ vanish because the changes of the quark currents are generators of the chiral flavour symmetry group. The RG equations (2.7.29), (2.7.29′) are linked together by the Wilson expansion and from this an RG equation for the Wilson coefficients can be derived. Eqn. (2.7.21) can be written in a more simple form, using only those indices which are absolutely necessary and this then helps to clarify the important steps in the following calculations;

$$[j, j] = \sum_{n,i} C_{n,i}(Q^2, g, \mathcal{M})\mathcal{O}_{n,i}(g, \mathcal{M}). \tag{2.7.21$'$}$$

Using Eqn. (2.7.29′) and substituting the above in Eqns. (2.7.28) and (2.7.29) gives the following:

$$\begin{aligned}
0 &= \left(\mathcal{M}\frac{\partial}{\partial \mathcal{M}} + \beta(g)\frac{\partial}{\partial g} - 2\gamma_\psi\right)\sum_{n,i} C_{n,i}(Q^2, g, \mathcal{M})\Gamma_{\psi\bar{\psi}\mathcal{O}_{n,i}}(g, \mathcal{M}) \\
&= \sum_{n,i} \Gamma_{\psi\bar{\psi}\mathcal{O}_{n,i}}(g, \mathcal{M})\left(\mathcal{M}\frac{\partial}{\partial \mathcal{M}} + \beta(g)\frac{\partial}{\partial g} - 2\gamma_\psi\right)C_{n,i}(Q^2, g, \mathcal{M}) \\
&\quad + \sum_{n,i} C_{n,i}(Q^2, g, \mathcal{M})\left(\mathcal{M}\frac{\partial}{\partial \mathcal{M}} + \beta(g)\frac{\partial}{\partial g}\right)\Gamma_{\psi\bar{\psi}\mathcal{O}_{n,i}}(g, \mathcal{M}) \\
&= \sum_{n,i,k} \Gamma_{\psi\bar{\psi}\mathcal{O}_{ni}}(g, \mathcal{M})\left[\left(\mathcal{M}\frac{\partial}{\partial \mathcal{M}} + \beta(g)\frac{\partial}{\partial g}\right)\delta_{ik} + \gamma_{n,ki}\right]C_{n,k}(Q^2, g, \mathcal{M}).
\end{aligned} \tag{2.7.31}$$

Due to Lorentz invariance, this equation must be fulfilled for each n separately:

$$0 = \sum_k \left[\left(\mathcal{M} \frac{\partial}{\partial \mathcal{M}} + \beta(g) \frac{\partial}{\partial g} \right) \delta_{ik} + \gamma_{n,ki} \right] C^{f,K}_{n,k}(Q^2, g, \mathcal{M}). \qquad (2.7.32)$$

As the Wilson coefficients are dimensionless, the associated scale transformation equation can be written:

$$\sum_k \left[\left(\lambda \frac{\partial}{\partial \lambda} + \beta(g) \frac{\partial}{\partial g} \right) \delta_{ik} + \gamma_{n,ki}(g) \right] C^{f,K}_{n,k}\left(\frac{Q^2}{\lambda}, g \right) = 0. \qquad (2.7.33)$$

This RG equation is a coupled system of linear, partial differential equations. As in Section 2.5.1, the equation an be solved by using the running coupling constant $\bar{g}(Q^2, g)$ (Eqn. (2.5.9))

$$\begin{aligned} C^{f,K}_{n,i}(Q^2, g) &= C^{f,K}_{n,k}(1, \bar{g}(Q^2)) \\ &\quad \times \left(\delta_{ki} + \int_g^{\bar{g}(Q^2)} \frac{dg'}{\beta(g')} \gamma_{n,ki}(g') \right. \\ &\quad \left. + \int_g^{\bar{g}(Q^2)} \frac{dg'}{\beta(g')} \int_g^{g'} \frac{dg''}{\beta(g'')} (\gamma_n(g')\gamma_n(g''))_{ki} + \ldots \right) \\ &=: \sum_k C^{f,K}_{n,k}(1, \bar{g}(Q^2)) \left[T_g \exp\left(\int_g^{\bar{g}(Q^2)} \frac{dg'}{\beta(g')} \gamma_n(g') \right) \right]_{ki} \end{aligned} \qquad (2.7.34)$$

in which T_g is an ordering of the non-commutative matrix multiplication of $\gamma_n(g)$ in analogy to Eqns. (2.2.20) and (2.2.21). Therefore:

$$M^f_{2,n}(Q^2) = \sum_{i,k} C^{f,A}_{n,k}(1, \bar{g}) \left[T_g \exp\left(\int \gamma_n / \beta \right) \right]_{ki} B^{f,A}_{n,i}. \qquad (2.7.35)$$

applies for the structure function moments. Their Q^2 dependence and thus breaking of scaling is determined by the anomalous dimensions of the Wilson operators and the RG function of QCD.

(d) Evaluation of Eqn. (2.7.35) using perturbative techniques is based on the power series expansions (2.5.26) and (2.5.27):

$$\beta(g) = -\beta_0 \frac{g^3}{16\pi^2} - \beta_1 \frac{g^5}{(16\pi^2)^2} + \ldots \qquad (2.5.26)$$

$$\gamma_n(g) = \gamma_n^{(0)} \frac{g^2}{16\pi^2} + \gamma_n^{(1)} \left(\frac{g^2}{16\pi^2} \right)^2 + \ldots \qquad (2.7.36)$$

The Wilson coefficients $C^{f,K}_n(1, \bar{g})$ are expanded around the parton model value:

$$\left.\begin{aligned} C^{f,K}_{n,NS}(1,\bar{g}) &= \delta^{f,K}_{NS}\left(1+\frac{\bar{g}^2}{16\pi^2}\,\varepsilon^{f,K}_{n,NS}+\ldots\right), \quad K=S,A \\ C^{f,L}_{n,NS}(1,\bar{g}) &= \delta^{f,L}_{NS}\left(0+\frac{\bar{g}^2}{16\pi^2}\,\varepsilon^{f,L}_{n,NS}+\ldots\right) \end{aligned}\right\} \text{non-singlet}$$

$$\left.\begin{aligned} C^{K}_{n,\psi}(1,\bar{g}) &= \delta^{K}_{\psi}\left(1+\frac{\bar{g}^2}{16\pi^2}\,\varepsilon^{f,K}_{n,\psi}+\ldots\right), \quad K=S,A \\ C^{L}_{n,\psi}(1,\bar{g}) &= \delta^{L}_{\psi}\left(0+\frac{\bar{g}^2}{16\pi^2}\,\varepsilon^{L}_{n,\psi}+\ldots\right). \\ C^{K}_{n,G}(1,\bar{g}) &= \delta^{K}_{G}\left(0+\frac{\bar{g}^2}{16\pi^2}\,\varepsilon^{K}_{n,G}+\ldots\right), \quad K=S,A,L \end{aligned}\right\} \text{singlet} \tag{2.7.37}$$

As $F_L=0$ in the parton model and the gluons do not couple directly to the flavour currents, the expansion of the corresponding quantities begins with the $\bar{g}^2$ terms.

Eqns. (2.7.35)–(2.7.37) form the starting point for calculating the *behaviour of structure function moments.* Calculations are made below *for the leading order*, i.e. using the first term of the above expansion. There is only one Wilson operator of order n, i.e. a single RG equation for the non-singlet structure functon moments and the parity-violating singlet structure functions $F_3(x, Q^2)$. Hence:

$$M^f_{2,n}(Q^2)=\int_0^1 dx x^{n-2}F^f_{2,NS}(x,Q^2)=\delta^{f,A}_{NS}B^f_{n,NS}\left(\log\frac{Q^2}{\Lambda^2}\right)^{-d^{NS}_n}, \quad f=1,\ldots,N_F^2-1, \tag{2.7.38}$$

$$M^f_{3,N}(Q^2)=\int_0^1 dx x^{n-3}F^f_3(x,Q^2)=\delta^{f,S}B^f_n\left(\log\frac{Q^2}{\Lambda^2}\right)^{-d^{NS}_n}, \quad f=0,1,\ldots,N_F^2-1,$$

with

$$d^{NS}_n=\frac{\gamma^{(0)}_n}{2\beta_0}.$$

For the singlet case $F^S_2(x, Q^2)$, $\gamma^{(0)}_n(g)$ is proportional to a constant 2×2 matrix:

$$\gamma^{(0)}_n=\begin{pmatrix}\gamma_{n,\psi\psi} & \gamma_{n,\psi G}\\ \psi_{n,G\psi} & \gamma_{n,GG}\end{pmatrix}, \tag{2.7.39}$$

which can be diagonalized:

$$U^{-1}\gamma^{(0)}_n U=\begin{pmatrix}\lambda_{n,-} & 0\\ 0 & \lambda_{n,+}\end{pmatrix}$$

with

$$\lambda_{n,\pm}=\tfrac{1}{2}\{\gamma_{n,\psi\psi}+\gamma_{n,GG}\pm\sqrt{[(\gamma_{n,\psi\psi}-\gamma_{n,GG})^2+4\gamma_{n,\psi G}\gamma_{n,G\psi}]}\} \tag{2.7.40}$$

and

$$U=\begin{pmatrix}\gamma_{n,\psi\psi}-\lambda_{n,+} & \gamma_{n,\psi\psi}-\lambda_{n,-}\\ \gamma_{n,\psi G} & \gamma_{n,\psi G}\end{pmatrix}\cdot\frac{1}{\lambda_{n,-}-\lambda_{n,+}}.$$

This decouples the two differential equations which can be solved using Eqn. (2.7.37):

$$\int_0^1 \mathrm{d}x x^{n-2}F_2^S(x,Q^2)=\delta_{\psi}^{S,2}\left[B_n^{S,-}\left(\log\frac{Q^2}{\Lambda^2}\right)^{-d_{n,-}}+B_n^{S,+}\left(\log\frac{Q^2}{\Lambda^2}\right)^{-d_{n,+}}\right] \tag{2.7.41}$$

with

$$d_{n,\pm}=\frac{\lambda_{n,\pm}}{2\beta_0}.$$

The QCD Feynman rules must be extended for the calculation of Green functions of composite fields (see Eqn. (2.3.22)) in order to calculate the anomalous dimensions of the operators $\mathcal{O}_{n,i}^{\{\mu_1\ldots\mu_n\}}(0)$ in order g^2. The following field points are assigned to the Wilson operators in Eqns. (2.7.25) and (2.7.26) (see [Gr 73] for further details):

$$\bar{\psi}(x)\left\{\gamma_{\mu_1}\frac{1-r\gamma_5}{2}\partial_{\mu_2}\ldots\partial_{\mu_n}\right\}\frac{\lambda^f}{2}\psi(x)\rightarrow$$

$$\rightarrow\left\{\gamma_{\mu_1}\frac{1-r\gamma_5}{2}p_{\mu_2}\ldots p_{\mu_n}\right\}\frac{\lambda^f}{2}\delta_{cc'}$$

$$\bar{\psi}(x)\left\{\gamma_{\mu_1}\frac{1-r\gamma_5}{2}\partial_{\mu_2}\ldots g\frac{\lambda^a}{2}A^a_{\mu_i}\ldots\partial_{\mu_n}\right\}\frac{\lambda^f}{2}\psi(x)\rightarrow$$

$$\rightarrow g\frac{\lambda^a_{cc'}}{2}\left\{\gamma_{\mu_1}\frac{1-r\gamma_5}{2}p_{\mu_2}\ldots\varepsilon_{\mu_i}\ldots p_{\mu_n}\right\}\frac{\lambda^f}{2}$$

$$\{F^\alpha_{\mu_1}(x)\,\partial_{\mu_2}\ldots\partial_{\mu_{n-1}}F_{\alpha\mu_n}(x)\}\rightarrow$$

$$\rightarrow\{g_{\mu\nu}k_{\mu_1}\ldots k_{\mu_n}+k^2 g_{\mu\mu_1}g_{\nu\mu_2}k_{\mu_3}\ldots k_{\mu_n}$$
$$-k_\mu g_{\nu\mu_1}k_{\mu_2}\ldots k_{\mu_n}-k_\nu g_{\mu\mu_1}k_{\mu_2}\ldots k_{\mu_n}\} \tag{2.7.42}$$

The following diagrams contribute in 1-loop approximation to the anomalous dimensions of the non-singlet operator:

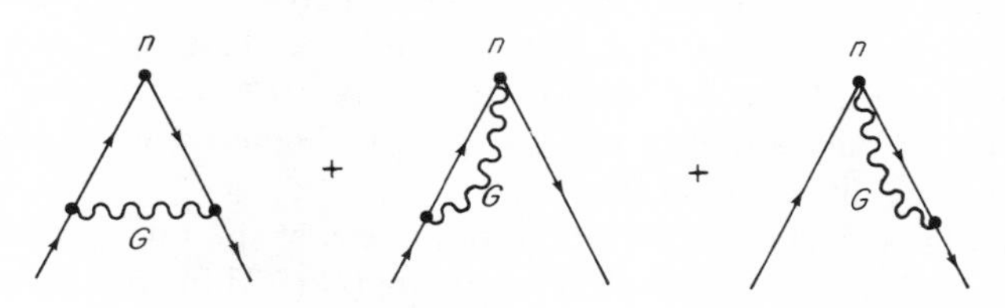

The calculation is similar to that for the gluon self-energy in Section 2.4.2. It was worked out initially by D. J. Gross and F. Wilzek [Gr 73] and gives the following formula for the $SU(3)_C$ group:

$$\gamma_{n,NS}^{(0)} = \frac{8}{3}\left(1 - \frac{2}{n(n+1)} + 4\sum_{2}^{n}\frac{1}{j}\right). \tag{2.7.43}$$

Hence:

$$d_n^{NS} = \frac{4}{3}\frac{1 - \dfrac{2}{n(n+1)} + 4\sum_2^n \dfrac{1}{j}}{11 - 2N_F/3}. \tag{2.7.44}$$

The 2×2 matrix elements for the singlet case have also been worked out in reference [Gr 73]:

$$\begin{aligned}
\gamma_{n,\psi\psi} &= \frac{8}{3}\left(1 - \frac{2}{n(n+1)} 4\sum_{2}^{n}\frac{1}{j}\right) = \gamma_{n,NS}^{(0)}, \\
\gamma_{n,\psi G} &= -4N_F \frac{n^2+n+2}{n(n+1)(n+2)}, \\
\gamma_{n,G\psi} &= -\frac{16}{3}\frac{n^2+n+2}{n(n^2-1)}, \\
\gamma_{n,GG} &= 6\left(\frac{1}{3} - \frac{4}{n(n-1)} - \frac{4}{(n+1)(n+2)} + 4\sum_{2}^{n}\frac{1}{j}\right) + \tfrac{4}{3}N_F.
\end{aligned} \tag{2.7.45}$$

Deviations from the scaling behaviour of the simple parton model for asymptotically free QCD are in the leading order powers in $\log Q^2/\Lambda^2$ whose exponents are represented by Eqn. (2.7.44), (2.7.40), and (2.7.45).

(e) All information on deep inelastic lepton–nucleon scattering in QCD given in the latter section was based on the assumption that $(Q^2, \nu) \gg M_{\mathcal{N}}^2$. Therefore, the calculations were carried out with the following approximations:

(1) vanishing nucleon mass;
(2) vanishing quark mass;
(3) 1-loop approximation;
(4) omission of operators with higher twist.

Experimental results are available for Q^2 values between $1\,\text{GeV}^2$ and $150\,\text{GeV}^2$, i.e. for values which are not necessarily big with respect to $M_{\mathcal{N}}^2$. Calculations have to be carried out beyond the leading order, for the principal reasons outlined in Section 2.5.2. As the strong fine structure constant $\bar{\alpha}_s$ is not very small in this Q^2 region, higher approximations may possibly give a significant contribution.

Kinematic effects connected with the finiteness of the nucleon mass can be considered by using the *Nachtmann moments* [Na 73] instead of the normal ones:

$$M_{2,n}(Q^2)=\int_0^1 dx x^{n-2}F_2(x, Q^2)$$

based on the fact that for $M_{\mathcal{N}} \neq 0$ the expression $(\xi P+q)^2$ in Eqn. (2.7.13) has its zero at

$$\xi=\frac{2x}{1+\surd(1+4x^2M_{\mathcal{N}}^2/Q^2)}, \tag{2.7.46}$$

and not at $\xi=x=Q^2/2\nu$. The moments are then calculated in accordance with the following prescriptions:

$$M_{2,n}(Q^2)=\int_0^1 dx x^{n-2}K_2(x, n, Q^2)F_2(x, Q^2),$$

$$M_{3,n}(Q^2)=\int_0^1 dx x^{n-1}K_3(x, n, Q^2)F_3(x, Q^2)$$

with

$$K_2(n, x, Q^2)=\frac{n^2+2n+3+3(n+1)\surd(1+4x^2M_{\mathcal{N}}^2/Q^2)+n(n+2)4x^2M_{\mathcal{N}}^2/Q^2}{(n+2)(n+3)[\frac{1}{2}+\frac{1}{2}\surd(1+4x^2M_{\mathcal{N}}^2/Q^2)]^{n+1}},$$

$$K_3(n, x, Q^2)=\frac{1+(n+1)\surd(1+4x^2M_{\mathcal{N}}^2/Q^2)}{(n+2)[\frac{1}{2}+\frac{1}{2}\surd(1+4x^2M_{\mathcal{N}}^2/Q^2)]^{n+1}}. \tag{2.7.47}$$

$\xi \to x$ and $K_{2,3} \to 1$ naturally apply for $M_{\mathcal{N}}^2/Q^2 \to 0$. The kinematic effects which occur when a heavy quark is produced from a light one can be considered in the same way [Ge 76, De 77].

Calculations of the *2-loop contributions in QCD* have been carried out independently by several authors for the *β-function* [Ca 74, Jo 74] (see Section 2.5.2(d)), for the Wilson operator *anomalous dimensions* [Fl 77, Go 79] and for the *Wilson coefficients* [Ba 78a, Fl 79a]. For the non-singlet case, results are given in the minimal subtraction scheme below. Summaries given by A. J. Buras [Bu 80] and J. Ellis [El 79a] and original research articles provide further details of both this and the singlet case.

In the 2-loop approximation, the running coupling constant emerges as:

$$\frac{\bar{g}^2(Q^2)}{16\pi^2}=\frac{1}{\beta_0 \log(Q^2/\Lambda^2)}-\frac{\beta_1}{\beta_0^3}\frac{\log\log(Q^2/\Lambda^2)}{[\log(Q^2/\Lambda^2)]^2} \tag{2.5.28}$$

with

$$\Lambda^2=\mathcal{M}^2 \exp\left(-\frac{16\pi^2}{\beta_0 g^2}-(\beta_1/\beta_0^2)\log\frac{\beta_0 g^2}{16\pi^2}\right). \tag{2.5.29}$$

The moments of the non-singlet structure functions $F_2^f(x, Q^2)$, $F_3^f(x, Q^2)$ have the following Q^2 dependence:

$$M_{i,n}^f(Q^2) \sim (\bar{g}^2(Q^2))^{d_n^{NS}}\left[1+\bar{g}^2(Q^2)\left(\frac{\gamma_n^{(1)}}{2\beta_0}-\frac{\beta_1^{(1)}}{\beta_0^{(0)}}d_n^{NS}+\varepsilon_n^{f,i}\right)\right]. \tag{2.7.48}$$

d_n^{NS} is given by Eqn. (2.7.44) and $\varepsilon_n^{f,i}$ by Eqn. (2.7.37). Numerical values for corrections to the anomalous dimensions and explicit expressions for $\varepsilon_n^{f,i}$ (which the authors write as $B_{K,n}^{NS}$) can be found e.g. in [Bu 80].

The minimal subtraction scheme simplifies the calculation of the two-loop graphs. However, other schemes, e.g. the MOM subtraction scheme, are better suited for comparison with experiment. W. Celmaster *et al.* give details of formulae for converting one scheme to another [Ce 79]. Calculations of the contributions from operators with higher twist have not yet been concluded but indications are that they are only important for $Q^2 < 5\ \text{GeV}^2$ [Ab 80].

(f) QCD information on structure function moments has been based on the Wilson expansion. Especially logarithmic corrections to the scaling behaviour of the simple parton model emerged from this. In the parton model, the structure functions are independent of Q^2, i.e. the probability of finding a parton with momentum fraction x in the nucleon does not depend on the resolution ($\sim Q^{-1}$) with which the nucleon is considered. This changes in QCD, in which interaction allows a quark to be converted to a quark and a gluon (Fig. 2.24(b)) and a gluon to become two gluons or a quark–antiquark pair. These fluctuations are observed with frequencies depending on the resolution [Ko 74].

Equations which enable direct statements on the Q^2 dependence of the structure functions are highly desirable. The solution (2.7.38) of the GL equation for non-singlet moments $M_{2,n}(Q^3)$ satisfies the differential equation

$$\frac{\mathrm{d}}{\mathrm{d}t} M_{2,n}^{f}(t) = -d_n^{NS} \quad \text{with } t = \log(Q^2/\Lambda^2).$$

On the other hand $\bar{\alpha}_s(Q^2) = 4\pi/\beta_0 t$ and $d_n^{NS} = \gamma_n^{(0)}/2\beta_0$, apply and thus:

$$\frac{\mathrm{d}}{\mathrm{d}t} M_{2,n}^{f}(t) = -\frac{\gamma_n^{(0)}}{8\pi} \bar{\alpha}_s(t) M_{2,n}^{f}(t).$$

Expressed by the structure function $F_2^f(x, t)$, this means

$$\frac{\mathrm{d}}{\mathrm{d}t} \int_0^1 \mathrm{d}x x^{n-2} F_2^f(x, t) = -\frac{\gamma_n^{(0)}}{8\pi} \bar{\alpha}_s(t) \int_0^1 \mathrm{d}x x^{n-2} F_2(x, t)$$

or with the quark distribution function $F_2^f(x, t) = xq^f(x, t)$ (see Eqn. (1.4.15):

$$\frac{\mathrm{d}}{\mathrm{d}t} \int_0^1 \mathrm{d}x x^{n-1} q^f(x, t) = -\frac{1}{8\pi} \gamma_n^{(0)} \bar{\alpha}_s(t) \int_0^1 \mathrm{d}x x^{n-1} q^f(x, t).$$

This equation is equivalent to an integro-differential equation for $q^f(x, t)$, the Altarelli–Parisi equation [Al 77]:

$$\frac{\partial}{\partial t} q^f(x, t) = \frac{\bar{\alpha}_s(t)}{2\pi} \int_0^1 \mathrm{d}y \int_0^1 \mathrm{d}z\, \delta(x - yz) q^f(y, t) P_{qq}(z)$$

with

$$\gamma_n^{(0)} = -4 \int_0^1 \mathrm{d}x x^{n-1} P_{qq}(x). \tag{2.7.50}$$

In the original variable Q^2, the AP equation is as follows:

$$\begin{aligned} Q^2 \frac{\partial q^f(x, Q^2)}{\partial Q^2} &= \frac{\bar{\alpha}_s(Q^2)}{2\pi} \int_0^1 \mathrm{d}y \int_0^1 \mathrm{d}z\, \delta(x - zy) q^f(y, Q^2) P_{qq}(z) \\ &= \frac{\bar{\alpha}_s(Q^2)}{2\pi} \int_x^1 \frac{\mathrm{d}y}{y} q^f(y, Q^2) P_{qq}(\tfrac{x}{y}). \end{aligned} \tag{2.7.51}$$

The branching or AP function $P_{qq}(z)$ is given by:

$$P_{qq}(z) = \frac{4}{3}\left(\frac{1+z^2}{(1-z)_+} + \tfrac{3}{2}\,\delta(1-z)\right), \tag{2.7.52}$$

in which

$$\int_0^1 \mathrm{d}z f(z) \frac{1}{(1-z)_+} := \int_0^1 \mathrm{d}z \frac{f(z) - f(1)}{1-z}, \quad \int_0^1 \mathrm{d}z f(z)\, \delta(1-z) := f(1).$$

A short calculation shows that their moments are in fact identical with $\gamma_n^{(0)}$ (Eqn. (2.7.43)):

$$\begin{aligned} \gamma_n^{(0)} &= -4 \cdot \frac{4}{3}\left(\int_0^1 \mathrm{d}x \frac{x^{n-1} - 1}{1-x}(1 + x^2) + \frac{3}{2}\int_0^1 \mathrm{d}x x^{n-1}\, \delta(1-x)\right) \\ &= \frac{16}{3}\left(\sum_{j=1}^{n-1} \frac{1}{j} + \sum_{j=3}^{n+1} \frac{1}{j} - \frac{3}{2}\right) = \frac{16}{3}\left(\tfrac{1}{2} + 2\sum_{j=2}^{n} \frac{1}{j} - \frac{1}{n(n+1)}\right). \end{aligned}$$

Section 2.7.3 shows that $P_{qq}(z)$ is proportional to the probability that a quark retains a fraction z of its original moentum after emitting a gluon.

Processes as shown in Fig. 2.24(c), i.e. quark pair creation by a gluon and gluon pair creation by a gluon also contribute to the singlet structure functions. This is expressed in the singlet equation by the fact that the gluon distribution function $G(x, Q^2)$ is coupled with the quark distribution function:

$$\begin{aligned} Q^2 \frac{\partial q^0(x, Q^2)}{\partial Q^2} &= \frac{\bar{\alpha}_s(Q^2)}{2\pi} \int_x^1 \frac{\mathrm{d}y}{y}\left[q^0(y, Q^2) P_{qq}\left(\frac{x}{y}\right)\right. \\ &\qquad \left. + G(y, Q^2) P_{qG}\left(\frac{x}{y}\right)\right], \end{aligned} \tag{2.7.53}$$

$$\begin{aligned} Q^2 \frac{\partial G(x, Q^2)}{\partial Q^2} &= \frac{\bar{\alpha}_s(Q^2)}{2\pi} \int_x^1 \frac{\mathrm{d}y}{y}\left[\sum_f (q^f(y, Q) + \bar{q}^f(y, Q) P_{Gq}\left(\frac{x}{y}\right)\right. \\ &\qquad \left. + G(y, Q^2) P_{GG}\left(\frac{x}{y}\right)\right] \end{aligned}$$

in which

$$P_{qG} = \tfrac{1}{2}[z^2 + (1-z^2)],$$
$$P_{Gq} = \frac{4}{3}\left(\frac{1+(1-z)^2}{z}\right), \tag{2.7.54}$$
$$P_{GG} = 6\left[\frac{z}{(1-z)_+} + \frac{1-z}{z} + z(1-z) + \left(\frac{11}{12} - \frac{N_F}{18}\right)\delta(1-z)\right]$$

are the corresponding branching functions.

The AP equations (2.7.51)–(2.7.54) describe the QCD results for the leading order in deep inelastic lepton scattering, in a concentrated form. They can be solved approximately by making suitable *ansätze* with free parameters for $q_i(x, Q^2)$, $\bar{q}_i(x, Q^2)$, $G(x, Q^2)$ and determining the parameters by substitution, to give phenomenologically useful, Q^2-dependent parton distribution functions [Bu 78].

(g) The QCD results can be compared with current standard experiments on inelastic lepton scattering at high energies and momentum transfers. These are

(1) $e\mathcal{N}$ scattering [Ri 75];
(2) $\mu\mathcal{N}$ scattering [Em 80, Au 82N]; and
(3) $\nu(\bar{\nu})$ scattering [De 79, De 79b].

Fig. 2.25 shows the structure functions $F_2(x, Q^2)$ for $e\mathcal{N}$ and $\nu\mathcal{N}$ scattering by way of an initial example. Deviations from the scaling behaviour of the

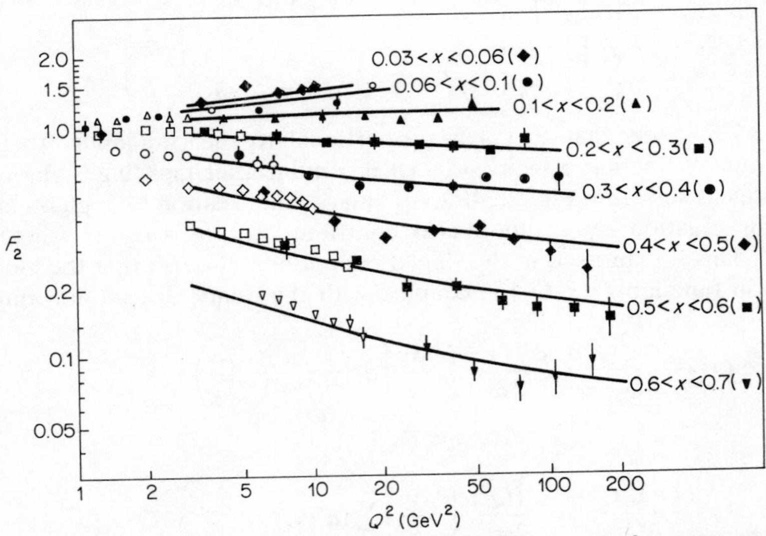

Figure 2.25 Comparison between structure function $F_2(x, Q^2)$ calculated in leading order with $\Lambda = 0.47$ GeV and the CDHS experiment (●, ▲, ■) and the SLAC experiment (○, △, □) $\cdot \frac{18}{5}$ [De 79]

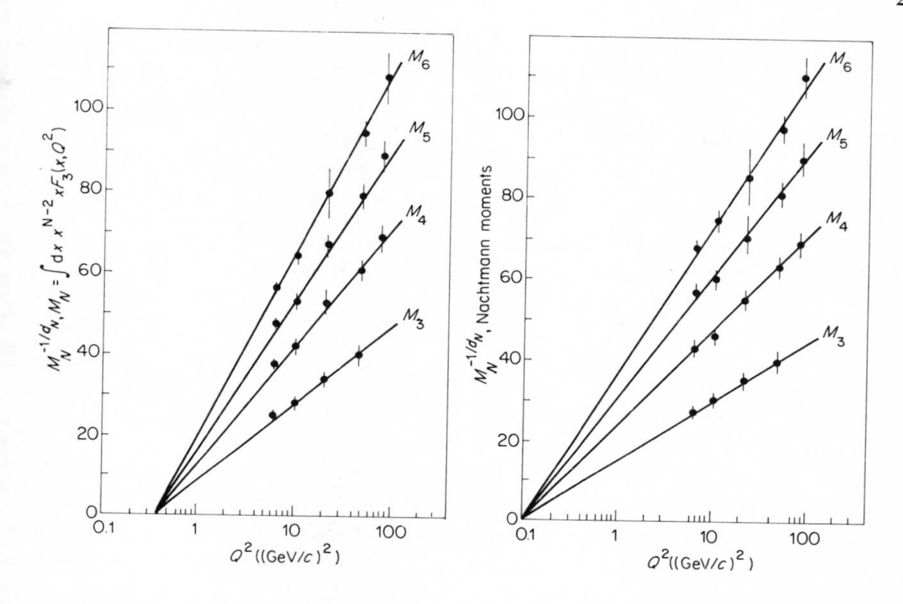

Figure 2.26 Normal and Nachtmann moments as a function of Q^2 for $n=3, 4, 5, 6$. The straight lines are fits to the data [De 79b]

parton model are clearly seen. As predicted by QCD, F_2 increases as a function of Q^2 for small values of x and decreases when x is large. A QCD fit to the data [Bu 78] is included in the figure with $\Lambda=(0.47\pm0.10)$ GeV, and this describes the Q^2 dependence of parton distributions in accordance with leading order AP equations.

Leading order QCD gives a behaviour like $[\log(Q^2/\Lambda^2)]^{-d_n^{NS}}$ (Eqns. (2.7.38), (2.7.44)) for non-singlet structure function moments. If the $(-d_n^{NS})$th root of M_n is calculated, this is then a linear function of $\log(Q^2/\Lambda^2)$ with a zero at $Q^2=\Lambda^2$:

$$[M_n(Q^2)]^{-1/d_n^{NS}} \sim \log Q^2/(\text{GeV})^2 - \log \Lambda^2/(\text{GeV})^2. \qquad (2.7.55)$$

Fig. 2.26 gives a corresponding analysis of the structure function $F_3(x, Q^2)$ in $\nu\mathcal{N}$ scattering for normal and Nachtmann moments and $Q^2 \geqslant 6.5$ (GeV)^2 [De 79b].

A fit to the data can be drawn up by straight lines with common zero position. However,m although only Q^2 values greater than 6.5 (GeV)^2 are used, the slopes and zeros of the normal ($\Lambda=0.60\pm0.15$ GeV) and the Nachtmann ($\Lambda=0.33\pm0.10$ GeV) moments are different.

Ratios of anomalous dimensions were also determined by the CDHS group and the BEBC/GGM group [Bo 78]. Table 2.4 compares these results with values calculated from (2.7.38) which are characteristic for the vector

Table 2.4 Comparison of ratios of anomalous dimensions in vector and scalar theories with experimental results

	CDHS		BEBC/GGM	Theory	
	$\int x^{n-2} x F_3 \, dx$	Nachtmann moments	Nachtmann moments	vector	scalar
d_5/d_3	1.58 ± 0.12	1.34 ± 0.12	1.50 ± 0.08	1.46	1.12
d_6/d_4	1.34 ± 0.07	1.18 ± 0.09	1.29 ± 0.06	1.29	1.06
d_6/d_3	1.76 ± 0.15	1.38 ± 0.15		1.62	1.14

nature of gluons and those for scalar gluons. Experiment favours vector gluons, but scalar theories are not excluded (one standard deviation) [Ab 82aN].

Out of the singlet moments (2.7.41), for which no general systematic determination has been performed, the second moment seems to be of special interest [Gl 79]. It drops with increasing Q^2 in QCD as a result of gluon self-interaction. It increases in vectorial gluon theories without this interaction. An analysis carried out by G. Glück and E. Reya [Gl 79] seems to indicate that experiment favours QCD.

Summing up this short survey on experimental analysis of QCD effects, it is clear that the fundamental concepts of gluonic radiative corrections in the parton model for deep inelastic lepton scattering are confirmed. However, agreement is merely semiquantitative presently; there is as yet no severe quantitative test of QCD which could be even remotely compared with QED tests (Section 2.1.2). The reasons for this are manifold:

(1) Experiments have not yet reached the necessary level of precision, especially when compared among each other.
(2) The comparison of theory and experiment has not yet been pursued in general to the required higher approximations of perturbative calculations.
(3) Theory itself is incomplete, especially as regards confinement, quark fragmentation, etc. on the one hand and the use of data with relatively low Q^2 values on the other, and hence the influence of finite nucleon mass, of operators with higher twist etc. are very difficult to estimate.

These uncertainties have an effect on the experimental determination of the asymptotic scale parameter Λ above all. As its logarithm enters in the measured quantities, errors are exponentiated. Therefore, the values for Λ obtained by elaboration of the various experiments vary between 100 MeV and 600 MeV. Using determinations mainly from analyses which are best suited to the above problems, gives the probable value of $\Lambda = 0.3 \pm 0.2$ [Ei 82N], which in that order of magnitude agrees well with the equation for the string constant and $\bar{\alpha}_s(Q^2)$ from hard gluon bremsstrahlung (Eqn. (2.7.5)) [Wo 82N].

2.7.3 Perturbative quantum chromodynamics

The operator product expansion and the RG equation are suitable methods for dealing with deep inelastic lepton–nucleon scattering. However, they cannot be used for other hard scattering processes such as the inclusive production of hadrons, the Drell–Yan process or the jets in e^+e^- annihilation. Therefore, it is important to provide rules for a direct summation of QCD perturbation series, which are suitable for calculating the processes mentioned above. These rules will below be heuristically derived using deep inelastic lepton scattering as an example and then the prescription for their general application will be formulated.

(a) A perturbative calculation of *QCD corrections of order* g^2 to the parton model results for deep inelastic electron–nucleon scattering is taken as the starting point. The relevant diagrams have already been shown in Figs. 2.24(b) and (d). For non-singlet structure functions, gluon bremsstrahlung is represented by (b) and internal radiative corrections by (d).

As can be read off from Eqns. (1.4.6) and (1.4.11), the calculation of the quark–lepton scattering yields for quark structure functions $\hat{F}_i(\hat{x}, Q^2)$:

$$\frac{1}{\hat{x}}\hat{F}_{2,i}(\hat{x}, Q^2) = Q_i^2\,\delta(1-\hat{x}) \quad \text{with } \hat{x} = \frac{Q^2}{2pq}, \qquad (2.7.56)$$
$$\hat{F}_{L,i}(\hat{x}, Q^2) = 0.$$

(Variables relating to partons are indicated by the superscript ˆ, see the kinematic information given in Fig. 2.27.) In the next order, the Feynman diagrams for real gluon bremsstrahlung (Fig. 2.27) give the invariant matrix

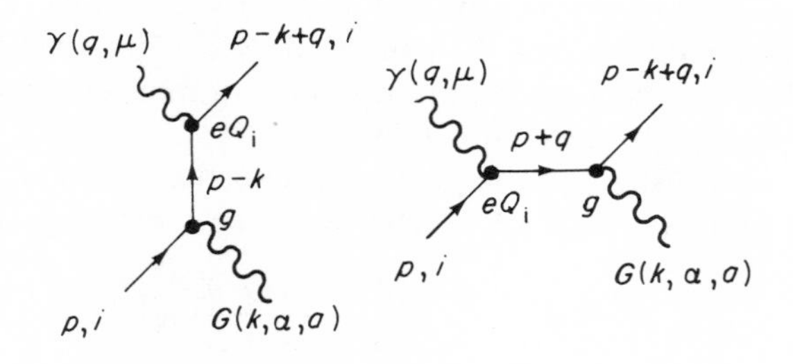

Figure 2.27 Feynman diagrams for $\gamma^* + q_i \to q_i + G$

$$\hat{s} = (p+q)^2 = 2\hat{\nu} - Q^2, \qquad Q^2 = -q^2, \qquad \hat{\nu} = pq$$

$$\hat{t} = (p-k)^2 = -2pk$$

$$\hat{u} = -\hat{t} - 2\hat{\nu}, \qquad qk = \frac{\hat{s}+\hat{t}}{2}$$

element:

$$\mathcal{M}_\mu = \frac{Q_i g}{(2\pi)^2} u(p)\left(\gamma^\alpha \frac{\lambda_a}{2} \frac{\not{p}-\not{k}}{\hat{t}} \gamma_\mu + \gamma_\mu \frac{\not{p}+\not{q}}{\hat{s}} \gamma^\alpha \frac{\lambda_a}{2}\right)\bar{u}(p-k+q)\varepsilon_\alpha(k, s_3).$$

Our interest lies in the unpolarized cross-section summed over gluon momentum and colours. With help of the polarization sum $\sum_{\text{spins}} \varepsilon^*_\alpha \varepsilon_\beta = -g_{\alpha\beta}$ (in Feynman gauge) the following equation is derived:

$$\mathcal{M}_{\mu\nu} = \sum_{\substack{\text{spins,}\\ \text{colours}}} \mathcal{M}_\mu \mathcal{M}^*_\nu = -\frac{Q_i^2 g^2}{(2\pi)^4} \cdot \tfrac{4}{3} \cdot \tfrac{1}{4} Sp\Big(\frac{1}{\hat{t}} \not{p}(\not{p}-\not{k})\gamma_\mu(\not{p}-\not{k}+\not{q})\gamma_\nu(\not{p}-\not{k})$$

$$+\frac{1}{\hat{s}\hat{t}} \not{p}(\not{p}-\not{k}-\not{q})\gamma_\mu(\not{p}-\not{k})(\not{p}+\not{q})\gamma_\nu + \frac{1}{\hat{s}\hat{t}} \gamma_\mu(\not{p}+\not{q}) \times(\not{p}-\not{k})\gamma_\nu(\not{p}-\not{k}+\not{q})$$

$$+\frac{1}{\hat{s}^2} \gamma_\mu(\not{p}+\not{q})(\not{p}-\not{k}+\not{q})(\not{p}+\not{q})\gamma_\nu\Big).$$

Projection of this tensor on $g^{\mu\nu}$ and $p^\mu p^\nu$ gives

$$g^{\mu\nu}\mathcal{M}_{\mu\nu} = \frac{Q_i^2 g^2}{(2\pi)^4} \frac{4}{3}\left(\frac{\hat{s}}{\hat{t}} + \frac{2Q^2 \hat{u}}{\hat{s}\hat{t}} + \frac{\hat{t}}{\hat{s}}\right),$$

$$p^\mu p^\nu \mathcal{M}_{\mu\nu} = \frac{Q_i^2 g^2}{(2\pi)^4} \frac{4}{3}\left(-\frac{\hat{u}}{2}\right). \tag{2.7.57}$$

When integrating with respect to $\hat{t}$ from $-2\hat{\nu} \leq \hat{t} \leq 0$, the $1/\hat{t}$ terms give a logarithmically divergent contribution to the quark structure functions at $\hat{t}=0$. In view of:

$$\hat{t} = -2p_0 k_0(1-\cos\theta), \quad \theta = \sphericalangle(\mathbf{p}, \mathbf{k}) \tag{2.7.58}$$

$\hat{t} \approx 0$ corresponds to $\theta \approx 0$, i.e. the radiation of gluons under small angles θ (almost collinear) or $k_0 \approx 0$, i.e. the radiation of soft gluons. The first case is known as mass singularity and is regularised by cutting off the $\hat{t}$ integration at $\hat{t} = -m^2$. The second case represents the infrared singularity caused by the fact that the gluons have zero mass.

Integration with respect to the phase space of the scattered particles gives the dominant term for large Q^2 of the quark structure function $g_{\mu\nu}\hat{W}^{\mu\nu}$:

$$g_{\mu\nu}\hat{W}^{\mu\nu} = -Q_i^2 \frac{\alpha_s}{2\pi} \frac{4}{3} \frac{\hat{s}^2 + 4Q^2\hat{\nu}}{2\hat{s}\hat{\nu}} \log\left(\frac{Q^2}{m^2}\right),$$

or, with

$$\hat{x} = Q^2/2\hat{\nu}:$$

$$g_{\mu\nu}\hat{W}^{\mu\nu} = -Q_i^2 \frac{\alpha_s}{2\pi} \frac{4}{3} \frac{1+\hat{x}^2}{1-\hat{x}} \log\left(\frac{Q^2}{m^2}\right).$$

Using Eqn. (2.7.7), the contribution of gluon bremsstrahlung to the quark structure functions can thus be derived:

$$\frac{1}{\hat{x}}\hat{F}_{2,i}(\hat{x},Q^2)=Q_i^2\frac{\alpha_s}{2\pi}\frac{4}{3}\frac{1+\hat{x}^2}{1-\hat{x}}\log\left(\frac{Q^2}{m^2}\right),$$
$$\hat{F}_{L,i}(\hat{x},Q^2)=Q_i^2\frac{\alpha_s}{2\pi}\frac{4}{3}2\hat{x}^2. \tag{2.7.59}$$

The expression for $\hat{F}_2$ is singular when $\hat{x}=1$. This is caused by the radiation of soft gluons and can be dealt with by dimensional regularization [Ma 75]: $(1-\hat{x})^{-1}\to(1-\hat{x})^{-1+\varepsilon}$. The contribution of the internal radiative corrections (Fig. 2.24(d)) is also infrared-singular and proportional to $(1/\varepsilon)\,\delta(1-\hat{x})$. The singularities of these two expressions cancel, i.e. their sum is finite and gives the final contribution in order g^2:

$$\frac{1}{\hat{x}}\hat{F}_{2,i}(\hat{x},Q^2)=Q_i^2\frac{\alpha_s}{2\pi}\frac{4}{3}\left(\frac{1+\hat{x}^2}{(1-\hat{x})_+}+\tfrac{3}{2}\,\delta(1-\hat{x})\right)\log\left(\frac{Q^2}{m^2}\right)$$
$$=Q_i^2\frac{\alpha_s}{2\pi}P_{qq}(\hat{x})\log\left(\frac{Q^2}{m^2}\right), \tag{2.7.60}$$

with the AP function $P_{qq}(x)$ as in Eqn. (2.7.52).

The Bloch–Nordsieck theorem of QED [Bl 37] states that this compensation of infrared singularities generally occurs when calculating inclusive cross-sections. However, there is still no general proof available to confirm the validity of this statement for QCD. If the result from the parton model in Eqn. (2.7.56) is now combined with the QCD calculations in order g^2 in Eqn. (2.7.60), the following formula emerges:

$$\frac{1}{\hat{x}}\hat{F}_{2,i}(\hat{x},Q^2)=Q_i^2\left[\delta(1-\hat{x})+\frac{\alpha_s}{2\pi}P_{qq}(\hat{x})\log\left(\frac{Q^2}{m^2}\right)\right]+O(g^4). \tag{2.7.61}$$

Obviously, the nucleon structure functions $F_2(x,Q^2)$ and their moments $M_{2,n}(Q^2)$ are the final aim of this analysis. These are worked out as in Section 1.4 by averaging the parton structure function $\hat{F}$ with the probability $f_i(z)$ of finding the corresponding parton with momentum fraction z in the nucleon. Using the fact that according to Eqn. (1.4.9)

$$p^\mu\approx zP^\mu_{\mathcal{N}},\quad \hat{x}=\frac{Q^2}{2pq}\approx\frac{Q^2}{2P_{\mathcal{N}}qz}=\frac{x}{z},$$

applies, then:

$$F_2(x,Q^2)=\sum_i\int_0^1 dz z f_i(z)\int_0^1 d\hat{x}\,\delta(x-z\hat{x})\hat{F}_{2,i}(\hat{x},Q^2)=\int_x^1 dz f_i(z)\hat{F}_{2,i}\left(\frac{x}{z},Q^2\right). \tag{2.7.62}$$

This multiplicative convolution can be split up into factors by calculating the moments:

$$\begin{aligned} M_{2,n}(Q^2) &= \int_0^1 dx x^{n-2} F_2(x, Q^2) \\ &= \sum_i \int_0^1 dx x^{n-2} \int_0^1 dz z f_i(z) \int_0^1 d\hat{x}\, \delta(x - z\hat{x}) \hat{F}_{2,i}(\hat{x}, Q^2) \\ &= \sum_i \int_0^1 dz f_i(z) z^{n-1} \int_0^1 d\hat{x} \hat{x}^{n-2} \hat{F}_{2,i}(\hat{x}, Q^2) \qquad (2.7.63) \\ &= \sum_i B_n^i \hat{M}_{2,n}^i(Q^2). \end{aligned}$$

Moments can be calculated from Eqn. (2.7.60), using Eqn. (2.7.50)

$$\begin{aligned} M_{2,n}(Q^2) &= B_n^i Q_i^2 \int_0^1 d\hat{x} \hat{x}^{n-1} \left[\delta(1-\hat{x}) + \frac{\alpha_s}{2\pi} P_{qq}(\hat{x}) \log\left(\frac{Q^2}{m^2}\right) \right] \\ &= B_n^i Q_i^2 \left[1 - \frac{\alpha_s}{8\pi} \gamma_n^{(0)} \log\left(\frac{Q^2}{m^2}\right) \right]. \qquad (2.7.64) \end{aligned}$$

This result still contains the mass singularity $\log(Q^2/m^2)$ and is, therefore, not directly usable. The Kinoshita–Lee–Nauenberg theorem [Ki 62, Le 64] guarantees that mass singularities occurring from emission of vector bosons in the initial state are compensated by those in the final state for QED at least. This means that the inclusion of gluon bremsstrahlung in quark–photon scattering and consideration of the quark distribution in the nucleon without gluons as in the simple parton model was somewhat inconsistent. The almost collinear direct emission of gluons from the nucleon gives a mass-singular parton distribution function $f_i(z, m^2)$. Mass singularities in the moments can be easily 'disposed of' by assuming the validity of the KLN theorem for QCD and using Eqn. (2.7.64). The equation can be rewritten as:

$$M_{2,n}(Q^2) = Q_i^2 B_n^i(m^2) \left[1 - \frac{\alpha_s}{8\pi} \gamma_n^{(0)} \log\left(\frac{\Lambda^2}{m^2}\right) \right] \left[1 - \frac{\alpha_s}{8\pi} \gamma_n^{(0)} \log\left(\frac{Q^2}{\Lambda^2}\right) + O(\alpha_s^2), \right.$$

and the last but one factor is absorbed in B_n^i

$$M_{2,n}(Q^2) = Q_i^2 B_n^i(\Lambda) \left[1 - \frac{\alpha_s}{8\pi} \gamma_n^{(0)} \log\left(\frac{Q^2}{\Lambda^2}\right) \right] \qquad (2.7.65)$$

thus giving a result free from any mass singularity. Similar calculations in order g^2 based on Fig. 2.24(c) give the Altarelli–Parisi functions $P_{Gq}(\hat{x})$, $P_{qG}(\hat{x})$, $P_{GG}(\hat{x})$ in agreement with Eqn. (2.7.54).

(b) Order g^2 calculations yield already some of the results obtained previously from the operator product formulation and the RG equation, summarized by the AP equations. Many Feynman diagrams occur in higher orders of g^2. Thus, the decisive question is which ones dominate in the

Bjorken limit. The answer to this was given by V. N. Gribov and L. N. Lipatov [Gr 72a] firstly for Abelian vector gluon theories and then extended to include QCD [Ll 78, Am 78]. They claimed that *generalized ladder diagrams* with renormalized propagators and vertices make the leading contributions (a few details only are given here—for further information, refer to the above quoted literature).

Unitarity relates the hadronic tensor $\hat{W}_{\mu\nu}$ to the amplitude for forward scattering of virtual photons off quarks (see Section 2.4.3(d)):

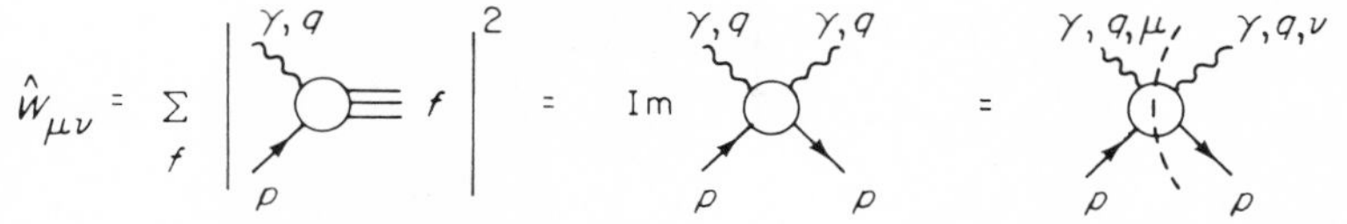

The ladder-like diagrams which have bare propagators and vertices (an example is shown in Fig. 2.28) can be investigated by iteration. Thus, the change in the corresponding amplitude which occurs when lengthening the ladder by one rung can be calculated as:

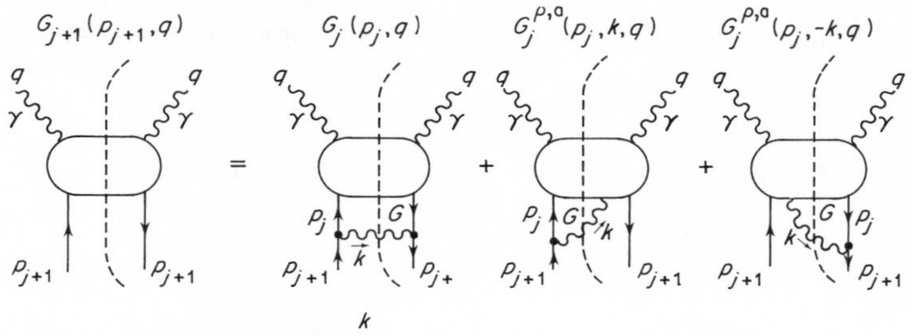

$$
\begin{aligned}
\hat{\sigma}_{j+1}(p_{j+1}, q) &= \sum_{\text{spins}} u(p_{j+1}) G_{j+1}(p_{j+1}, q) \bar{u}(p_{j+1}) \\
&= \mathrm{Tr}\, \not{p}_{j+1} G_{j+1}(p_{j+1}, q) \\
&= (2\pi)^{-3} \int \frac{\mathrm{d}^3 k}{2k_0} \sum_{\text{spins}} u(p_{j+1}) \\
&\quad \times \left(\not{\varepsilon}(k) \frac{\lambda_a}{2} g \frac{\not{p}_j}{t_j} G_j(p_j, q) \frac{\not{p}_j}{t_j} \not{\varepsilon}^*(k) \frac{\lambda_a}{2} g \right. \\
&\quad \left. + 2\gamma_\rho \frac{\lambda_a}{2} g \frac{\not{p}_j}{t_j} G_j^{\rho,a}(p_j, k, q) \right) \bar{u}(p_{j+1}),
\end{aligned}
$$

or, with Eqn. (2.7.57):

$$
\begin{aligned}
\mathrm{Tr}\, \not{p}_{j+1} G_{j+1}(p_{j+1}, q) = (2\pi)^{-3} \int \frac{\mathrm{d}^3 k}{2k_0} \mathrm{Tr}\bigg(& g^2 \cdot \tfrac{4}{3} \cdot 2 \frac{\not{p}_j \not{p}_{j+1} \not{p}_j}{t_j^2} G_j(p_j, q) \\
& + 2g \frac{\lambda_a}{2} \frac{\not{p}_{j+1} \gamma_\rho \not{p}_j}{t_j} G_j^{\rho,a}(p_j, k, q) \bigg).
\end{aligned}
$$

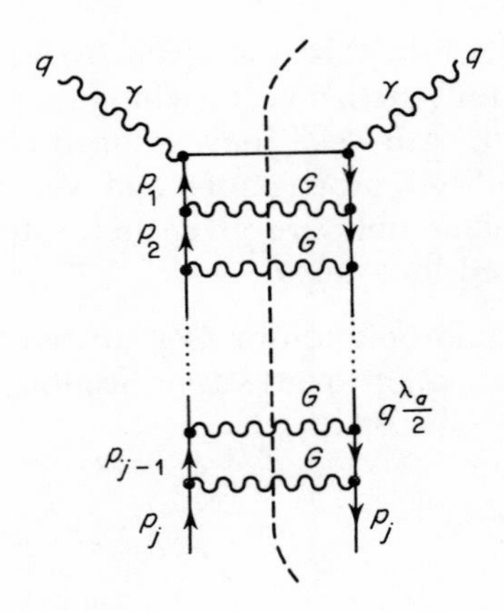

Figure 2.28 Example of a ladder diagram for virtual photon–quark scattering ($t_j = p_j^2$; $q^2 \leq t_1 \leq t_2 \leq \cdots \leq t_j \leq 0$)

The dominating contributions are once again made by the emission of almost collinear gluons; hence, the following approximations can be used:

$$\not{p}_j \approx z_j \not{p}_{j+1}, \quad \not{k} \approx (1-z_j)\not{p}_{j+1},$$

$$\not{p}_j \not{p}_{j+1} \not{p}_j = \not{k} \not{p}_{j+1} \not{k} = 2kp_{j+1}\not{k} = -t_j \not{k} \approx -t_j \frac{1-z_j}{z_j} \not{p}_j, \quad (k^2=0),$$

$$\not{p}_{j+1} \gamma_\rho \not{p}_j \approx -2 \frac{1}{1-z_j} \not{p}_j k_\rho.$$

In view of the latter relationship, only the divergence of $G_j^{\rho,a}(p_j, k, q)$ is needed and this is related to $G_j(p_j, q)$ by a Ward identity (see Section 2.4.2):

$$k_\rho G_j^{\rho,a}(p_j, k, q) = g \frac{\lambda_a}{2} G_j(p_j, p). \tag{2.7.66}$$

Using all the above and $d^3k/2k_0 = -\pi \, dt_j \, dz_j$, results in:

$$\mathrm{Tr}\, \not{p}_{j+1} G_{j+1}(p_{j+1}, q) = \int_{q^2}^{-m^2} \frac{dt_j}{t_j} \int_0^1 \frac{dz_j}{z_j} \frac{g^2}{8\pi^2} \frac{4}{3} \frac{(1-z_j)^2 + 2z_j}{1-z_j} \mathrm{Tr}\, \not{p}_j G_j(p_j, q),$$

or, for the cross-section for forward scattering of quarks on virtual photons,

$$\hat{\sigma}_{j+1}(p_{j+1}, q) = \int_{-Q^2}^{-m^2} d \log t_j \int_0^1 \frac{dz_j}{z_j} \frac{g^2}{8\pi^2} \frac{4}{3} \frac{1+z_j^2}{1-z_j} \hat{\sigma}_j(p_j, q) \quad \text{with } p_j = z_j p_{j+1}.$$

Once again, mass singularities have been regularized by a cut-off parameter $-m^2$. The transition to dressed vertices and propagators takes place by replacing g^2 by the running coupling constant $\bar{g}^2(t_j) = 16\pi^2/\beta_0 \log(-t_j/\Lambda^2)$; the infrared singularities are also compensated in this way, so that P_{qq} can be given by Eqn. (2.7.52) [Ll 78]:

$$\hat{\sigma}_{j+1}(p_{j+1}, q) = \frac{2}{\beta_0} \int_{-Q^2}^{-m^2} \frac{dt_j}{t_j \log(-t_j/\Lambda^2)} \int_0^1 \frac{dz_j}{z_j} P_{qq}(z_j) \hat{\sigma}_j(p_j, q)$$
$$= \frac{2}{\beta_0} \int_{-Q^2}^{-m^2} d \log \log(-t_j/\Lambda^2) \int_0^1 \frac{dz_j}{z_j} P_{qq}(z_j) \hat{\sigma}_j(p_j, q). \tag{2.7.67}$$

This is the desired iteration formula for the dominant terms.

Calculations were carried out in Feynman gauge. With suitable gauges [Li 75, Do 80a], one can achieve that only ladder diagrams make any contributions, i.e. only the first of the three diagrams given must be calculated.

The zero-order term is the parton model contribution $\hat{\sigma}_0 = \delta(1 - Q^2/2p_1 q) Q_i^2$. This can be used to cary out iteration and $\hat{\sigma}_j$ can be represented explicitly by:

$$\hat{\sigma}_j(p, q) = \left(\frac{2}{\beta_0}\right)^j \int_{-Q^2}^{-m^2} d \log \log(-t_j/\Lambda^2) \int_{t_1}^{-m^2} d \log \log(-t_2/\Lambda^2) \ldots$$
$$\times \int_{t_{j-1}}^{-m^2} d \log \log(-t_j/\Lambda^2) \int_0^1 \frac{dz_j}{z_j} P_{qq}(z_j) \int_0^1 \frac{dz_{j-1}}{z_{j-1}} P_{qq}(z_{j-1}) \ldots$$
$$\times \int_0^1 \frac{dz_1}{z_1} P_{zz}(z_1)\, \delta\!\left(1 - \frac{\xi}{z_1 \ldots z_{j-1} z_j}\right) Q_i^2$$

with

$$\xi = \frac{-q^2}{2pq} = \frac{Q^2}{2pq} \approx \frac{Q^2}{2p_j q} z_j \approx \ldots = \frac{Q^2}{2p_1 q} z_j z_{j-1} \ldots z_1. \tag{2.7.68}$$

The j-fold t integrals give

$$\frac{1}{j!}\left(\log \frac{\log(Q^2/\Lambda^2)}{\log(m^2/\Lambda^2)}\right)^j.$$

The z integrals can again be decoupled by calculation of moments:

$$\hat{M}_n^{(j)}(Q^2) = \int_0^1 d\xi \xi^{n-1} \hat{\sigma}_j(\xi, Q^2) = \frac{1}{j!}\left[\frac{2}{\beta_0} \log\left(\frac{\log Q^2/\Lambda^2}{\log m^2/\Lambda^2}\right) \int_0^1 dy y^{n-1} P_{qq}(y)\right]^j Q_i^2$$
$$= \frac{1}{j!}\left[-d_n^{NS} \log \frac{\log Q^2/\Lambda^2}{\log m^2/\Lambda^2}\right]^j Q_i^2 \tag{2.7.69}$$

in which $d_n^{NS} = \gamma_n^{(0)}/2\beta_0$ and Eqn. (2.7.50) have been used. Summation over the number of rungs gives the result

$$\sum_{j=0}^{\infty} \hat{M}_n^{(j)}(Q^2) = Q_i^2 \sum_{j=0}^{\infty} \frac{1}{j!}\left(-d_n^{NS} \log \frac{\log(Q^2/\Lambda^2)}{\log(m^2/\Lambda^2)}\right)^j$$
$$= Q_i^2 [\log(m^2/\Lambda^2)]^{d_n^{NS}} [\log(Q^2/\Lambda^2)]^{-d_n^{NS}} \tag{2.7.70}$$

for the ladder diagrams. This expression factorizes into the mass singularity and a regular part. The nucleon structure function moments follow from this and application of Eqn. (2.7.63):

$$
\begin{aligned}
M_{2,n}(Q^2) &= \sum_i Q_i^2 B_n^i \left(\frac{1}{\log(m^2/\Lambda^2)}\right)^{-d_n^{NS}} \left[\log\left(\frac{Q^2}{\Lambda^2}\right)\right]^{-d_n^{NS}} \\
&= \sum_i Q_i^2 B_n^i(\Lambda) \left[\log\left(\frac{Q^2}{\Lambda^2}\right)\right]^{-d_n^{NS}} .
\end{aligned}
\tag{2.7.71}
$$

This result is the same as that obtained with the RG equation and the Wilson expansion in Eqns. (2.7.24) and (2.7.38). From this, it is clear that the summation of generalized ladder diagrams in conjunction with the running strong coupling constant gives the leading terms for the Bjorken limit, for deep inelastic lepton–nucleon scattering at least. The dominant process is the collinear radiation of hard gluons. The associated mass singularities factorize and are compensated by those from initial state gluon radiation in accordance with the KLN theorem. Formally, this corresponds in Eqn. (2.7.71) to the replacement of the bare Wilson coefficients B_n^f, which also occur in Eqn. (2.7.19), by the renormalized mass singularity-free $B_n^f(\Lambda)$ from Eqn. (2.7.22). This picture of hard QCD processes which now seems plausible can be extended to cover other reactions and higher-order calculations.

(c) The above idea is, in fact, the starting point for applying QCD to many hard scattering processes. *Perturbative QCD* justifies the universality of the parton model and permits calculations of radiative corrections for different processes [Do 80a].

The results given in Eqns. (2.7.62) and (2.7.68) for the structure functions $F_2^{e\mathcal{N}}(x, Q^2)$:

$$
\begin{aligned}
F_2^{e\mathcal{N}}(x, Q^2) &= \sum_i Q_i^2 \int_0^1 d\xi\, \delta\left(\xi - \frac{x}{z}\right) \int_0^1 dz f_i(z) \\
&\times \sum_{j=0}^{\infty} \frac{1}{j!} \left(\frac{2}{\beta_0} \log \frac{\log(Q^2/\Lambda^2)}{\log(m^2/\Lambda^2)}\right)^j \int_0^1 \frac{dz_j}{z_j} P_{qq}(z_j) \ldots \\
&\times \int_0^1 \frac{dz_1}{z_1} P_{qq}(z_1)\, \delta\left(1 - \frac{\xi}{z_1 \ldots z_j}\right)
\end{aligned}
\tag{2.7.72}
$$

have two interpretations:

(1) $F_2^{e\mathcal{N}}(x, Q^2)$ is the convolution of the bare parton distribution $f_i(z)$ with the parton structure function calculated in ladder approximation.
(2) $F_2^{e\mathcal{N}}(x, Q^2)$ is the convolution of the Q^2-dependent parton distribution function $q(x, Q^2)$ calculated by ladder summation with the bare parton structure function.

The second aspect can be generalized to cover other hard scattering processes [El 79]. Thus, Eqn. (2.7.72) can be written as:

$$F_2^{e,\mathcal{N}}(x, Q^2) = \sum_i \int_x^1 \frac{dz}{z} zq_i(z, Q^2)Q_i^2\, \delta\left(1 - \frac{x}{z}\right) \tag{2.7.73}$$

with

$$q_i(z, Q^2) = f_i(z) \sum_{j=0}^{\infty} \frac{1}{j!} \left(\frac{2}{\beta_0} \log \frac{\log(Q^2/\Lambda^2)}{\log(m^2/\Lambda^2)}\right)^j$$
$$\times \int_0^z \frac{dz_1}{z_1} P_{qq}(z_1) \int_0^{z_1} \frac{dz_2}{z_2} P_{qq}(z_2) \dots \int_0^{z_{j-1}} \frac{dz_j}{z_j} P_{qq}\left(\frac{z_j}{z_{j-1}}\right). \tag{2.7.74}$$

The Altarelli–Parisi equation (2.7.51) is satisfied by the quark distribution function $q_i(z, Q^2)$ as can be seen by diffeentiation of Eqn. (2.7.74) with respect to Q^2.

The rules for dealing with hard scattering processes in the leading order of QCD can thus be formulated as follows: based on the parton model, the process must be split up into parton subprocesses. The cross-section is then composed of the parton model cross-sections $\hat{\sigma}$ for parton–lepton or parton–parton scattering and universal, Q^2-dependent parton distribution functions $q_i(x, Q^2)$ (i represents quarks and antiquarks), $G(x, Q^2)$ or parton fragmentation functions $D_i^h(z, Q^2)$. This can be illustrated by means of two examples:

(1) Inclusive hadron production in e^+e^- annihilation: $e^+e^- \to h(P) + X$ [Ge 78]:

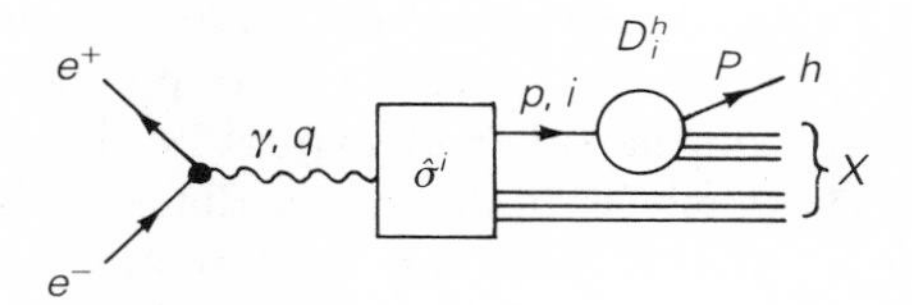

With

$$z = \frac{2Pq}{Q^2}, \quad \hat{z} = \frac{2pq}{Q^2}, \quad \xi = \frac{z}{\hat{z}},$$

the inclusive cross-section is:

$$\sigma_n(z, Q^2) = \sum_i \int_0^1 d\hat{z} \int_0^1 d\xi\, \delta(z - \xi\hat{z})\hat{\sigma}^i(\hat{z})\xi D_i^h(\xi, Q^2) \tag{2.7.75}$$

with

$$\hat{\sigma}^i(z) = e^2 Q_i^2\, \delta(1 - \hat{z}).$$

The fragmentation functions $D_i^h(\xi, Q^2)$ satisfy the 'crossed AP equations', e.g. in the non-singlet case:

$$Q^2 \frac{\partial D_i^h(\xi, Q^2)}{\partial Q^2} = \frac{\bar{\alpha}_s(Q^2)}{2\pi} \int_\xi^1 \frac{dy}{y} P_{qq}\left(\frac{\xi}{y}\right) D_i^h(y, Q^2). \tag{2.7.76}$$

(2) Drell–Yan process $h_1h_2 \to \mu^+\mu^- + X$ [Dr 70, Po 77, Sa 78]:

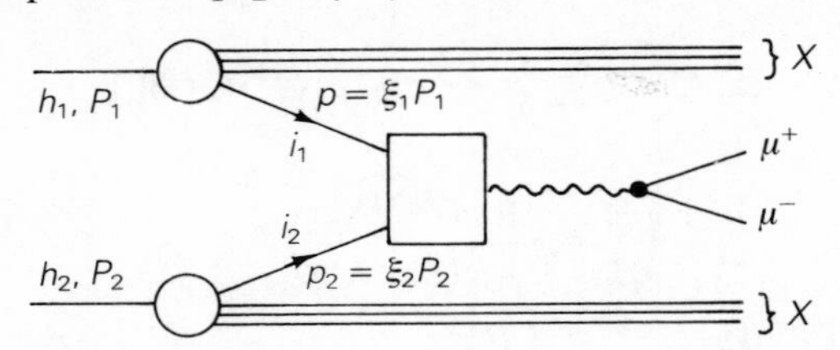

with

$$x_{1,2} = \frac{Q^2}{2P_{1,2}q}, \quad \xi_{1,2} = \frac{Q^2}{2p_{1,2}q}, \quad x_{12} = \frac{2P_1qP_2q}{Q^2P_1P_2} \quad \xi_{12} = \frac{2p_1qp_2q}{Q^2p_1p_2}$$

one has

$$\sigma_{h_1h_2}(x_1, x_2, x_{12}, Q^2)$$
$$= \sum_{i_1,i_2} \int dz_1\, d\xi_1\, dz_2\, d\xi_2\, \delta(x_1 - z_1\xi_1)\, \delta(x_2 - z_2\xi_2)$$
$$\times \hat{\sigma}^{i_1i_2}(z_1, z_2, z_{12}) \cdot \xi_1 q_{i_1}^{h_1}(\xi_1, Q^2)\xi_2 q_{i_2}^{h_2}(\xi_2, Q^2),$$
$$\hat{\sigma}^{i_1i_2}(z_1, z_2, z_{12}) = \tfrac{1}{3}\, \delta_{i_1i_2} e^2 Q_{i_1}^2\, \delta(1 - z_1)\, \delta(1 - z_2)\, \delta(z_{12}^{-1} - z_1 - z_2 - z_1z_2). \quad (2.7.77)$$

Quark distributions in the nucleon measured in deep inelastic lepton scattering can be used to make absolute predictions for the case in which h_1 and h_2 are nucleons [Gl 78]. If an $SU(3)_F$ symmetrical sea distribution is assumed, then these are a factor of 2–3 lower than the experimental results available [An 77, Ka 78, Co 82N].

As in the case of deep inelastic lepton scattering, it is important that higher approximations be calculated, as on the one hand $\bar{\alpha}_s$ is not very small compared with one and on the other, Λ cannot be fixed until second-order effects are known. The analysis of Feynman diagrams gives the following picture [El 79c]: the cross-sections for hard scattering processes can be obtained from universal parton distributions and fragmentation functions calculated in the next order and parton cross-sections calculated in α_s order. In this case, gluons can be 'seen' as partons in the hadrons during the inclusive hadron production and Drell–Yan process.

(d) Description of quark and gluon jets in e^+e^- annihilation [El 76a, De 78, Sc 79] acquires physical significance once infrared and mass singularities in QCD have been dealt with by perturbative techniques. Taking parton model diagram (a), vertex correction (b), and self-energy diagram (c) as a starting point:

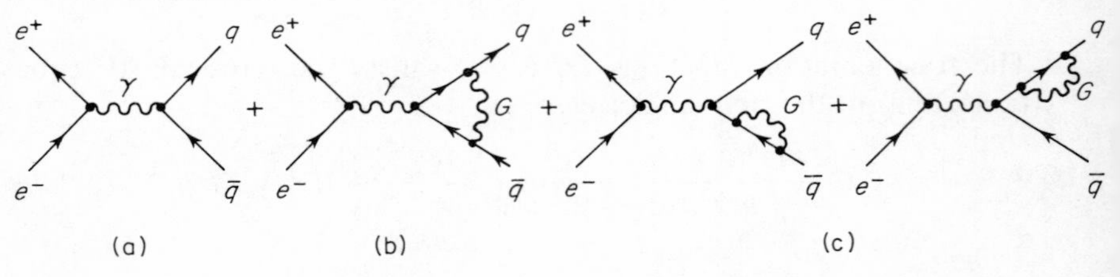

the resultant 'cross-section' still contains an infrared singularity after renormalization of UV divergences, which can be regularized by introduction of a gluon mass μ:

$$\frac{d\sigma}{d\Omega}=\left\{1+\frac{4\bar{\alpha}_s}{3\pi}\left[-\frac{1}{2}\left(\log\frac{s}{\mu^2}\right)^2+\frac{3}{2}\log\frac{s}{\mu^2}+\frac{\pi^2}{6}-\frac{7}{4}\right]\right\}\left(\frac{d\sigma}{d\Omega}\right)_{\text{parton}} \tag{2.7.78}$$

with

$$\left(\frac{d\sigma}{d\Omega}\right)_{\text{parton}}=\frac{\alpha^2}{4s}N_C\sum_f Q_f^2(1+\cos^2\vartheta);\quad s=E_{\text{CMS}}^2;\quad N_C=3\quad\text{for } SU(3)_C.$$

According to the Bloch–Nordsieck theorem only the inclusive cross-section is physically meaningful. This includes the radiation (d) of soft gluons (infrared singularity) and the collinear radiation of hard gluons (mass singularity) in the following way: $d\sigma(\sqrt{s},\vartheta;\varepsilon,\delta)$ is the cross-section for the

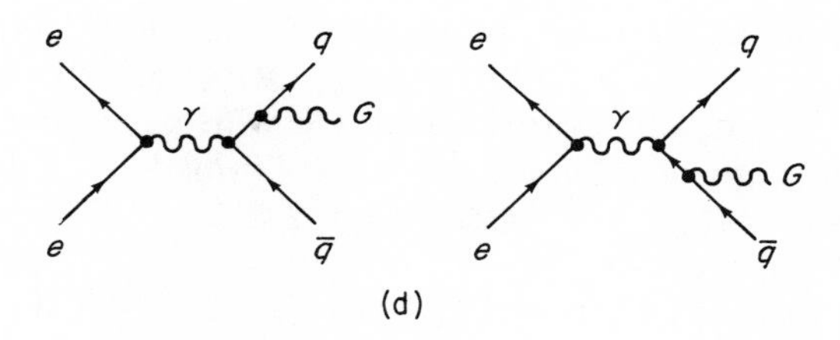

(d)

events in which the fraction $1-\varepsilon$, $\varepsilon\ll 1$, of $\sqrt{s}$ is emitted in a pair of opposing cones with half-angle aperture $\delta(\delta\ll 1)$:

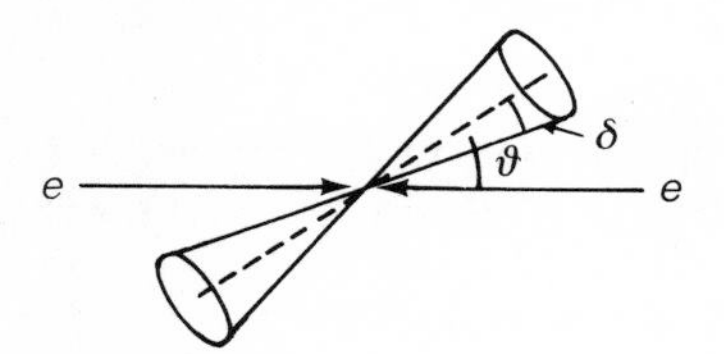

Soft gluon bremsstrahlung ($E_G<\varepsilon\sqrt{s}$) gives:

$$\left(\frac{d\sigma}{d\Omega}\right)_{\text{soft}}=\frac{\bar{\alpha}_s}{\pi}C_2^F\left[2(\log 2\varepsilon)^2+2\log 2\varepsilon\,\log\frac{s}{\mu^2}+\frac{1}{2}\left(\log\frac{s}{\mu^2}\right)^2-\frac{\pi^2}{6}\right]\left(\frac{d\sigma}{d\Omega}\right)_{\text{parton}}.$$

Radiation of hard, almost collinear gluons ($E_G>\varepsilon\sqrt{s}$, $\theta_G<\delta$) gives:

$$\left(\frac{d\sigma}{d\Omega}\right)_{\text{hard}}=\frac{\bar{\alpha}_s}{\pi}C_2^F\left(-3\log\delta-\frac{3}{2}\log\frac{s}{\mu^2}-2(\log 2\varepsilon)^2-4\log\delta\log 2\varepsilon\right.$$
$$\left.-2\log\frac{s}{\mu^2}\log 2\varepsilon+\frac{17}{4}-\frac{\pi^2}{3}\right)\left(\frac{d\sigma}{d\Omega}\right)_{\text{parton}}$$

with C_F^2 as in Eqn. (1.2.7). The sum of these contributions is free from singularities and thus gives the QCD corrections to the two-jet cross-section [St 77]:

$$\left(\frac{d\sigma}{d\Omega}\right)^{2\text{ jet}}_{\text{phys}} = \left(\frac{d\sigma}{d\Omega}\right)^{2\text{ jet}}_{\text{parton}} + \left(\frac{d\sigma}{d\Omega}\right)_{\text{virt}} + \left(\frac{d\sigma}{d\Omega}\right)_{\text{soft}} + \left(\frac{d\sigma}{d\Omega}\right)_{\text{hard}}$$
$$= \left[1 - \frac{4\bar{\alpha}_s}{\pi} C_2^F\left(\frac{3}{4}\log\delta + \log\delta\log 2\varepsilon + \frac{\pi^2}{12} - \frac{5}{8} + O(\varepsilon, \delta)\right)\right]$$
$$\times \left(\frac{d\sigma}{d\Omega}\right)^{2\text{ jet}}_{\text{parton}}. \qquad (2.7.79)$$

Diagrams (a)–(d) also determine the total cross-section for e^+e^- annihilation in hadrons in $\bar{\alpha}_s$ order. The result is:

$$\sigma(e^+e^- \to \text{hadrons}) = \frac{4\pi\alpha^2}{3s} N_C \sum_f Q_f^2\left(1 + \frac{3\bar{\alpha}_s(s)}{4\pi} C_2^F + O(\bar{\alpha}_s^2)\right). \qquad (2.7.80)$$

Thus, the parton model value is reached from above with increasing s Current accuracy is not sufficient to test the QCD correction term, but both sign and order of magnitude are compatible with experimental results (see Fig. 1.14). This state of affairs has been confirmed by a new comparison between theory, $\Delta R_{\text{theor}} = 3\sum Q_f^2\alpha_s/\pi \simeq 0.21$, and experiment, $\Delta R_{\text{exp}} = 0.30 \pm 0.16$ [Wi 80, Wo 82N].

The ratio of Eqns. (2.7.79) and (2.7.80) measures the fraction $f(\sqrt{s}, \varepsilon, \delta)$ of two-jet events:

$$f(\sqrt{s}, \varepsilon, \delta) = 1 - \frac{4\bar{\alpha}_s}{3\pi}\left(\log\delta(4\log 2\varepsilon + 3) + \frac{\pi^2}{3} - \frac{7}{4} + O(\varepsilon, \delta)\right). \qquad (2.7.81)$$

This expression approaches one with increasing energy as a result of the asymptotic freedom of QCD. With $\bar{\alpha}_s = 4\pi/(11 - \frac{2}{3}N_F)\log(Q^2/\Lambda^2)$ this equation gives:

$$\delta(\sqrt{s}, \varepsilon, f) = \left(\frac{\sqrt{s}}{\Lambda}\right)^{(33-2N_F)/8(4\log 2\varepsilon+3)(1-f)} \exp\left(-\frac{\frac{\pi^2}{3} - \frac{7}{4}}{4\log 2\varepsilon + 3}\right) \sim (\sqrt{s})^{-d(f,\varepsilon)}, \qquad (2.7.82)$$

when solved with respect to the jet opening angle δ. This angle drops like the power $d(f, \varepsilon)$. When $N_F = 3$, i.e. N_F is equal to the number of light quark flavour degrees of freedom and $\varepsilon = 0.1$, the result is $d = 0.98(1-f)$.

The broadening linked with quark and gluon hadronization is superimposed on the perturbative opening angle of the jet. This non-perturbative part can be worked out with help of a model using the fragmentation functions [Fi 78, Ho 79, Al 80]. A rough estimate based on the average transverse momentum of the hadrons gives: $\delta_{np} \approx \langle n\rangle(0.3\text{–}0.5)\ \text{GeV}/E_{\text{jet}}$, in which $\langle n\rangle$ is the average particle multiplicity in the jet. For 15 GeV jet energy, this gives a value of $\delta_{np} \approx 15°$ approximately [Wo 80]. On the other hand, hard gluon bremsstrahlung (2.7.82) gives an angle aperture of 1° for $\sqrt{s} = 30$ GeV, $\varepsilon = 0.2$, and $f = 0.7$ with $\Lambda = 0.5$ GeV. This means that PETRA/PEP energies are probably not big enough for providing a solid test of QCD radiative corrections of two-jet phenomena.

Leading order QCD radiative corrections can also be calculated for jets of n quarks, n, antiquarks and m gluons [Sm 79, Ei 78]. If δ_i represents the maximum permitted half-angle aperture of jet i, ε_i represents the maximum permitted ratio between the total energy of the non-jet particles and the total energy of jet i and ε represents the same ratio as above, but applied to average jet energy, then:

$$\begin{aligned} d\sigma(2n, m) = \Big[1 &- \frac{\bar{\alpha}_s}{\pi} C_2^F \sum_{i=1}^{2n} 2 \log \delta_i (\log \varepsilon_i + \tfrac{3}{4}) \\ &- \frac{\bar{\alpha}_s}{\pi} C_2 \sum_{i=1}^{m} 2 \log \delta_i \left(\log \varepsilon_i + \frac{11}{12} - \frac{N_F}{6C_2^F}\right) \\ &- \frac{\bar{\alpha}_s}{\pi} (A \log \varepsilon + O(1))\Big] d\sigma(2n, m)_{\text{parton}}. \end{aligned} \tag{2.7.83}$$

A is a process and angle-dependent quantity. The value of the quadratic Casimir operator in the gluon representation C_2 determines the gluon jet angle aperture. In view of $C_2 = 3$, $C_2^F = \frac{4}{3}$ from (Eqns. (1.2.7), (1.2.8)), this is $\frac{9}{4}$ times greater than that of the quark jet. Because of their greater colour charges, gluons generate stronger bremsstrahlung than quarks. In practice, this means that at similar energy levels, the gluon jets are less jet-like than the quark ones. The result for $q\bar{q}G$ jets is given by way of a concrete example for the application of Eqn. (2.7.83):

$$\begin{aligned} d\sigma(2, 1) = \Big\{1 - \frac{\bar{\alpha}_s}{\pi} \Big[&\tfrac{8}{3} \log \delta_q (\log \varepsilon_q + \log \varepsilon_{\bar{q}} + \tfrac{3}{2}) + \tfrac{2}{3} \log \varepsilon \log \sin\left(\frac{\theta_{q\bar{q}}}{2}\right) \\ &+ 6 \log \delta_G \left(\log \varepsilon_G + \frac{11}{12} - \frac{N_F}{18}\right) \\ &- 3 \log \varepsilon \log \sin\left(\frac{\theta_{qG}}{2}\right) \sin\left(\frac{\theta_{\bar{q}G}}{2}\right)\Big]\Big\} d\sigma(2, 1)_{\text{parton}}. \end{aligned} \tag{2.7.84}$$

These results conclude this survey of perturbative QCD. Methods for dealing with hard scattering processes beyond the leading order have not yet been firmly established. The same applies for theoretical research into the effects of higher order and processes which occur during the hadronization of quark and gluon jets. A consistent picture of these processes, including the higher orders in $\bar{\alpha}_s$, is important to the recent phenomenological discussions stimulated by QCD.

2.7.4 Quarkonia

Hadron spectroscopy for light hadrons in particular cannot yet be clarified in the light of the current status of the confinement problem. However, simple

QCD concepts have formed the basis for the treatment of both meson and baryon spectroscopy which had detailed results on hadron masses and decays [No 78, Gr 79, Kr 79, Ei 80, Bu 81aN]. Section 1.1.3 described some calculations carried out in the realm of non-relativistic quantum mechanics. Interaction between quarks at short distances takes place mainly through the one-gluon exchange potential $V_G(r)$ containing the running coupling constant $\bar{\alpha}_s$

$$V_G(r) = -\frac{\bar{\alpha}_s}{r}\begin{pmatrix}4/3\\2/3\end{pmatrix} \begin{matrix}\text{meson } q\bar{q}\\ \text{baryon } qqq\end{matrix}, \tag{1.1.14}$$

and by the confinement potential for greater distances:

$$V_C(r) = \kappa r. \tag{1.1.13}$$

Only the simplest case will be considered here, using the total potential $V(r) = V_G + V_C$ which then produces the spectrum shown in Fig. 1.4. If the spin dependence of the interaction given by the one-gluon exchange is also taken into account, then fine structure spliting, which will be discussed below, is the result [Ap 75a, De 75a]. Finally, gluon dynamics in the form of quark–antiquark annihilation in gluons is applied to heavy meson decays [Ap 75].

(a) In atomic physics, relativistic effects give spin-dependent potentials and also a *fine and hyperfine structure of the spectrum*. As the expectation value of the velocity in the quarkonium system is not very small compared with the velocity of light c, a considerable degree of fine structure splitting is expected. This can be calculated for the short-range component V_G of the potential, in the same way as for the positronium and muonium system [Be 57], by expanding the one-gluon exchange graph as far as c^{-2} order. The long-range component of the potential V_C is best regarded as spin-independent. Hence, the following spin-dependent terms occur in the Hamilton operator (m_i, $\mathbf{p}_i$, $\mathbf{S}_i \equiv$ mass, momentum, spin of quark i, $\mathbf{r}_{ij} \equiv \mathbf{r}_i - \mathbf{r}_j$):

$$H_{SS} = \sum_{i>j} \frac{2}{3m_i m_j} \mathbf{S}_i \cdot \mathbf{S}_j \,\Delta V_G(r_{ij}) \quad \text{spin–spin interaction} \tag{2.7.85}$$

$$H_{LS} = \sum_{i>j} \frac{1}{r_{ij}} \frac{\mathrm{d}(3V_G - V_C)}{2\,\mathrm{d}r_{ij}} \left(-\frac{\mathbf{r}_{ij} \times (\mathbf{p}_j \mathbf{S}_i - \mathbf{p}_i \mathbf{S}_j)}{m_i m_j} - \frac{1}{2}\frac{\mathbf{r}_{ij} \times \mathbf{p}_j \cdot \mathbf{S}_j}{m_j^2} + \frac{1}{2}\frac{\mathbf{r}_{ij} \times \mathbf{p}_i \cdot \mathbf{S}_i}{m_i^2} \right) \quad \text{spin–orbit interaction} \tag{2.7.86}$$

$$H_T = \sum_{i>j} \left(\frac{\mathrm{d}^2 V_G}{\mathrm{d}(r_{ij})^2} - \frac{1}{r_{ij}} \frac{\mathrm{d}V_G}{\mathrm{d}r_{ij}} \right) \left(\frac{\mathbf{S}_i \mathbf{S}_j}{3m_i m_j} - \frac{(\mathbf{S}_i \mathbf{r}_{ij})(\mathbf{S}_j \mathbf{r}_{ij})}{m_i m_j (r_{ij})^2} \right) \quad \text{tensor interaction} \tag{2.7.87}$$

These additional terms to the Hamilton operator (1.1.11) are treated in first-order perturbation theory. Only H_{SS} makes a contribution for S-wave states (pseudoscalar and vector mesons, nucleons and Δ resonances). In view of:

$$\langle \mathbf{S}_1\mathbf{S}_2\rangle = \tfrac{1}{2}\langle \mathbf{S}^2 - \mathbf{S}_1^2 - \mathbf{S}_2^2\rangle = \begin{pmatrix} 1/4 \\ -3/4 \end{pmatrix} \text{ for } \begin{pmatrix}\text{triplet}\\ \text{singlet}\end{pmatrix} \text{ states}$$

the relevant mass splitting is

$$\langle H_{SS}\rangle = \frac{2}{3m_i m_j}\begin{pmatrix} 1/4 \\ -3/4 \end{pmatrix}\langle \Delta V_G\rangle.$$

As a result of $\Delta(r^{-1}) = -4\pi\,\delta^{(3)}(\mathbf{r})$,

$$\langle H_{SS}\rangle = \frac{8\pi\bar{\alpha}_s}{9m_i m_j}\begin{pmatrix} 1 \\ -3 \end{pmatrix}|\psi(0)|^2 \quad \text{for } \begin{pmatrix}\text{triplet}\\ \text{singlet}\end{pmatrix} \text{ states} \tag{2.7.88}$$

applies with a Coulomb-like $V_G(r)$ (Eqn. (1.1.14)). This means that the ortho-states are heavier than the para-states, as in the positronium system; in fact $M_\rho > M_\pi$, $M_{K^*} > M_K$, $M_{D^*} > M_D$, $M_{J/\psi} > M_{\eta_c}$, ... As the quark–quark potential is also attractive in QCD, $M_\Delta > M_N$, ... follows for the baryons.

The van Royen–Weisskopf formula [Ro 67] expresses the leptonic width of vector mesons by the modulus squared of the wavefunction for relative motion at the origin

$$\Gamma_{V\to e^+e^-} = 16\pi\alpha^2 Q_f^2 \frac{|\psi_V(0)|^2}{M_V^2}. \tag{2.7.89}$$

It is used to estimate the mass difference $J/\psi(1^{--}) - \eta_c(0^{-+})$ for the charmonium system. When $|\psi(0)|^2$ from (2.7.89) is substituted in (2.7.88), the following formula is obtained:

$$\langle H_{SS}\rangle = \frac{2\bar{\alpha}_s}{9}\frac{\Gamma_{V\to e^+e^-}}{\alpha^2 Q_f^2}\left(\frac{M_V}{2m_c}\right)^2\begin{pmatrix} 1 \\ -3 \end{pmatrix}. \tag{2.7.90}$$

Using the measured width $\Gamma_{J/\psi\to e^+e^-} = (4.7\pm 0.7)$ keV and $\alpha_s = 0.4$ results in:

$$\begin{aligned} M_{J/\psi} - M_{\eta_c} &= 70 \text{ MeV}\left(\frac{M_\psi}{2m_c}\right)^2 \\ M_{\psi'} - M_{\eta_c'} &= 35 \text{ MeV}\left(\frac{M_{\psi'}}{2m_c}\right)^2. \end{aligned} \tag{2.7.91}$$

If $m_c = 1.4$ GeV (see Eqn. (1.1.15)), $M_{\eta_c} = 3.0$ GeV, $M_{\eta_c'} = 3.6$ GeV are the predicted values.

A resonance is observed at 2.98 GeV in the inclusive photon spectrum $\psi' \to \gamma + X$ [Hi 80, Pa 80b, Pa 82]. The estimate would be confirmed if it were the $0^{-+} - c\bar{c}$ particle. The P-wave triplet states (2^{++}, 1^{++}, 0^{++}) split up by the spin–orbit and tensor force. The experimental results in the charmonium system agree well with the calculations using the parameters (1.1.15).

The one-gluon exchange with running coupling constant $\bar{g}_s(Q^2)$ gives no Coulomb potential, strictly speaking. This plus the fact that the level of current knowledge of QCD does not cover potential calculations for moderate distances, can be used to apply potentials which are slightly modified in comparison with Eqns. (1.1.13) and (1.1.14). These can be chosen to give good results for the spectroscopy of both charmonium and bottomonium at the same time [Kr 79, Bu 81aN].

(b) One remarkable property of mesons with hidden strangeness ($\phi = s\bar{s}$), hidden charm (J/ψ, $\psi' = c\bar{c}$) and hidden bottom (Υ, Υ', $\Upsilon'' = b\bar{b}$), which have a mass below or only slightly above the threshold for proction of meson pairs with corresponding flavour quantum numbers, is their long lifetime. The *Zweig rule* provides a phenomenological explanation of this [Zw 64]. It permits only those reactions in which the valence quarks in the initial states are also present in the final state. This selection rule goes way beyond flavour symmetry. A discussion of the ϕ meson decay will be used to illustrate the Zweig rule.

Meson-mass formulae show that the ϕ meson is an almost pure $s\bar{s}$ state (ideal vector meson mixing). The decay $\phi \to 3\pi$ is permitted by $SU(3)_F$ but forbidden by the Zweig rule as the 3π final state contains no s quarks. Experimentally, $\Gamma_{\phi\to 3\pi} = 0.6$ MeV which represents a suppression by a factor of 50 (this estimate is based on the following: $\Gamma_{\omega\to 3\pi} = 9$ MeV, but the phase space for $\phi \to 3\pi$ is four times greater than that for $\omega \to 3\pi$). The $\phi \to K^+K^-$ and $\phi \to K^0\bar{K}^0$ decays are allowed by the Zweig rule, as the s quarks in the ϕ's re-emerge in the K mesons. The width $\Gamma_{\phi\to K\bar{K}} = 3.4$ MeV is only apparently small, as the associated phase space is very small in view of $M_\phi - 2M_K = 30$ MeV.

Applied to the charmonium states, the Zweig rule means that strong and radiative decays for particles with a mass of less than $2M_D = 3.73$ GeV are suppressed. However, the Zweig suppression factor is of the order 1000 for the J/ψ meson with $\Gamma(J/\psi \to \text{hadron}) = 60$ keV and the ψ' with $\Gamma(\psi' \to \text{light hadron}) = 40$ keV, i.e. it is much greater than for the ϕ mesons.

The Zweig rule is an important element of quark phenomenology and has always been confirmed by experiment. It has a plausible explanation in QCD and moreover QCD can provide quantitative estimates of the suppression factor, viz.: in QCD quarks interact by exchange of flavour-neutral gluons. Reactions in which quarks in the initial state do not occur in the final state can only proceed after annihilation into a suitable number of gluons. If the initial state is massive, then the process becomes more suppressed the greater the number n of hard gluons required, due to the fact that the running coupling constant is small for high momentum transfers. $[\bar{\alpha}_s(Q^2)]^n$ is a measure for the Zweig suppression factor.

Quarkonium states with even charge conjugation parity π_c decay into hadrons via at least two gluons. The decay width of these particles can be calculated by analogy with 2γ decay, which from the Feynman graph in Fig.

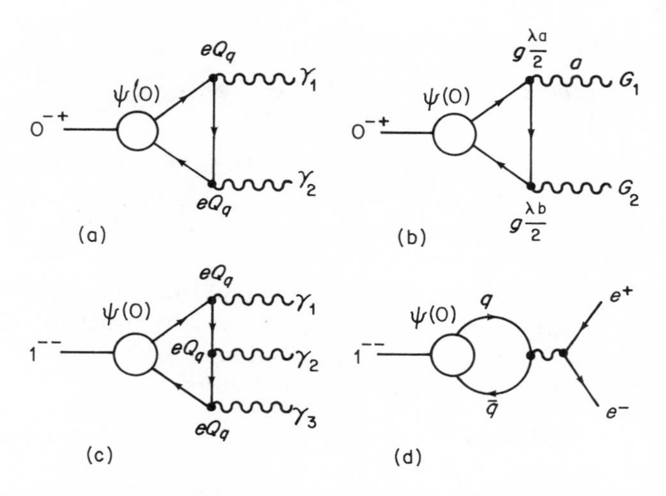

Figure 2.29 Diagrams for the decay of quarkonium states with $\pi_c = 1$ in (a) 2 photons, (b) 2 gluons, with $\pi_c = -1$ in (c) 3 photons and (d) e^+e^-. As photons and gluons satisfy Bose statistics, diagrams are summed with permuted final states

2.29(a) can be represented by:

$$\Gamma(0^{-+} \to 2\gamma) = 12\pi\alpha^2 Q_f^4 \frac{|\psi(0)|^2}{m_f^2}. \tag{2.7.92}$$

This formula is based on the assumption that the kinetic energy of the quarks in the bound state is small in comparison with their rest energy. In this case, the decay width Γ can be calculated from the cross-section $\sigma(q\bar{q} \to 2\gamma)$ at threshold and the incoming current composed of relative velocity v and particle density $|\psi(0)|^2$:

$$\Gamma = \lim_{v \to 0} v\, |\psi(0)|^2 \bar{\sigma}(q\bar{q} \to 2\gamma)(v).$$

For small velocities, the cross-section averaged over quark spin is:

$$\bar{\sigma}(q\bar{q} \to 2\gamma) = \tfrac{1}{4}\sigma^{\text{sing}}(q\bar{q} \to 2\gamma) = \frac{\pi\alpha^2 Q_f^4}{m_f^2 v}.$$

In the pseudoscalar meson, the spins form a singlet state and hence the expression σ^{sing} is used. Moreover, a summation with respect to the colour degrees of freedom of the quarks must be carried out and thus the colour wavefunction $1/\sqrt{(3)}\, \delta_{cc'}$ (Eqn. (1.1.8)) must be used. The overall result is then Eqn. (2.7.92).

For the decay into two gluons $0^{-+} \to G + G$, the quark–gluon couplings $g\lambda_a/2$ must be used in the Feynman diagram (Fig. 2.29(b)) instead of the electric charges of the quarks and summation carried out with respect to the eight gluon colour degrees of freedom:

$$\sum_{a,b}\left|\sum_{c,c',c''}\frac{\delta_{cc'}}{\sqrt{3}}\frac{(\lambda_a)_{c'c''}}{2}\frac{(\lambda_b)_{c''c}}{2}\right|^2=\sum_{a,b}\frac{1}{3\cdot 16}|Sp\,\lambda_a\lambda_b|^2=\tfrac{2}{3}.$$

The following formula then results for the decay width:

$$\Gamma(0^{-+}\to 2G)=\Gamma(0^{-+}\to \text{hadrons})=\frac{8\pi}{3}\frac{\bar{\alpha}_s}{m_f^2}|\psi(0)|^2. \tag{2.7.93}$$

Decays of states with $\pi_c=-1$ into three photons, three gluons or two gluons and one photon are calculated in a similar way by using the Ore–Powell formula for fermion–antifermion annihilation into 3γ [Or 49] and then multiplying with the corresponding colour factors:

$$\left.\begin{aligned}\Gamma(1^{--}\to 3\gamma)&=\tfrac{4}{3}\alpha^3Q_f^6\\ \Gamma(1^{--}\to 3G)&=\tfrac{10}{81}\bar{\alpha}_s^3\\ \Gamma(1^{--}\to 2G\gamma)&=\tfrac{8}{9}\alpha Q_f^2\bar{\alpha}_s^2\end{aligned}\right\}\times 4(\pi^2-9)\frac{|\psi(0)|^2}{m_f^2}. \tag{2.7.94}$$

The decay of flavour-neutral 1^{--} mesons into e^+e^- pairs can be calculated from the diagram in Fig. 2.29(d), to give the Royen–Weisskopf formula (2.7.89).

The first step in the dynamical mechanism for the hadronic decays of quarkonium states is annihilation into gluons and the second is gluon hadronization. For heavy $q\bar{q}$ systems such as bottomonium, toponium etc., the gluons hadronize in the form of gluon jets. Vector mesons such as Υ convert to three gluons, i.e. three jets in view of $\pi_c=-1$, and pseudoscalar mesons convert to two gluons, i.e. two jets. Off-resonance, two quark-jets occur. Hence the topology of hadronic events on the Υ resonances and in the continuum is expected to be different and this is in fact confirmed by experiment [Be 79b]. Moreover, experimental results on thrust axis distribution favour gluons with spin 1 more than those with spin 0 [Ko 79b].

Experimental determination of leptonic and hadronic decay widths for 1^{--} quarkonia states offers a further possibility of determining the strong fine-structure constant. With $\Gamma(1^{--}\to 3G)\equiv\Gamma(1^{--}\to\text{hadrons})$, $M_V\approx 2m_f$ division of Eqn. (2.7.89) by Eqn. (2.7.94) results in:

$$B=\frac{\Gamma(V\to e^+e^-)}{\Gamma(V\to\text{hadrons})}=\frac{81\pi}{10(\pi^2-9)}\frac{\alpha^2}{\bar{\alpha}_s^3}Q_f^2. \tag{2.7.95}$$

Solution of this equation with respect to $\bar{\alpha}_s$ gives the values listed in the '$\bar{\alpha}_s$ uncorrected' column of Table 2.5. The $\bar{\alpha}_s$ (uncorrected) values in the ψ region differ from the $\bar{\alpha}_s=0.35-0.40$ (Eqn. (1.1.15)) obtained from the potential. G. Parisi and R. Petronzio [Pa 80c] connect this with non-perturbative effects. They have attempted to describe this very roughly by giving the gluons an effective mass $m_G\approx 800$ MeV the size of a half glue ball

Table 2.5 Determination of the running coupling constant ($\bar{\alpha}_s$) from decays of quarkonium states

V	B_{exp}	$\bar{\alpha}_s$ uncorrected	$\bar{\alpha}_s$ corrected
J/ψ	0.08 [Pa 82]	0.21	0.36
ψ'	0.04 [Pa 82]	0.26	0.37
Υ	$\sim 0.03 \pm 0.02$ [LE 81]	0.18 ± 0.04	0.19 ± 0.04

mass. This reduces the theoretical width Γ ($V \rightarrow$ hadrons) in Eqn. (2.7.94) by a phase-space factor of $f(2m_G/m_V) = 0.2$; 0.35; 0.9 for $V = J/\psi$; ψ'; Υ and thus gives the corrected values for $\bar{\alpha}_s$ in Table 2.5.

This represents the last example for QCD application to experimental data. To sum up, QCD certainly appears to provide a consistent qualitative explanation of the multifaceted phenomena in hadron physics, in the form of a unified theory. However, in the light of all knowledge currently available, only a semiquantitative comparison is now possible. For example, if the question as to what extent the basic concept in QCD, the running coupling constant, has been confirmed by experiment is considered in conjunction with the various specifications for $\bar{\alpha}_s(Q^2)$ which have been discussed in this book, then the result is as shown in Table 2.6. An average momentum transfer has been chosen as the characteristic Q^2 range for a special process. As emphasized many times, comparisons with higher-order calculations are essential for a more accurate assessment of momentum, but this is still a long way off. Information on the experimental error of $\bar{\alpha}_s$ is not very systematic—part of it is based on various experimental estimates of the systematic error but part of the determination of $\bar{\alpha}_s$ is not based on systematic variation of the model parameters. The theoretical value results

Table 2.6

	Reference	Q (GeV)	$\bar{\alpha}_s(Q^2)$ exp.	$\bar{\alpha}_s(Q^2)$ theoretical $\Lambda = 0.25$	$\Lambda = 0.35$
Charmonium potential	Eqn. (1.1.15)	1.5	0.3–0.4		
Decay widths J/ψ, ψ'	Table 2.5	1.5	~ 0.35	0.28	0.34
Charmonium hyperfine structure	Eqn. (2.7.91)	1.5	~ 0.4		
Decay widths Υ	Table 2.5	4.5	0.19 ± 0.04	0.20	0.23
Scaling violation in deep inelastic lepton scattering	Fig. 2.25	7	0.25 ± 0.03	0.20	0.22
Gluon brems-strahlung	Eqn. (2.7.5)	10	0.17 ± 0.05	0.18	0.20
$\sigma_{tot}(e^+e^- \rightarrow \text{hadrons})$	Eqn. (2.7.80)	10	0.26 ± 0.13		

from a second-order calculation (Eqn. (2.5.29)), with $\Lambda = 0.25$ or 0.35 GeV. As $\bar{\alpha}_s$ is only a logarithmic function of Λ, the determination of the asymptotic scale parameter has an uncertainty factor ≈ 2. A Λ of this order then agrees with theoretical calculations of confinement (Eqn. (2.6.124)).

This short review is intended to give an idea of the status of the experimental confirmation of QCD. Other experimental comparisons, e.g. the Drell–Yan process, transverse momentum distributions in jets, inclusive hadron production, gluonic corrections for weak decay, etc. which provide a similar or lower degree of accuracy, can be brought in to provide further substantiation.

However, final confirmation of QCD still requires more extensive and more accurate measurements and more advanced theoretical research, in particular. The objective of our discussions on QCD was to present the basic facts needed for this.

CHAPTER 3

Gauge Theory for the Electroweak Interaction

There are good reasons for assuming that the electroweak interaction also takes the form of a non-Abelian gauge theory. If an attempt is made to solve the problem of the bad high energy behaviour of the Fermi model by introducing massive vector bosons, the necessity for a unification of the electromagnetic and weak interactions shows up. Moreover, a coupling structure which compares with that of a non-Abelian gauge theory emerges (Section 3.1). Any attempt to formulate a gauge theory for the electroweak interaction is confronted with the problem that the intermediate vector bosons have mass because of the short range of the weak interaction (Section 1.3.1), in contrast to the gauge fields of a pure gauge theory. The Higgs–Kibble mechanism for dynamic mass generation (Section 3.2) provides a solution to this problem. From this, the only known form of a renormalizable field theory for the interaction of massive vector bosons is derived.

Phenomenological analysis of electromagnetic and weak processes carried out so far shows the universality of the electroweak interaction in the form of an $SU(2)_W \times U(1)$ symmetry (Section 1.3). The corresponding gauge theory, the Glashow–Salam–Weinberg theory, makes interesting predictions for the masses of the vector bosons $W^{\pm}$, Z^0 and for scattering processes at high energies. On the other hand, the Fermi model results are reproduced for energies which are small in comparison with the vector boson masses (Section 3.3).

The concluding part of our review of gauge theories for elementary particle interactions will consider attempts to describe the strong, electromagnetic, and weak interactions by means of a grand unified field theory (Section 3.4).

3.1 Unification of the electromagnetic and weak interactions

Section 1.3 described the Fermi model for the weak interaction of elementary particles and compared it for various reactions with the experimental results. This gave a good quantitative description of the phenomena occurring at presently available energies. However, the discussion of the neutrino cross-sections showed that these violate the S-matrix unitarity at a centre of mass energy of 500 GeV. The problem was alleviated by the introduction of massive intermediate vector bosons, but this is only the first step towards the formulation of a consistent theory for the weak interaction.

Theories involving massive vector bosons are in general not renormalizable. Only models with a sufficient number of particles and special coupling constants can provide adequate compensation for the ultraviolet divergences of the individual Feynman diagrams. This section is devoted to following the consequences of such compensation mechanisms on tree graphs [Ll 74]. The significance of this heuristic investigation is that it leads to renormalizable theories for the electroweak interaction.

3.1.1 Foundation of the unification of electromagnetic and weak interactions based on the high energy behaviour of cross-sections

The initial consequence of the demand for an acceptable behaviour at high energies by mutual compensation of interaction processes, is that the weak and electromagnetic interactions are unified. If the hypothesis for mediating the weak interaction by heavy vector bosons is to be taken seriously, then scattering processes such as $e^+e^- \to W^+W^-$, $W^+W^- \to W^+W^-$ etc. must also be considered. The production of W-pairs in e^+e^- annihilation is described in lowest order by the e.m. (a) and weak (b) Feynman diagrams below:

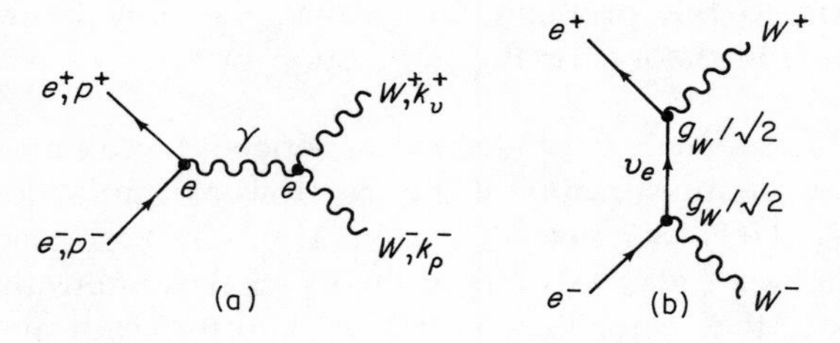

The γW^+W^- vertex is required for calculating the e.m. diagram (a). This can be obtained from the Lagrangian density $\mathscr{L}_0$ of free, massive W bosons:

$$\mathscr{L}_0 = -\tfrac{1}{2}(\partial_\mu W^*_\nu - \partial_\nu W^*_\mu)(\partial^\mu W^\nu - \partial^\nu W^\mu) + M^2_{\mathrm{W}} W^*_\mu W^\mu, \quad W \equiv W^+,\ W^* = W^-,$$

according to the principle of minimal gauge-invariant coupling $\partial_\mu \to D_\mu = \partial_\mu - \mathrm{i}eA_\mu$ (Section 2.1.3(c)) and results in:

$$= \frac{(2\pi)^4}{\mathrm{i}} e[-g_{\mu\nu}k^+_\rho + g_{\nu\rho}(k^+ - k^-)_\mu + g_{\rho\mu}k^-_\nu] \times \delta^{(4)}(k^+ + k^- + k), \tag{3.1.1}$$

(the additional arrows on the graph represent momentum flux).

The polarization vectors ε^μ for the massive W bosons take the following form for a particle propagating in the z direction ($k^\mu=(k^0, 0, 0, k)$, $k^0=\sqrt{(M_W^2+k^2)}$):

$$\varepsilon^\mu_{T_1}=(0,1,0,0), \quad \varepsilon^\mu_{T_2}=(0,0,1,0),$$
$$\varepsilon^\mu_L=(k/M_W, 0, 0, k_0/M_W). \tag{3.1.2}$$

Both transversal polarization vectors ε^μ_T are energy-independent. Conversely, the longitudinal polarization vector ε^μ_L is proportional to the four-momentum for large k:

$$\varepsilon^\mu_L=\frac{k^\mu}{M_W}+O\left(\frac{M_W}{k}\right). \tag{3.1.3}$$

Hence the invariant matrix element can now be calculated from the e.m. (a) and weak (b) Feynman graphs:

$$\begin{aligned}\mathcal{M}=&-(2\pi)^{-2}\frac{e^2}{(p^++p^-)^2}\bar{v}_{e^+}[(\not{k}^+-\not{k}^-)(\varepsilon(k^+)\cdot\varepsilon(k^-))\\&+(\varepsilon(k^+)\cdot k^-)\not{\varepsilon}(k^-)-(\varepsilon(k^-)\cdot k^+)\not{\varepsilon}(k^+)]u_{e^-}\\&-(2\pi)^{-2}\frac{g_W^2}{4(p^+-k^+)^2}\bar{v}_{e^+}\not{\varepsilon}(k^+)(\not{p}^+-\not{k}^+)\not{\varepsilon}(k^-)(1-\gamma_5)u_{e^-};\end{aligned}$$

$$p^++p^-=k^++k^-.$$

For high energies ($\sqrt{s}\gg M_W$) and longitudinal polarization of the $W^\pm$ bosons ($\varepsilon^\mu_L=k^\mu/M_W$), the ν_e exchange diagram:

$$\mathcal{M}=-\frac{(2\pi)^{-2}}{4M_W^2}\frac{g_W^2}{(p^+-k^+)^2}\bar{v}_{e^+}\not{k}^+(\not{p}^+-\not{k}^+)\not{k}^-(1-\gamma_5)u_{e^-}$$

dominates.

Using the Dirac equation ($\not{p}^-u_{e^-}=M_eu_{e^-}$; $\bar{v}_{e^+}\not{p}^+=-\bar{v}_{e^+}M_e$) and leaving out terms proportional to the lepton mass M_e, this can be simplified to:

$$\mathcal{M}=-\frac{(2\pi)^{-2}}{M_W^2}\frac{g_W^2}{4}\bar{v}_{e^+}\not{k}^+(1-\gamma_5)u_{e^-}. \tag{3.1.4}$$

The cross-section derived from this diverges for high energies: $\sigma\sim s$, and thus there is no compensation between the electromagnetic matrix element with minimal coupling and the weak matrix element.

Partial compensation can be achieved if the $W^\pm$ bosons have a gauge-invariant coupling to the photon corresponding to an anomalous magnetic moment κ, i.e. when the expression

$$\frac{(2\pi)^4}{i}e\kappa^2(+g_{\mu\nu}k_\rho-g_{\rho\mu}k_\nu)\,\delta^{(4)}(k^++k^-+k) \tag{3.1,3'}$$

is added to the vertex. A γW^+W^- vertex which is symmetrical in the three vector bosons is obtained for $\kappa=1$. The contribution of this term to the invariant matrix element for $e^+e^- \to W^+W^-$ is

$$\mathcal{M} = -(2\pi)^{-2}\frac{e^2\kappa^2}{(p^+ + p^-)^2}\bar{v}_{e^+}[(\varepsilon(k^+)\cdot k)\not{\varepsilon}(k^-) - (\varepsilon(k^-)\cdot k)\not{\varepsilon}(k^+)]u_{e^-}.$$

From this, the following term emerges for longitudinal polarization and high energies:

$$\mathcal{M} = \frac{(2\pi)^{-2}}{M_W^2} e^2\kappa^2 \bar{v}_{e^+}\not{k}^+ u_{e^-}. \tag{3.1.4'}$$

This term can compensate the vector part in Eqn. (3.1.4) for $e\kappa = g_W/2$, but not the axial vector part, as the electromagnetic interaction conserves parity.

The desired good high energy behaviour correlates the e.m. and weak coupling. However, it can only be achieved by starting with minimal coupling and then introducing additional suitable reaction mechanisms. Neutral vector bosons Z^0 and/or heavy neutral fermions E^0 fall into the latter category and these give rise to the following additional Feynman diagrams:

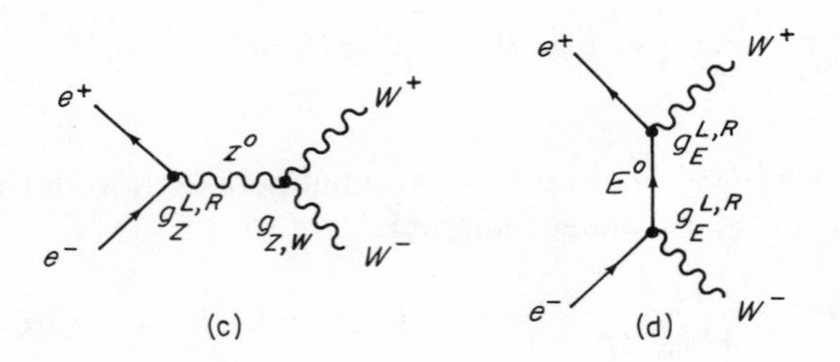

At high energies, graphs (c) and (d) must compensate the leading terms in the e.m. (a) and weak (b) graphs (Eqns. (3.1.4), (3.1.4')). This is the case when the following relationship exists between the e.m. coupling constant e, the W^- boson coupling g_W, and the Z^0 coupling constants g_Z^L, g_Z^R, $g_{Z,W}$ and the E^0 coupling constants g_E^L, g_E^R:

$$e^2\kappa^2 = \frac{g_W^2}{2} + g_{Z,W}g_Z^L + (g_E^L)^2 = g_{Z,W}g_Z^R + (g_E^R)^2. \tag{3.1.5}$$

Eqn. (3.1.5) states that the coupling strength for e.m. interaction and the various weak interaction processes has the same magnitude. This means that e.m. and weak reactions are equally strong at these high energies and can therefore no longer be discussed separately. Extension of the 1-boson exchange model into a weak interaction theory thus requires combination with the e.m. interaction! As we have seen, the small strength of the weak interaction at low energies is due to the fact that the exchanged vector bosons are very heavy.

The possibilities on which a combined electroweak theory can be built are summarized below. The following conditions are essential [Bj 76]:

(1) either a neutral vector boson Z^0, which does not provide just a left-handed coupling;
(2) or a neutral, heavy electron E^0, which does not provide just a left-handed coupling;
(3) or a double-charged electron E^{++}, which does not provide just a left-handed coupling;
(4) or any combination of the above options.

3.1.2 Coupling structure in models with good high energy behaviour

The possibilities for compensation between electromagnetic and weak processes as outlined in the above section will now be systematically built up and thereby the restriction to the electron–neutrino system will be lifted.

(a) The vector bosons involved are denoted by

$$W_a = (W^+, W^-, Z^0, \gamma, \ldots), \quad a = 1, \ldots, N.$$

In accordance with Eqn. (2.3.20), the W_a propagators are given by:

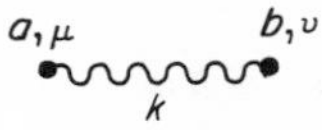

$$\overset{a,\mu}{\bullet}\!\!\sim\!\!\underset{k}{\sim}\!\!\sim\!\!\overset{b,\nu}{\bullet} = \frac{\mathrm{i}}{(2\pi)^4}\frac{-g_{\mu\nu}+k_\mu k_\nu/M_a^2}{k^2-M_a^2+\mathrm{i}\varepsilon}\delta_{ab}. \tag{3.1.6}$$

A derivative coupling with dimensionless coupling constants F_{abc} (see Section 2.4.1(c)) is chosen for the coupling of three vector bosons:

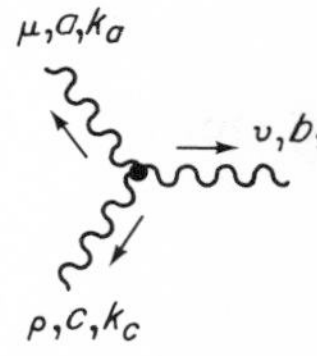

$$\begin{aligned}(\mu,a,k_a;\ \nu,b,k_b;\ \rho,c,k_c) = \frac{(2\pi)^4}{\mathrm{i}}&[F_{abc}g_{\mu\nu}(k_a-k_b)_\rho \\ &+F_{bca}g_{\nu\rho}(k_b-k_c)_\mu \\ &+F_{cab}g_{\rho\mu}(k_c-k_a)_\nu]\,\delta^{(4)}(k_a+k_b+k_c)\end{aligned} \tag{3.1.6'}$$

Fermions can also be collected:

$$\psi_i = (\nu_e, e, \nu_\mu, \mu, \ldots, u_r, d_r, s_r, c_r, \ldots), \quad i = 1, \ldots, n,$$

with r, g, b the colour index of the quarks. The fermion propagators are

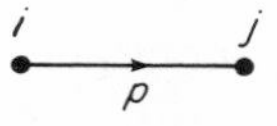

$$\overset{i}{\bullet}\!\!\xrightarrow[p]{}\!\!\overset{j}{\bullet} = \frac{\mathrm{i}}{(2\pi)^4}\frac{\not{p}+M_i}{p^2-M_i^2+\mathrm{i}\varepsilon}\delta_{ij}. \tag{3.1.7}$$

(see Eqn. (1.2.39)). The fermion vector boson coupling constants are called L_{ij}^a and R_{ij}^a for left- and right-handed coupling respectively:

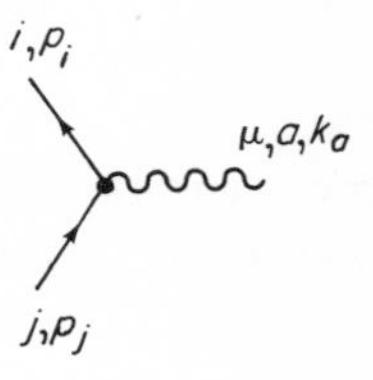

$$(i,p_i;\ j,p_j;\ \mu,a,k_a) = \frac{(2\pi)^4}{\mathrm{i}}(L_{ij}^a\gamma_{\mu L}+R_{ij}^a\gamma_{\mu R})\,\delta^{(4)}(p_i+p_j-k_a). \tag{3.1.7'}$$

(b) The propagators and vertices defined in Eqns. (3.1.6) and (3.1.7) are used to calculate the fermion–antifermion annihilation in vector boson pairs in lowest approximation. The following diagrams contribute to this:

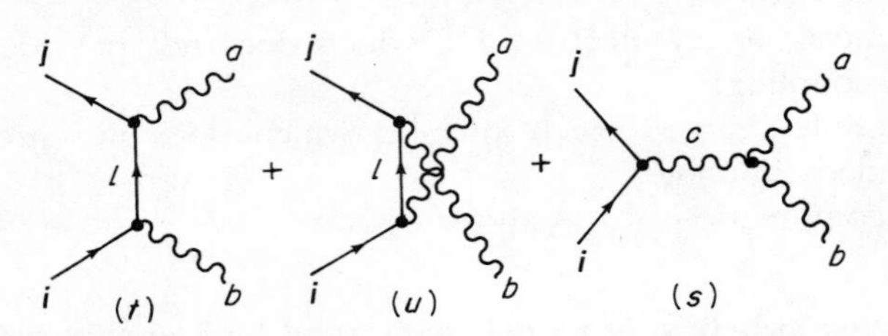

The analytical expression:

$$-(2\pi)^{-2}\sum_{l}\bar{v}_j\not{\varepsilon}_{bL}\frac{\not{p}_i-\not{k}_a+M_l}{(p_i-k_a)^2-M_l^2}\not{\varepsilon}_{aL}u_iL_{il}^{a}L_{lj}^{b}\quad \text{with } \not{\varepsilon}_{bL}=\varepsilon_b^{\mu}\gamma_{\mu L}$$

corresponds to the t-channel diagram. (As left- and right-handed couplings can be considered separately, only the left-handed ones are calculated here.) As in Section 3.1.1, the longitudinal polarization states of the vector bosons are critical at high energies as $\varepsilon_L^{\mu}\sim k^{\mu}/M$. Thus, the leading order becomes:

$$-(2\pi)^{-2}\sum_{l}L_{il}^{a}L_{lj}^{b}\bar{v}_j\not{k}_{bL}u_i/M_aM_b. \tag{3.1.8}$$

The result for the u-channel diagram is obtained by exchanging a with b:

$$(2\pi)^{-2}\sum_{l}L_{il}^{b}L_{lj}^{a}\bar{v}_j\not{k}_{bL}u_i/M_aM_b. \tag{3.1.8$'$}$$

Finally, the contribution made by the s-channel diagram is:

$$-(2\pi)^{-2}\sum_{c}\bar{v}_j\gamma_{\mu L}u_i\frac{-g^{\mu\nu}+(p_i+p_j)^{\mu}(p_i+p_j)^{\nu}/M_c^2}{(p_i+p_j)^2-M_c^2+i\varepsilon}L_{ij}^{c}$$
$$\times[F_{abc}(\varepsilon_a\varepsilon_b)(k_a-k_b)_{\nu}+F_{bca}(\varepsilon_a(k_b+k_c))\varepsilon_{b,\nu}-F_{cab}((k_a+k_c)\varepsilon_b)\varepsilon_{a,\nu}].$$

At high energies, the leading terms become

$$\frac{(2\pi)^{-2}}{M_aM_b}\sum_{c}L_{ij}^{c}\bar{v}_j\{\not{k}_a(F_{abc}-F_{cab})+\not{k}_bF_{bca}\}u_i. \tag{3.1.8$''$}$$

for longitudinally polarized vector bosons. Compensation of the terms in the t-, u-, and s-channel matrix elements, which would give increasing cross-sections at high energies, demands that:

$$F_{abc}=F_{cab} \tag{3.1.9}$$

$$\sum_{l}(L_{il}^{a}L_{lj}^{b}-L_{il}^{b}L_{lj}^{a})=\sum_{c}L_{ij}^{c}F_{bca}. \tag{3.1.10}$$

It follows from this that F_{abc} is totally antisymmetrical and that very severe conditions must be imposed on the fermion–vector boson coupling constants. This becomes even more apparent if L^a_{ij} is interpreted as matrix L^a with row index i and column index j:

$$[L^a, L^b] = F_{abc} L^c. \tag{3.1.11}$$

A similar result is obtained for right-handed couplings:

$$[R^a, R^b] = F_{abc} R^c. \tag{3.1.11$'$}$$

Further information on the structure of the vector boson coupling constants F_{abc} can be obtained from the Jacobi identity:

$$[[L^a, L^b], L^c] + [[L^b, L^c], L^a] + [[L^c, L^a], L^b] = 0$$

and application of Eqn. (3.1.11). The result is:

$$\sum_l (F_{abe}F_{ecd} + F_{bce}F_{ead} + F_{cae}F_{ebd}) = 0 \tag{3.1.12}$$

or, if matrix notation $(F^a)_{bc} = F_{abc}$ is incorporated:

$$[F^a, F^b] = F_{abc} F^c. \tag{3.1.13}$$

A direct group theoretical interpretation can be given for this analysis of compensation conditions for the matrix elements of fermion–antifermion annihilation into vector boson pairs. The coupling constants F_{abc} are structure constants of a Lie group and the matrices F^a are the basis elements which span the adjoint representation of the accompanying Lie algebra. Matrices L^a, R^a of the fermion–vector boson couplings are representation matrices of the Lie algebra of this internal symmetry group.

(c) The couplings used so far are not sufficient for obtaining good results for the high energy behaviour of vector boson–vector boson scattering $W_a + W_b \rightarrow W_c + W_d$. If the sum of the matrix elements of the three Feynman diagrams:

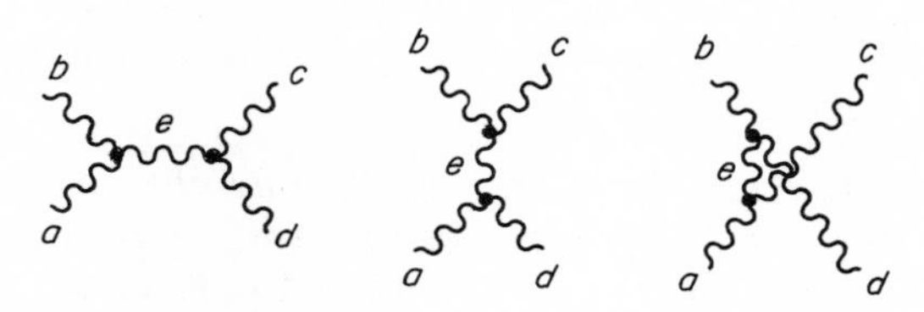

is calculated for longitudinally polarized W's at high energies, together with the vertices from Eqn. (3.1.6′), then the following is obtained for equal W boson masses:

$$\mathcal{M} = -\frac{(2\pi)^{-2}}{4M^4} \sum_e \left(F_{abe}F_{ecd} \frac{s^2}{s - M_e^2}(t-u) + F_{bce}F_{eda} \frac{t^2}{t - M_e^2}(s-u) \right.$$

$$\left. + F_{bde}F_{eca} \frac{u^2}{u - M_e^2}(s-t) \right), \tag{3.1.14}$$

$$s = (k_a + k_b)^2, \quad t = (k_a - k_c)^2, \quad u = (k_a - k_d)^2.$$

This would give a cross-section divergent like s^3. Compensation is achieved by a four-boson coupling of the following form:

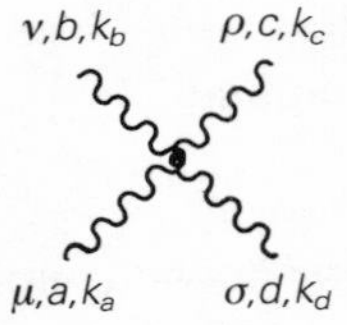

ν, b, k_b ρ, c, k_c

μ, a, k_a σ, d, k_d

$$= -\frac{(2\pi)^4}{\mathrm{i}} \sum_e [F_{abe}F_{ecd}(g_{\mu\sigma}g_{\nu\rho} - g_{\mu\rho}g_{\nu\sigma}) + F_{bce}F_{eda}(g_{\mu\nu}g_{\rho\sigma} - g_{\mu\rho}g_{\nu\sigma}) + F_{bde}F_{eca}(g_{\mu\nu}g_{\rho\sigma} - g_{\mu\sigma}g_{\nu\rho})] \times \delta^{(4)}(k_a + k_b + k_c + k_d). \quad (3.1.15)$$

For longitudinal polarization at high energies, the additional coupling makes the following contribution to $W_a + W_b \to W_c + W_d$:

$$\mathcal{M} = \frac{(2\pi)^{-2}}{4M^4} \sum_e [F_{abe}F_{ecd}s(t-u) + F_{bce}F_{eda}t(s-u) + F_{bde}F_{eca}u(s-t)]. \quad (3.1.16)$$

Hence, the term which increases with s^2 is compensated in $\mathcal{M}$. Terms which are still linear in s remain for $M_a \neq 0$, i.e. for massive vector bosons, and these can be removed by the introduction of scalar particles with suitable couplings proportional to the W mass.

After the above compensation of leading terms in fermion–antifermion annihilation into vector boson pairs, some expressions $M_f \bar{v}_i u_j$ remain which violate unitarity at high energies. Thus scalar particles must also be coupled with fermions too, with a strength which is proportional to the fermion mass.

These aspects of unitarity of the S-matrix calculated in tree-graph approximation can be summarized as follows:

> the vector bosons of the theory must have Yang–Mills type couplings and the fermions are coupled with vector bosons like in a non-Abelian gauge theory. Effects which are proportional to the fermion or vector boson mass must be removed by introducing scalar fields. The coupling of scalar fields to vector bosons and fermions is proportional to the masses M_a or M_f.

An attempt to describe these tree graphs by the classical Lagrangian function for a non-Abelian gauge theory now seems reasonable. If this Lagrangian function is to lead to a renormalizable quantum field theory, then the scalar particles must be specifically related to the vector boson mass. This contains a mechanism for dynamic mass generation which will be explained in the following section.

3.2 Gauge theories with spontaneously broken symmetry

The gauge fields of all gauge field theories considered so far have zero mass—at least as long as the non-renormalized Lagrangian functions, i.e. the tree graph approximation for these theories, is considered. On the other hand, apart from the photon, only massive vector particles are expected in a theory of the electroweak interaction, because the Fermi model could not be reproduced at low energies otherwise. Thus, mass terms must be introduced without disturbing gauge invariance; from gauge invariance, the Ward identities (Section 2.4.3) follow which are needed for the proof of renormalizability and unitarity as outlined in Section 2.3.3(b). For example, a mass term $\frac{1}{2}M^2 A_\mu A^\mu$ must not be added to the Lagrangian density as this is not gauge-invariant. The only known mechanism for introducing masses without violating gauge invariance is the Higgs mechanism which causes spontaneous symmetry breaking. In this, gauge fields are coupled with additional, postulated scalar fields—'Higgs fields'—the field equations of which have unsymmetrical solutions, despite their gauge symmetry. In Section 3.1.2(c) hints to this type of scalar fields have been found.

A Lagrangian function which is invariant under a group of transformations G is used to define more accurately what is meant by *spontaneous symmetry breaking.** There are two alternatives for the ground states Ω, i.e. the states with lowest energy:

(1) Ω is unique; in this case, Ω must be invariant under group transformations from G.
(2) Ω consists of a set of degenerate states which transform into one another under the group G; if an Ω is chosen as the ground state of the theory then the latter is not G-invariant and the G-symmetry is said to be spontaneously broken.

Systems with a finite number of degrees of freedom usually possess a unique ground state. Spontaneous symmetry breaking is much more likely to occur in theories involving an infinite number of degrees of freedom. The phenomenon is independent of its application for mass generation in gauge field theories of distinct physical significance [He 67, Na 61, Ha 62]. We know that chiral symmetry is approximately spontaneously broken [We 68, Da 69, Le 72b]. Another well-known example is the infinitely extended ferromagnet below the Curie temperature [He 28]: although the interaction of the magnetic moments is rotation-symmetrical, in the ground state, all moments are aligned in an arbitrary, but fixed direction which causes a spontaneous magnetization $\mathbf{M} \neq \mathbf{0}$. Conversely, rotation symmetry is not spontaneously broken in the phase above the Curie temperature: $\mathbf{M} = \mathbf{0}$.

The example incorporates two characteristic points which occur in all the modes described below:

(1) A symmetrical phase and a phase with spontaneous symmetry breaking can (separated by a phase transition) occur in *one and the same* system depending on the boundary conditions (the temperature of the ferromagnet, in the example).
(2) In contrast to the symmetrical phase, the one with spontaneous symmetry breaking is characterized by a non-vanishing order parameter;

* The phrase was coined by M. Baker and S. L. Glashow [Ba 62].

this is the magnetization **M** for the ferromagnet or corresponding quantities for other models. These are vacuum expectation values in quantized models.

The Higgs mechanism has already been discussed on a classical level using the example of the Nielsen–Olesen model during discussions on the chromoelectric Meissner effect in QCD (Section 2.6.2). After a preparatory section on the breaking of a gauge symmetry of the first kind (Section 3.2.1), the whole concept will be extended to include non-Abelian local gauge symmetry (Section 3.2.2). The Higgs mechanism is used to explain dynamic mass generation in the Glashow–Salam–Weinberg model for the electroweak interaction (Section 3.3).

The main question is whether Higgs fields are fundamental fields with corresponding, physically detectable particles or not [We 76a, Su 79, Di 79, Pa 80a]. The dynamical explanation of the Landau–Ginzburg model for superconduction with help of the BCS theory [Ba 57] (cf. Section 2.6.2) can be considered as a parallel for the latter case.

3.2.1 Spontaneous symmetry breaking of a gauge symmetry of the first kind

Because of the importance of spontaneous symmetry breaking its dynamics is first described using a gauge symmetry of the first kind as an example.

(a) The *Goldstone model* [Go 61] describes the self-interaction of a complex field $\phi(x)$. Its Lagrangian density which is invariant under global phase transformations

$$\phi(x) \to \mathrm{e}^{-\mathrm{i}\theta}\phi(x), \quad \phi^\dagger(x) \to \phi^\dagger(x)\mathrm{e}^{\mathrm{i}\theta}, \quad \theta = \text{constant}, \tag{3.2.1}$$

is given by

$$\mathscr{L} = (\partial^\mu \phi)^\dagger\, \partial_\mu \phi - V(\phi)$$

with

$$V(\phi) = \frac{\lambda}{4}\left(|\phi|^2 - \frac{\varepsilon\mu^2}{\lambda}\right)^2 = \frac{\lambda}{4}|\phi|^4 - \frac{\varepsilon\mu^2}{2}|\phi|^2 + \frac{\mu^4}{4\lambda}. \tag{3.2.2}$$

The coupling constant λ is positive, otherwise the energy would not be bounded from below. On the other hand, there are two possibilities for the sign ε of μ^2 (>0) in Eqn. (3.2.2).

(1) $\varepsilon = -1$: the $\varepsilon\mu^2$ term is a mass term; the potential $V(\phi)$ has a unique, classical vacuum (= field configuration with lowest energy): $\phi(x) \equiv 0$.

(2) $\varepsilon = +1$: in this case, Eqn. (3.2.1) describes a phase with spontaneously broken symmetry; all fields (= 'order parameter', see above)

$$\phi(x) = v\mathrm{e}^{-\mathrm{i}\theta}, \quad v = \sqrt{\left(\frac{\mu^2}{\lambda}\right)}, \quad \begin{array}{l} \theta = \text{constant} \\ 0 \leqslant \theta < 2\pi \end{array} \tag{3.2.3}$$

describe configurations of lowest energy (see Fig. 3.1). The phase θ of the field is thus arbitrary and can be changed by a gauge transformation of the first kind (Eqn. (3.2.1)).

The classical (degenerate) ground state, i.e. the field configurations (3.2.3) are used as a basis for the perturbative treatment of the self-interaction of the field ϕ for $\varepsilon=+1$. The fields η and ξ are introduced by

$$\phi(x)=\left(v+\frac{1}{\sqrt{2}}\,\eta(x)\right)\exp[-\mathrm{i}(\theta+\xi(x)/\sqrt{2}\,v)] \tag{3.2.4}$$

in which $\eta(x)$ and $\xi(x)$ are real fields describing small deviations of the modulus and the phase of the ϕ-field from the ground state (3.2.3), respectively. The ground state is realized for $\eta(x)\equiv\xi(x)\equiv 0$. Hence, the Lagrangian density (3.2.2) reads:

$$\begin{aligned}\mathscr{L}=\tfrac{1}{2}(\partial_\mu\xi)\,\partial^\mu\xi+\tfrac{1}{2}(\partial_\mu\eta)\,\partial^\mu\eta-\tfrac{1}{2}\mu^2\eta^2-\frac{\lambda v}{2\sqrt{2}}\,\eta^3-\frac{\lambda}{16}\,\eta^4\\ +\text{coupling terms between } \xi \text{ and } \eta \text{ field}\end{aligned} \tag{3.2.5}$$

and the results obtained are as follows:

(1) $\eta(x)$ is a scalar field with mass μ; it describes the radial oscillations of the field around the vacuum value v (see Fig. 3.1). The finite mass results from the curvature of the potential which does not vanish along the radial direction.

(2) $\xi(x)$ is a zero mass, scalar field, the 'Goldstone' field. It is massless as a result of the vanishing curvature along the potential minima; the field (3.2.4) does not oscillate classically in this direction. $\xi(x)$ describes the associated normal mode with eigenfrequency zero (see Fig. 3.1).

The Goldstone model exhibits a property generally associated with theories with spontaneous breaking of a continuous gauge symmetry of the first kind, namely the occurrence of zero mass, scalar field excitations.

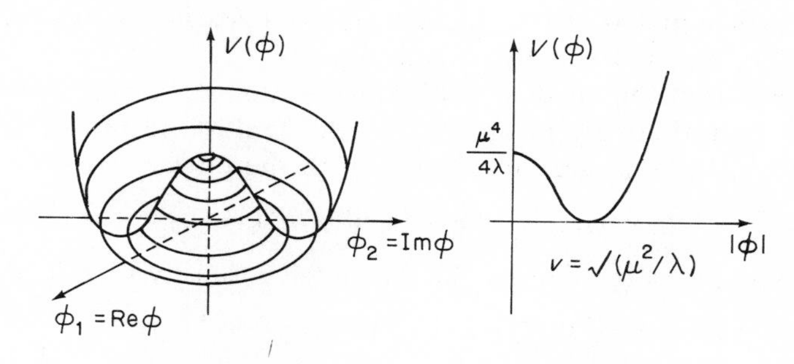

Figure 3.1 Potential of the Goldstone model in the phase with spontaneous symmetry breaking $\varepsilon=+1$

(b) The gauge group of the Goldstone model is the commutative $U(1)$ group of phase transformations (3.2.1). Its *extension to non-Abelian groups* G ($\dim G = N$) is obvious. A real multiplet of scalar fields $\phi^i(x)$ which is transformed as follows under a real representation of the group G, is considered:

$$\begin{aligned} \phi^i(x) &\to (\exp(-\mathrm{i}\theta^a T_a))^i_j \phi^j(x), \quad \theta^a = \text{constant}, \\ \delta\phi^i(x) &= -\mathrm{i}(T_a)^i_j \phi^j(x)\, \delta\theta^a \end{aligned} \tag{3.2.6}$$

(cf. Eqns. (1.2.62)–(1.2.63)). Complex representations can be dealt with in an analogous way. The potential $V(\phi)$ of the Lagrangian density:

$$\mathcal{L} = \tfrac{1}{2}(\partial_\mu \phi_i)(\partial^\mu \phi^i) - V(\phi) \tag{3.2.7}$$

is required to be invariant under the gauge transformations (3.2.6):

$$0 = \delta V = \frac{\partial V}{\partial \phi^i}\, \delta\phi^i = -\mathrm{i}\frac{\partial V}{\partial \phi^i}(T_a)^i_j \phi^j\, \delta\theta^a. \tag{3.2.8}$$

In view of the free choice of $\delta\theta^a$, this means:

$$\frac{\partial V}{\partial \phi^i}(T_a)^i_j \phi^j = 0, \quad a = 1, \ldots, N. \tag{3.2.9}$$

Differentiation with respect to ϕ^k gives:

$$\frac{\partial^2 V}{\partial \phi^i\, \partial \phi^k}(T_a)^i_j \phi^j + \frac{\partial V}{\partial \phi^i}(T_a)^i_k = 0. \tag{3.2.10}$$

As the example of the Goldstone model showed, spontaneous symmetry breaking occurs when $V(\phi)$ has a minimum for $\phi^i(x) \equiv v^i \neq 0$. Because of the invariance of $V(\phi)$ under the transformations (3.2.6), $\exp(-\mathrm{i}\theta^a T_a)v$ is then also a minimum:

$$\left.\frac{\partial V}{\partial \phi^i}\right|_{\phi^i \equiv v^i} = 0, \quad \left.\frac{\partial^2 V}{\partial \phi^i\, \partial \phi^k}\right|_{\phi^i \equiv v^i} = (M^2)_{ik} \geqslant 0. \tag{3.2.11}$$

The symmetrical matrix $(M^2)_{ik}$ formed from the second derivatives of V in the minimum, has non-negative eigenvalues.

Deviations from the vacuum field configuration $\phi^i(x) = v^i = \text{constant}$ are considered perturbatively in analogy to the Goldstone model, and hence fields $\eta^i(x) = \phi^i(x) - v^i$ are introduced. With the power series

$$V(\phi) = V_0 + \tfrac{1}{2}(M^2)_{ik}\eta^i(x)\eta^k(x) + \ldots \tag{3.2.12}$$

for the potential energy density the Lagrangian density (3.2.7) is:

$$\mathcal{L} = \tfrac{1}{2}(\partial_\mu \eta_i(x))\, \partial^\mu \eta^i(x) - \tfrac{1}{2}(M^2)_{ik}\eta^i(x)\eta^k(x) + \ldots. \tag{3.2.13}$$

This gives the physical interpretation of $(M^2)_{ik}$: $(M^2)_{ik}$ is the matrix of mass squares for the fields $\eta^i(x)$. Eqn. (3.2.10) is specialized to $\phi^i(x) \equiv v^i$ in order to obtain information on the eigenvalues of M^2:

$$(M^2)_{ik}(T_a)^i_j v^j = 0, \quad a = 1, \ldots, N. \tag{3.2.14}$$

In general, v *is invariant under a subgroup* H *of* G. The generators of the Lie algebra of H are written as: $\hat{T}_{\hat{a}}$, $\hat{a} = 1, \ldots, \hat{N}$. The invariance of v under transformations from H means that:

$$(\hat{T}_{\hat{a}})^i_j v^j = 0, \quad \hat{a} = 1, \ldots, \hat{N}. \tag{3.2.15}$$

As a result, $\hat{N}$ of the N Eqns. (3.2.14) are trivially satisfied. However, the remaining $N - \hat{N}$ equations give eigenvalues zero for M^2. The result of this consideration is the *Goldstone theorem* [Go 61, Go 62, Bl 63]: the matrix of mass squares M^2 has $(N - \hat{N})$ eigenvalues zero. The associated field excitations are called Goldstone fields and result from spontaneous symmetry breaking. The remaining eigenvalues of M^2 are positive.

(c) The σ *model* [Ge 60, Ba 70] for spontaneously broken, chiral $SU(2) \times SU(2)$ symmetry is given as an *example* with physical application. The group $SU(2) \times SU(2)$ is six-dimensional ($N = 6$). It is locally isomorphic to the group $G = SO(4)$. The fields ϕ^i should form 4-component, real vectors which are transformed according to the $SO(4)$ group. The Lagrangian density can be written as (cf. Eqn. (3.2.2)):

$$\mathcal{L} = \tfrac{1}{2}(\partial_\mu \phi_i)(\partial^\mu \phi^i) - V(\phi), \quad V(\phi) = \frac{\lambda}{4}\left(\phi_i\phi^i - \frac{\mu^2}{\lambda}\right)^2 \tag{3.2.16}$$

The minima of $V(\phi)$ are characterized by

$$\phi_i\phi^i = \frac{\mu^2}{\lambda} = v^2 > 0 \tag{3.2.17}$$

i.e. the fields which minimalize V form a 3-dimensional sphere with radius v in 4-dimensional Euclidean ϕ-space. From these, the vector

$$\phi_0 = (0, 0, 0, v)$$

is chosen. This is invariant under the orthogonal transformations of the first three coordinates, i.e. under the subgroup $H = SO(3)$. Hence $N = 6$, $\hat{N} = 3$, and $N - \hat{N} = 3$.

According to the Goldstone theorem, there are three zero mass fields $\boldsymbol{\pi} = (\pi^1, \pi^2, \pi^3)$, the quanta of which can be interpreted as pions in the σ model and one massive field σ, which appears in the quantized version as σ meson. If $\phi^i = (\boldsymbol{\pi}, \sigma + v)$, the Lagrangian density (3.2.16) is expressed as:

$$\mathcal{L} = \tfrac{1}{2}(\partial_\mu \sigma)\,\partial^\mu \sigma - \tfrac{1}{2}2\mu^2\sigma^2 + \tfrac{1}{2}(\partial_\mu \boldsymbol{\pi})\,\partial^\mu \boldsymbol{\pi} - \frac{\lambda}{4}(\boldsymbol{\pi}^2 + \sigma^2)^2 - \lambda v \sigma(\boldsymbol{\pi}^2 + \sigma^2). \tag{3.2.18}$$

The mass of the σ meson is $\sqrt{2}\mu^2$. The fact that the physical pions are not completely massless like Goldstone bosons can be ascribed to an explicit symmetry breaking of the chiral $SU(2)\times SU(2)$ group. According to the PCAC hypothesis [Ad 68], the π mass represents a measure for the explicit breaking of the spontaneously broken chiral symmetry.

(d) So far, the models have been considered as classical field theories. They form the lowest order of a semiclassical approximation for the corresponding quantized theories according to Section 2.3.2(e). It will now be shown that the *Goldstone theorem is also valid for the quantized, non-Abelian Goldstone model.* One method of proving this is to analyse the matrix elements of the charge operator which generates the internal, spontaneously broken symmetry [Gi 64, Be 74a]. Instead the path integral quantization (Section 2.3.2) is used [Jo 64] and the generating functional of the disconnected Green functions is considered:

$$T\{j\}=Z\{j\}/Z\{0\}$$

with

$$Z\{j\}=\int \mathscr{D}[\phi]\exp\left(\mathrm{i}\int \mathrm{d}^4x(\mathscr{L}+j_i(x)\phi^i(x))\right) \tag{3.2.19}$$

and $\mathscr{L}$ from Eqn. (3.2.7). An infinitesimal gauge transformation (of the first kind) is carried out in the partition integral $Z\{j\}$, in the same way as for the derivation of the Ward identities in Sections 2.3.2(f) and 2.3.3(c): $\phi^i(x)\to\phi^i(x)+\delta\phi^i(x)$, $\delta\phi^i(x)$ according to Eqn. (3.2.6). The formula:

$$\int \mathrm{d}^4x j_i(x)(T_a)^i_k\frac{\delta T\{j\}}{\mathrm{i}\,\delta j_k(x)}=0$$

is obtained as a result of the gauge invariance of $\mathscr{L}$ and $\mathscr{D}[\phi]$, or, using the generating functional of connected Green functions $\phi\{j\}=\log T\{j\}$ (see Eqn. (2.3.68)):

$$\int \mathrm{d}^4x j_i(x)(T_a)^i_k\frac{\delta \phi\{j\}}{\mathrm{i}\,\delta j_k(x)}=0. \tag{3.2.20}$$

A Legendre transformation gives the generating functional of the vertices (see Eqns. (2.3.69)–(2.3.70)). This has the following form:

$$\Gamma\{\varphi\}=-\mathrm{i}\int \mathrm{d}^4x j_i(x)\varphi^i(x)+\phi\{j\}$$
$$\varphi^i(x)=\frac{\delta\phi\{j\}}{\mathrm{i}\,\delta j_i(x)},\quad j_i(x)=\mathrm{i}\frac{\delta\Gamma\{\varphi\}}{\delta\varphi^i(x)}, \tag{3.2.21}$$

and Eqn. (3.2.20) is changed to:

$$\int \mathrm{d}^4x\frac{\delta\Gamma\{\varphi\}}{\delta\varphi^i(x)}(T_a)^i_k\varphi^k(x)=0. \tag{3.2.22}$$

In the ground state, the field remains constant but differs from zero for spontaneous symmetry breaking, i.e.

$$v^i = \langle 0|\, \phi^i(x)\, |0\rangle = \frac{\delta\phi\{j\}}{\mathrm{i}\,\delta j_i(x)}\bigg|_{j=0} \neq 0.$$

The inverse propagator of the ϕ field then follows from Eqn. (2.3.72):

$$\frac{\delta^2\Gamma\{\varphi\}}{\delta\varphi^i(x)\,\delta\varphi^k(y)}\bigg|_{\varphi=v} = -(D'^{-1})_{ik}(x-y). \tag{3.2.23}$$

If Eqn. (3.2.22) is differentiated with respect to φ^k, if φ^i is put equal to v^i and if the transformation to momentum space is performed, then:

$$(D'^{-1})_{li}(p=0)(T_a)^i_k v^k = 0, \tag{3.2.24}$$

is obtained with Eqn. (3.2.23), i.e. the same formula as in the classical approximation (3.2.14) which was used as the starting point for the proof of the Goldstone theorem. This means that zero mass Goldstone fields also occur in quantized theories with gauge symmetry of the first kind and mass generation caused by spontaneous symmetry breaking.*

3.2.2 Spontaneous symmetry breaking of a gauge symmetry of the second kind

(a) The result of spontaneous symmetry breaking in a theory with gauge symmetry of the second kind in which the gauge fields are minimally coupled to Higgs fields ϕ with a Goldstone potential $V(\phi)$, is that part of the gauge fields become massive. No zero mass Goldstone bosons occur in this case.

This mass generation mechanism will now be considered using the Higgs–Kibble model [Hi 64, En 64, Gu 64, Ki 67, Be 74a]. The Lagrangian density of this model describes a non-Abelian gauge theory, with Higgs fields as the charged fields:

$$\begin{gathered}\mathscr{L} = -\tfrac{1}{4}F^a_{\mu\nu}F^{\mu\nu}_a + \tfrac{1}{2}(D_\mu\phi_i)D^\mu\phi^i - V(\phi),\\ V(\phi) = \frac{\lambda}{4}\left(\phi_i\phi^i - \frac{\varepsilon\mu^2}{\lambda}\right)^2, \quad D_\mu = \partial_\mu + \mathrm{i}\bar{g}A^a_\mu T_a.\end{gathered} \tag{3.2.25}$$

Eqn. (3.2.25) is the generalization of the Nielsen–Olesen string discussed in Section 2.6 (see Eqn. (2.6.24)) to non-Abelian gauge groups. Only the group $SO(n)$ is considered here, otherwise the group theoretical formalism has to be developed too far (Reference [Ki 67] gives further details on the group theory of symmetry breaking.) The subgroup H of the symmetry which is preserved is $SO(n-1)$ (see Eqn. (3.2.15)).

* The limit $\varphi^i \to v^i$ was interchanged with the integration $\int \mathrm{d}^4x$ when deriving Eqn. (3.2.24) from Eqn. (3.2.22); this might be not allowed if singularities arise thereby. Restrictions on the validity of the Goldstone theorem could be imposed as a result of this [Jo 64].

In our example, the n scalar fields ϕ^i are transformed according to the defining representation and the $N=n(n-1)/2$ gauge fields according to the adjoint representation of the group $SO(n)$ (compare Eqn. (2.2.7)). $\mathscr{L}$ is invariant under these gauge transformations of the second kind. The classical ground state, i.e. the state with lowest total energy, is represented by all constant fields ϕ^i with $\phi_i\phi^i=\mu^2/\lambda=v$. As in the previous examples, small excitations of the ground state are parameterized by:

$$\phi^i(x)=\exp(-\mathrm{i}\check{T}_{\check{a}}\xi^{\check{a}}(x)/v)\begin{pmatrix}0\\ \vdots\\ 0\\ v+\eta(x)\end{pmatrix},\quad \check{a}=1,\ldots,N-\hat{N},$$

with (3.2.26)

$$\check{T}_{\check{a}}\begin{pmatrix}0\\ \vdots\\ 0\\ v\end{pmatrix}\neq 0\quad\text{for } \check{a}=1,\ldots,N-\hat{N}$$

(see Eqns. (3.2.4), (3.2.15)). $N-\hat{N}=n(n-1)/2-(n-1)(n-2)/2=n-1$ applies for the group $SO(n)$. In contrast with the models considered up to now the Goldstone fields $\xi^{\check{a}}(x)$ can be regarded as parameters of gauge transformations of the second kind $\exp(-\mathrm{i}\check{T}_{\check{a}}\xi^{\check{a}}(x)/v)$ and thus have no physical significance. As a result of the gauge symmetry of the Lagrangian density (3.2.25), the Goldstone degrees of freedom can be gauged away by carrying out the gauge transformations $\exp(\mathrm{i}\check{T}_{\check{a}}\xi^{\check{a}}(x)/v)$. The Higgs field $\phi^i(x)$ in Eqn. (3.2.26) is then given by:

$$\phi^i(x)=\begin{pmatrix}0\\ \vdots\\ 0\\ v+\eta(x)\end{pmatrix}. \qquad (3.2.26')$$

So far, gauges have always been fixed by specifying the form of the gauge field, e.g. $A^a_\mu(x)$ so that $\partial^\mu A^a_\mu(x)=0$. However, other fields which are not transformed trivially can be used for gauge fixing, as well (see p. 110): the requirement that $\phi^i(x)$ has the form Eqn. (3.2.26′) fixes a gauge, which is in fact a 'unitary gauge' because the unphysical ξ degrees of freedom do not occur and therefore the unitarity of the S-matrix is evident (see Section 2.4.3(d)). In the gauge Eqn. (3.2.26′), the Lagrangian density (3.2.25) is as follows:

$$\mathscr{L}=-\tfrac{1}{4}F^a_{\mu\nu}F^{\mu\nu}_a+\tfrac{1}{2}M^2_{\check{a}\check{a}'}A^{\check{a}}_\mu A^{\mu,\check{a}'}+\tfrac{1}{2}(\partial_\mu\eta)\,\partial^\mu\eta-\mu^2\eta^2$$
$$+\text{coupling terms between 3 and 4 fields} \qquad (3.2.27)$$

with

$$M^2_{\check{a}\check{a}'} = g^2 v^2 (\check{T}_{\check{a}} \check{T}_{\check{a}'})_{nn} = \text{matrix of mass squares.}$$

From this one concludes:

(1) the $N - \hat{N}$ $(= n-1$ for $SO(n))$ ξ fields are gauged away. No Goldstone bosons appear when spontaneous symmetry breaking occurs in gauge theories;
(2) the η field is a massive $(M = \mu\sqrt{2})$, neutral, scalar field; its quanta are called Higgs particles;
(3) $N - \hat{N}$ $(= n-1$ for $SO(n))$ vector fields have become massive $(M = gv)$ and thus have longitudinal, physical polarization states too, in contrast to the zero mass vector fields (compare Eqns. (2.3.17)–(2.3.20));
(4) the remaining $\hat{N}$ $(= (n-1)(n-2)/2$ for $SO(n))$ vector fields have zero mass.

The comparison of the degrees of freedom described by $\mathscr{L}$ according to Eqn. (3.2.25) or—gauge equivalent—Eqn. (3.2.28), gives the following picture:

	$\mathscr{L}$ as in Eqn. (3.2.25)		$\mathscr{L}$ as in Eqn. (3.2.28)	
Gauge fields $M = 0$, physical polarization states	$N = n(n-1)/2$	$2N$	$\hat{N} = (n-1)(n-2)/2$	$2\hat{N}$
Gauge fields $M \neq 0$, physical polarization states	0	0	$N - \hat{N} = n - 1$ ↑	$3(N - \hat{N})$
Higgs fields	$1 + (n-1) =$ ↓	n		1
		(+) $2N + n$		(+) $2N + n$

$\hat{N} = \dim H$, $H = SO(n-1)$.

The transmutation of Goldstone modes into longitudinal polarization states of massive vector bosons is known as the Higgs mechanism.

(b) The gauge field theory defined by Eqn. (3.2.25) can be quantized and renormalized according to Sections 2.3 and 2.4. This is certainly true for $\varepsilon = -1$, but also applies in the phase considered here, i.e. $\varepsilon = +1$ with spontaneous symmetry breaking ('Higgs phase', see Section 2.6.2) [tH 71, Le 74a]. Since Eqn. (3.2.28) represents only a gauge-equivalent formulation of the same theory, the Higgs mechanism can be used to formulate a *renormalizable theory with massive vector fields.* The consistent quantization and evaluation beyond the tree graph approximation can therefore be carried out by: (1) fixing a gauge (see below); and (2) introducing Faddeev–Popov fields according to the quantization procedure outlined in Sections 2.3 and 2.4.

In this connection, the meaning of the concept 'spontaneous symmetry breaking' established at the beginning of this section must be stressed once again. The Lagrangian density (3.2.28) is gauge-invariant! The currents coupled to vector fields are conserved. Only the gauge transformation law, expressed in the η field, becomes inhomogeneous.

(c) The quantization of gauge field theories with spontaneous symmetry breaking requires *gauge fixing* (Section 2.3.3(a)). The unitary gauge used in Eqn. (3.2.28) gives the canonical form (2.3.20) of the vector boson propagators and unphysical scalar bosons are not present. Unitarity of the S-matrix is manifest but Green functions cannot be renormalized in this gauge. Divergences do not cancel until the S-matrix is calculated [We 72, Le 72c, Ap 72]. Moreover, the definition of the finite part of the S-matrix may not be unique under certain circumstances [Ja 72, Ba 72a]. Thus, a renormalizable gauge (in the sense of Section 2.4.1(b)) is appropriate, just as it was for the case without symmetry breaking.

In principle, Eqns. (2.3.124) and (2.3.129″) can be used as a gauge fixing term, but this is often unsuitable. More frequently used are generalized forms, viz.: (1) the Fradkin–Tyutin gauges [Fr 74]; and (2) the 't Hooft gauges [tH 71].

As a 't Hooft gauge will be used in the next section, the Higgs model (3.2.25) for the Abelian gauge group $SO(2)$ will now be considered in this gauge. When $\phi=(\phi_1, v+\eta)$, the gauge-invariant Lagrangian density (3.2.25) is as follows:

$$\begin{aligned}\mathscr{L} = &-\tfrac{1}{4}F_{\mu\nu}F^{\mu\nu}+\frac{g^2v^2}{2}A_\mu A^\mu+\tfrac{1}{2}(\partial_\mu\eta)\,\partial^\mu\eta-\mu^2\eta^2\\ &+\tfrac{1}{2}(\partial_\mu\phi_1)\,\partial^\mu\phi_1+gv(\partial^\mu A_\mu)\phi_1\\ &+\text{coupling terms between three and four fields.}\end{aligned}\tag{3.2.29}$$

In the 't Hooft gauge, the gauge fixing part of the Lagrangian density, $\mathscr{L}_{\text{fix}}$, is chosen so that all bilinear mixed terms in the fields disappear, in the example $gv(\partial^\mu A_\mu)\phi_1$:

$$\begin{aligned}C[A_\mu(x)] &= \sqrt{(\xi)}\,\partial^\mu A_\mu(x)+\frac{1}{\sqrt{\xi}}M\phi_1(x), \quad M=gv,\\ \mathscr{L}_{\text{fix}} &= -\tfrac{1}{2}(C[A_\mu(x)])^2=-\frac{\xi}{2}\left(\partial^\mu A_\mu(x)+\frac{1}{\xi}M\phi_1(x)\right)^2.\end{aligned}\tag{3.2.30}$$

Thus the mass matrix of $\mathscr{L}+\mathscr{L}_{\text{fix}}$ is diagonal in the fields. The ϕ_1 propagator is given as:

$$\frac{\mathrm{i}}{(2\pi)^4}\frac{1}{p^2-M^2/\xi}.\tag{3.2.31}$$

When $\xi \to 0$, the ϕ_1 mass $M/\sqrt{\xi}$ is infinite and hence the ϕ_1 field should decouple from the physical degrees of freedom in this limit. The gauge field propagator has a form already given in Section 2.4.1(b):

$$\frac{\mathrm{i}}{(2\pi)^4}\frac{-g_{\mu\nu}+k_\mu k_\nu(\xi-1)/(\xi k^2-M^2)}{k^2-M^2+\mathrm{i}\varepsilon}. \tag{3.2.32}$$

For $\xi = 0$, it agrees with the propagator (2.3.20), so that the 't Hooft gauge (3.2.30) has a unitary limit $\xi = 0$, whereas it is renormalizable for $\xi \neq 0$ (see Section 2.4.1(c)). The proof of the unitarity of the S-matrix simplifies according to Section 2.4.3(e).

The Lagrangian density for the Faddeev–Popov ghost fields $\mathscr{L}_{\text{ghost}}$ is calculated from $C^a[A_\mu(x)]$ according to Eqns. (2.3.126) and (2.3.127).

(d) A non-Abelian gauge theory for spinor and scalar fields which interact with each other via massive, intermediate vector bosons, is a suitable dynamic concept for describing the electroweak phenomena (Section 3.1). The Higgs mechanism can be used for constructing a renormalizable theory of this kind: the scalar fields in the form of Higgs fields generate the masses of the gauge bosons by spontaneous symmetry breaking. The construction of the general, gauge-invariant part of the *Lagrangian function* of such a theory is outlined below:

(1) a gauge group G is chosen; as the gauge fields are always transformed according to the adjoint representation of G, the number of gauge fields $W_a = (W^+, W^-, Z^0, \gamma, \ldots)$ can be derived from it. The gauge field couplings are given by the structure constants f_{abc} $(=-\mathrm{i}F_{abc}/g$ in comparison with Section 3.1);

(2) representations L_a, R_a, and P_a of G are chosen, under which left- and right-handed fermions $\psi_i = (\nu_e, e, \nu_\mu, \mu, \ldots, u, d, s, c, \ldots)$ and scalars ϕ^r are transformed, respectively:

$$\begin{aligned}&[L_a, L_b] = \mathrm{i}f_{abc}L^c, \quad [R_a, R_b] = \mathrm{i}f_{abc}R^c,\\ &[P_a, P_b] = \mathrm{i}f_{abc}P^c.\end{aligned} \tag{3.2.33}$$

When choosing the representations, care must be taken that no γ_5 anomalies can occur (see Section 2.4.3(f));

(3) the general form of the gauge-invariant, local, renormalizable Lagrangian density, which also includes self-couplings $V(\phi)$ of the scalar fields and a coupling between scalars and fermions, is then given by:*

$$\begin{aligned}\mathscr{L} = &-\tfrac{1}{4}F^a_{\mu\nu}(x)F^{\mu\nu}_a(x) + \bar{\psi}_i(x)(\mathrm{i}\gamma^\mu D^{ij}_\mu - M^{ij})\psi_j(x)\\ &+\tfrac{1}{2}(D^{rs}_\mu\phi_s(x))(D^\mu_{rs'}\phi^{s'}(x)) - V(\phi(x))\\ &+g\bar{\psi}_i(x)\phi^r(x)\left(X^{ij}_{L,r}\frac{1-\gamma_5}{2}+X^{ij}_{R,r}\frac{1+\gamma_5}{2}\right)\psi_j(x)\end{aligned} \tag{3.2.34}$$

* For the choice of the coupling constants, see the note given at the end of Section 2.2.2(c) (Eqn. (2.2.37)).

with

$$F^a_{\mu\nu}(x) = \partial_\mu W^a_\nu(x) - \partial_\nu W^a_\mu(x) - gf^{abc} W^b_\mu(x) W^c_\nu(x),$$

$$D^{ij}_\mu \psi_j(x) = \left[\delta^{ij}\,\partial_\mu + igW^a_\mu(x)\left(L^{ij}_a \frac{1-\gamma_5}{2} + R^{ij}_a \frac{1+\gamma_5}{2}\right)\right]\psi_j(x), \quad (3.2.34')$$

$$D^{rs}_\mu \phi_s(x) = (\delta^{rs}\,\partial_\mu + igW^a_\mu(x) P^{rs}_a)\phi_s(x).$$

$V(\phi)$ is a polynomial of the fourth degree at most in ϕ (see Eqn. (3.2.25)). Higher powers may not occur in the fields in the other coupling terms, too. Otherwise, coupling constants with negative mass dimension destroy the renormalizability of the theory (see Section 2.4.1(c) and Table 2.2). The G symmetry can be spontaneously broken and a Higgs–Kibble mechanism introduced by a suitable choice of $V(\phi)$. The gauge bosons of the symmetry group $H \subset G$, which leave a classical vacuum vector ϕ_0 invariant remain massless and the remainder become massive (see Section 3.2.2(a)). The gauge theory is said to be broken spontaneously $G \to H$ by the 'Higgs structure'. The $\bar{\psi}\phi\psi$ coupling term is gauge-invariant if (in matrix notation):

$$[L_a, X^r_L] = X_{L,s} P^{sr}_a \quad [R_a, X^r_R] = X_{R,s} P^{sr}_a$$

applies.

The phenomenological analysis of the electroweak interaction in Sections 1.3 and 3.1 has shown that the gauge group $G = SU(2)_W \times U(1)$ of the weak isospin and weak hypercharge, and hence four gauge fields W^+_μ, W^-_μ, Z_μ, A_μ are needed as the basis for a 'minimal' theory. More complicated models based on the above scheme must contain this structure.

3.3 The Glashow–Salam–Weinberg theory

The preceding sections have given plausibility to the reasons why the extension of the Fermi model to a renormalizable theory of the electroweak interaction results in a spontaneously broken, non-Abelian gauge theory. The following section gives details of the standard model for the electroweak interaction which was drawn up for leptons by S. L. Glashow [Gl 61], S. Weinberg [We 67], and A. Salam [Sa 68] and extended to include hadronic degrees of freedom by S. L. Glashow, J. Iliopoulos, and L. Maiani [Gl 70]. It is a minimal model as far as the number of required fields is concerned [GSW].

3.3.1 The Lagrangian function for the GSW theory

The structure of the phenomenological Fermi model as described in Section 1.3 is the basis for the construction of the GSW theory. This is characterized by the representation of weak interaction universality as a symmetry under weak isospin $SU(2)_W$ and weak hypercharge $U(1)$ transformations. Thus, the GSW theory describes the dynamics of electroweak interaction as a gauge theory with the group $SU(2)_W \times U(1)$ as a spontaneously broken, non-Abelian gauge group.

(a) The gauge fields are four *vector fields* which span the adjoint representation of this minimal gauge group $SU(2)_W \times U(1)$. The gauge fields belonging to I_W^a, $a=1,2,3$ are called $W_\mu^a(x)$, and that belonging to the weak hypercharge Y_W is $B_\mu(x)$. Thus the Lagrangian density is:

$$\mathscr{L}_E = -\tfrac{1}{4}(\partial_\mu W_\nu^a - \partial_\nu W_\mu^a + g_2 \varepsilon_{abc} W_\mu^b W_\nu^c)^2 - \tfrac{1}{4}(\partial_\mu B_\nu - \partial_\nu B_\mu)^2, \qquad (3.3.1)$$

where $\varepsilon_{abc} \equiv$ isospin structure constants (1.2.1). As the group is not simple, but rather a product of $SU(2)$ and $U(1)$, electromagnetic and weak interactions are combined by two coupling constants: g_2 for weak isospin $SU(2)_W$, g_1 for weak hypercharge $U(1)$—see p. 74.

(b) *Fermions* can be arranged in particle pairs (i numbers the pairs):

$$\{(\nu_e, e^-), (\nu_\mu, \mu^-), (\nu_\tau, \tau^-), \ldots, (u, d'), (c, s'), (t, b'), \ldots\} = \{(a^i, b^i)\} = \{\psi_f\}.$$

For leptons, this is suggested by the individually conserved lepton numbers. For quarks, it can be regarded as a classification based on mass: (u, d) are the light quarks, (c, s) the moderately heavy ones, etc. The dashes on the quarks d', s', b' indicate that Cabibbo-transformed fields (Eqn. (1.3.20)) are used. The left-handed fields $\psi_L(x) = \frac{1}{2}(1-\gamma_5)\psi(x)$ in these pairs can be incorporated into the fundamental two-dimensional representation of weak isospin $\psi_L^i(x) = (a_L^i(x), b_L^i(x))$, $I_W^a = \tau^a/2$ ($\tau^a \equiv$ Pauli matrices). Each right-handed field $\psi_R(x) = \frac{1}{2}(1+\gamma_5)\psi(x)$ is put in a one-dimensional trivial representation. Weak hypercharge is arranged such that the Gell-Mann–Nishijima formula applies for the fermion electric charges Q:

$$Q = I_W^{(0)} + \frac{Y_W}{2}. \qquad (3.3.2)$$

Hence the Lagrangian density of the fermions is:

$$\mathscr{L}_F = \sum_i \Big[\bar{\psi}_L^i \gamma^\mu \Big(i\,\partial_\mu + g_2 \frac{\tau^a}{2} W_\mu^a - g_1 \frac{y_i}{2} B_\mu \Big) \psi_L^i + \bar{a}_R^i \gamma^\mu \Big(i\,\partial_\mu + g_1 \frac{\alpha_i}{2} B_\mu \Big) a_R^i + \bar{b}_R^i \gamma^\mu \Big(i\,\partial_\mu + g_1 \frac{\beta_i}{2} B_\mu \Big) b_R^i \Big]. \qquad (3.3.3)$$

The Gell-Mann–Nishijima formula is satisfied if the following relationship exists between the weak hypercharges of right- and left-handed fields:

$$\alpha_i = 1 + y_i \quad \beta_i = -1 + y_i. \qquad (3.3.4)$$

(c) Three of the gauge bosons (two charged, one neutral) must be massive and the photon must obviously have zero mass. To achieve this, the *Higgs mechanism* for spontaneous symmetry breaking of the $SU(2)_W \times U(1)$ symmetry must be applied in such a way that the electromagnetic gauge group $U(1)_{em}$ remains conserved as a residual symmetry (see Section 3.2.2). Two complex scalar fields $\phi(x) = (\phi^+(x), \phi^0(x))$, which form a weak isospin doublet and couple gauge invariantly to the vector bosons, are chosen for this purpose:

$$\mathcal{L}_H = \left|\left(\partial_\mu - \mathrm{i}g_2 \frac{\tau^a}{2} W_\mu^a + \mathrm{i}g_1 \tfrac{1}{2} B_\mu\right)\phi\right|^2 - \frac{\lambda}{4}\left(\phi^\dagger\phi - \frac{v^2}{2}\right)^2$$
$$+ \sum_{i,j} (g_{ij} \bar{a}_R^i \tilde{\phi}^* \psi_L^j - \tilde{g}_{ij} \bar{b}_R^i \phi^* \psi_L^j + hc). \tag{3.3.5}$$

Here $\tilde{\phi}(x)$ is the charge-conjugated Higgs field $\tilde{\phi} = (-\phi^{0*}, \phi^-)$. The classical Lagrangian density $\mathcal{L}_{GSW}$ in the GSW theory is then the sum of these individual terms (Eqns. (3.3.1), (3.3.3), and (3.3.5)).

$$\mathcal{L}_{GSW} = \mathcal{L}_E + \mathcal{L}_F + \mathcal{L}_H. \tag{3.3.6}$$

3.3.2 Spontaneous symmetry breaking in the GSW theory

In the GSW theory, the scalar field self-interaction $V(\phi)$ is chosen so that spontaneous symmetry breaking and Higgs mechanism take place as specified in Section 3.2. From amongst the vectors for which $V(\phi)$ assumes its minimum value: $|\phi|^2 = v^2/2$, a ϕ_0 is chosen so that the remaining symmetry is that of electromagnetic gauge transformations $U(1)_{em}$, generated by the electric charge:

$$Q\phi_0 = (I_W^{(0)} + Y_W/2)\phi_0 = \begin{pmatrix} 1 & 0 \\ 0 & 0 \end{pmatrix}\begin{pmatrix} \phi_{01} \\ \phi_{02} \end{pmatrix} = 0.$$

Therefore the classical vacuum:

$$\phi_0 = \begin{pmatrix} 0 \\ \frac{1}{\sqrt{2}} v \end{pmatrix}$$

is the starting point for the spontaneous symmetry breaking $SU(2)_W \times U(1) \to U(1)_{em}$. The following can then be written for $\phi(x)$:

$$\phi(x) = \begin{pmatrix} \phi^+(x) \\ \frac{1}{\sqrt{2}}(v + \eta(x) + \mathrm{i}\chi(x)) \end{pmatrix}; \quad \phi^-(x) = (\phi^+(x))^\dagger. \tag{3.3.7}$$

In the fields $\phi^+(x)$, $\phi^-(x)$, $\eta(x)$, $\chi(x)$, $W_\mu^\pm(x) = (1/\sqrt{2})(W_\mu^{(1)}(x) \mp \mathrm{i}W_\mu^{(2)}(x))$,

$W^0_\mu(x) = W^{(3)}_\mu(x)$, the Higgs field Lagrangian density $\mathscr{L}_H$ can be written as

$$\mathscr{L}_H = \left| v\frac{g_2}{2}W^+_\mu + \frac{g_2}{2}W^+_\mu(\eta + \mathrm{i}\chi) + \mathrm{i}\,\partial_\mu\phi^+ + \frac{g_2}{2}\phi^+W^0_\mu - \frac{g_1}{2}\phi^+B_\mu\right|^2$$

$$+\frac{1}{2}\left|\mathrm{i}\,\partial_\mu\eta - \frac{g_2}{2}vW^0_\mu - v\frac{g_1}{2}B_\mu - \left(\frac{g_2}{2}W^0_\mu + \frac{g_1}{2}B_\mu\right)(\eta + \mathrm{i}\chi) - \partial_\mu\chi + g_2W^-_\mu\phi^+\right|^2$$

$$-\frac{\lambda}{4}[v^2\eta^2 - 2v\eta(\phi^+\phi^- + \tfrac{1}{2}\eta^2 + \tfrac{1}{2}\chi^2) + (\phi^+\phi^- + \tfrac{1}{2}\eta^2 + \tfrac{1}{2}\chi^2)^2]$$

$$-\sum_{f,f'}\left(\frac{v}{\sqrt{2}}g_{ff'}\bar{\psi}_f\psi_{f'} + \frac{g_{ff'}}{\sqrt{2}}(\bar{\psi}_f\psi_{f'}\eta + \mathrm{i}\bar{\psi}_f\gamma_5\psi_{f'}\chi)\right)$$

$$+\frac{1}{2}\sum_{i,j}(g_{ij}\bar{a}^i_Rb_L\phi^+ + \tilde{g}_{ij}\bar{b}^i_Ra^i_L\phi^- + \mathrm{hc}). \tag{3.3.8}$$

The fields $\phi^+(x)$, $\phi^-(x)$, $\chi(x)$ are non-physical degrees of freedom and can be gauged away after transition to the unitary gauge by means of Eqn. (3.2.26). In unitary gauge, $\mathscr{L}_H$ is obtained as a special case $\phi^\pm \equiv 0$, $\chi \equiv 0$ of Eqn. (3.3.8):

$$(\mathscr{L}_H)_{\mathrm{unitary}} = \tfrac{1}{4}v^2g_2^2W^+_\mu W^{-\mu} + \tfrac{1}{8}v^2(g_2W^0_\mu + g_1B_\mu)^2$$

$$+\tfrac{1}{2}(\partial_\mu\eta)^2 - \frac{\lambda}{4}v^2\eta^2 + \text{higher order terms}$$

$$-\sum_{f,f'}(vg_{ff'}\bar{\psi}_f\psi_{f'}/\sqrt{2} + \tfrac{1}{2}g_{ff'}\bar{\psi}_f\psi_{f'}\eta/\sqrt{2}). \tag{3.3.8$'$}$$

The properties of the physical states emerge from this as follows:

(1) The scalar, neutral Higgs field $\eta(x)$ has the mass:

$$M_H = v\sqrt{\frac{\lambda}{2}}. \tag{3.3.9}$$

(2) Masses of the vector fields result from the bilinear form:

$$\tfrac{1}{4}v^2g_2^2W^+_\mu W^{-\mu} + \tfrac{1}{8}v^2(g_2W^0_\mu + g_1B_\mu)^2.$$

The charged vector bosons $W^\pm$ have the mass:

$$M_{W^\pm} = vg_2/2. \tag{3.3.10}$$

In view of the term $g_1g_2W^0_\mu B^\mu$, the mass square matrix $(M^0)^2$ of neutral vector bosons (see Eqn. (3.2.27)) must be diagonalized:

$$(M^0)^2 = \frac{v^2}{4}\begin{pmatrix} g_1^2 & g_1g_2 \\ g_1g_2 & g_2^2 \end{pmatrix}. \tag{3.3.11}$$

The determinant of $(M^0)^2$ vanishes: $\det(M^0)^2 = 0$. Therefore one of the two eigenvalues is equal to zero which reflects the fact that a neutral vector field—the photon—remains massless due to the unbroken $U(1)_{\text{em}}$ gauge invariance. The sum of the eigenvalues is $v^2(g_1^2+g_2^2)/4$, and hence the mass of the heavy neutral vector boson Z is:

$$M_Z = \frac{v}{2}\sqrt{(g_1^2+g_2^2)}. \tag{3.3.12}$$

(3) In order to diagonalize the matrix of mass squares new fields are introduced by an orthogonal transformation:

$$\begin{aligned} Z_\mu &= \cos\theta_W \cdot W_\mu^0 + \sin\theta_W \cdot B_\mu; & B_\mu &= \cos\theta_W \cdot A_\mu + \sin\theta_W \cdot Z_\mu \\ A_\mu &= -\sin\theta_W \cdot W_\mu^0 + \cos\theta_W \cdot B_\mu; & W_\mu^0 &= -\sin\theta_W \cdot A_\mu + \cos\theta_W \cdot Z_\mu. \end{aligned} \tag{3.3.13}$$

The Weinberg angle θ_W is determined so that $(M^0)^2$ is diagonal. This requires:

$$g_2 \sin\theta_W = g_1 \cos\theta_W, \quad \sin\theta_W = \frac{g_1}{\sqrt{(g_1^2+g_2^2)}}. \tag{3.3.14}$$

(4) A mass term for fermions is obtained from the original fermion–Higgs coupling. Obviously the coupling constants $g_{ff'}$ must be chosen diagonal and such that the experimental masses emerge from $vg_f^H/\sqrt{2}$:

$$M_f = \frac{v}{\sqrt{2}} g_f^H \quad \text{with } g_{ff'} = \delta_{ff'} g_f^H. \tag{3.3.15}$$

(5) The Higgs field $\eta(x)$ is coupled to the fermions by $g_f^H = M_f/v$, i.e. proportional to the fermion masses.

For the fields $W_\mu^\pm$, A_μ, Z_μ, $\mathscr{L}_{GSW}$ can now be written ($c_W = \cos\theta_W$, $s_W \equiv \sin\theta_W$):

$$\begin{aligned} \mathscr{L}_{GSW} = {} & \sum_f [\bar{\psi}_f(\mathrm{i}\,\not{\partial} - M_f)\psi_f - e\bar{\psi}_f\gamma^\mu Q_f\psi_f A_\mu] \\ & + \sum_i \frac{g_2}{\sqrt{2}}(\bar{a}_L^i\gamma^\mu b_L'^i W_\mu^+ + \bar{b}_L'^i\gamma^\mu a_L^i W_\mu^-) \\ & + \sum_i \frac{g_2}{c_W}(\bar{\psi}_L^i\gamma^\mu I_i^{(0)}\psi_L^i - s_W^2\bar{\psi}^i Q_i\gamma^\mu\psi^i)Z_\mu \\ & - \tfrac{1}{4}|\partial_\mu A_\nu - \partial_\nu A_\mu - \mathrm{i}e(W_\mu^- W_\nu^+ - W_\mu^+ W_\nu^-)|^2 \\ & - \tfrac{1}{2}|\partial_\mu W_\nu^+ - \partial_\nu W_\mu^+ - \mathrm{i}e(W_\mu^+ A_\nu - W_\nu^+ A_\mu) + \mathrm{i}gc_W(W_\mu^+ Z_\nu - W_\nu^+ Z_\mu)|^2 \\ & - \tfrac{1}{4}|\partial_\mu Z_\nu - \partial_\nu Z_\mu + \mathrm{i}gc_W(W_\mu^- W_\nu^+ - W_\nu^- W_\mu^+)|^2 \\ & - \tfrac{1}{2}M_H^2\eta^2 - \frac{g_2 M_H^2}{2M_W}\eta(\phi^+\phi^- + \tfrac{1}{2}|\eta + \mathrm{i}\chi|^2) \\ & - \frac{g^2 M_H^2}{8M_W^2}(\phi^+\phi^- + \tfrac{1}{2}|\eta + \mathrm{i}\chi|^2)^2 \end{aligned}$$

$$+\left|\partial_\mu\phi^+ - \mathrm{i}g\frac{c_W^2 - s_W^2}{2c_W}Z_\mu\phi^+ + \mathrm{i}eA_\mu\phi^+ - \mathrm{i}M_W W_\mu^+ - \frac{\mathrm{i}}{2}g_2 W_\mu^+(\eta + \mathrm{i}\chi)\right|^2$$

$$+\frac{1}{2}\left|\partial_\mu(\eta + \mathrm{i}\chi) - \mathrm{i}g_2 W_\mu^-\phi^+ + \mathrm{i}M_Z Z_\mu + \frac{\mathrm{i}g_2}{2c_W}Z_\mu(\eta + \mathrm{i}\chi)\right|^2$$

$$-\sum_f \frac{g_2}{2}\frac{M_f}{M_W}(\bar{\psi}^f\psi^f\eta + \mathrm{i}\bar{\psi}^f\gamma_5\psi^f\chi)$$

$$-\sum_i \frac{g_2}{\sqrt{2}}\left(\frac{M_b^i}{M_W}\bar{a}_R^i b_L^i\phi^+ + \frac{M_a^i}{M_W}\bar{b}_R^i a_L^i\phi^- + \mathrm{hc}\right). \tag{3.3.16}$$

The free fermion Lagrangian density and the coupling of the fermion to the photon field A_μ are represented in the first line. The elementary charge $e = \sqrt{(4\pi\alpha)}$ results in:

$$e = g_2 \sin\theta_W = g_1 \cos\theta_W, \quad \alpha = \frac{e^2}{4\pi} \approx \frac{1}{137}. \tag{3.3.17}$$

The second line gives the universal coupling for left-handed fermions to charged gauge bosons and the third the coupling of fermions to the neutral gauge boson Z. These expressions can obviously reproduce at small energies and in leading order of g_2 the Fermi model. Thus, the following can be identified:

$$4\frac{G_F}{\sqrt{2}} = \frac{g_2^2}{2M_W^2} = \frac{2}{v^2}. \tag{3.3.18}$$

This gives $v \approx 250$ GeV and with Eqns. (3.3.10) and (3.3.12):

$$M_W = \frac{37.3\ \text{GeV}}{\sin\theta_W}, \quad M_Z = \frac{M_W}{\cos\theta_W} = \frac{37.3\ \text{GeV}}{\sin\theta_W \cos\theta_W}. \tag{3.3.19}$$

$M_W = 78$ GeV, $M_Z = 89$ GeV is obtained for $\sin^2\theta_W = 0.23$. Hence the gauge boson mass can be calculated from the low energy parameters G_F, α, $\sin^2\theta_W$. The two gauge coupling constants can also be determined from Eqn. (3.3.17):

$$\begin{aligned} g_1 &= 0.36, & g_2 &= 0.63, \\ \alpha_1 &= \frac{g_1^2}{4\pi} = 0.010, & \alpha_2 &= \frac{g_2^2}{4\pi} = 0.032. \end{aligned} \tag{3.3.17$'$}$$

The other expressions in $\mathscr{L}_{GSW}$ describe gauge boson couplings and the mass and couplings of Higgs particles. The parameter λ cannot be determined from available data so that no value can be given for the Higgs particle mass. On the other hand, v and M_f fix the coupling g_f^H of Higgs particles with fermions:

$$g_f^H = \frac{M_f}{v} \approx \frac{M_f}{250\ \text{GeV}}. \tag{3.3.20}$$

Therefore, with the exception of λ, all parameters can be determined in the GSW model.

Eqn. (3.3.16) can be used for $\mathscr{L}_{GSW}$ in the tree graph approximation calculations. The unitary gauge can be applied, i.e. $\phi^{\pm}=\chi=0$. As with all non-Abelian gauge theories, a gauge fixing term and Faddeev–Popov ghost fields must be introduced if quantization is to be consistent. A 't Hooft gauge of the following form is chosen for this purpose (see Section 3.2.2(c)):

$$\mathscr{L}_{\text{fix}}=-\frac{\xi_A}{2}(\partial^\mu A_\mu)^2-\xi_W\left|\partial^\mu W^+_\mu-\mathrm{i}\frac{M_W}{\xi_W}\phi^+\right|^2-\frac{\xi_Z}{2}\left(\partial^\mu Z_\mu-\frac{M_Z}{\xi_Z}\chi\right)^2. \quad (3.3.21)$$

This gives a renormalizable theory for $\xi_a\neq 0$, $a=A, W, Z$, the Feynman gauge $\xi_a=1$ is an advantage. The gauge (3.3.21) approaches the unitary gauge when $\xi_a\to 0$.

When $\xi_a=1$, the vector boson propagators take the following form:

$$D^a_{\mu\nu}(k)=-\mathrm{i}(2\pi)^4 g_{\mu\nu}/(k^2-M_a^2), \quad a=W, Z, \gamma. \quad (3.3.22)$$

The propagators for the $\phi^{\pm}$ and χ fields also contain the gauge boson mass:

$$\Delta^a_F(q)=\mathrm{i}(2\pi)^{-4}/(q^2-M_a^2). \quad (3.3.22')$$

Denoting the ghost fields by $u^{\pm}$, u_Z, u_A, the Faddeev–Popov Lagrangian density for $\xi_a=1$ is:

$$\begin{aligned}\mathscr{L}_{FP}=&-\bar{u}^+(\partial^\mu\,\partial_\mu+M_W^2)u^+ +\mathrm{i}g_2c_W\bar{u}^+\,\partial^\mu(Z_\mu u^+)-\mathrm{i}e\bar{u}^+\,\partial^\mu(A_\mu u^+)\\&-\mathrm{i}g_2\bar{u}^+\,\partial^\mu(W^+(u_Zc_W-u_As_W))\\&-M_W\bar{u}^+\left(g_2(c_W^2-s_W^2)/2c_Wu_Z\phi^+-eu_A\phi^+ +\frac{g_2}{2}u^+(\eta+\mathrm{i}\chi)\right)\\&+(u^+\to u^-,\, W^+\to W^-,\, i\to -i)\\&-\bar{u}_Z(\partial^\mu\,\partial_\mu+M_Z^2)u_Z-\mathrm{i}g_2c_W\bar{u}_Z\,\partial^\mu(u^+W^-_\mu-u^-W^+_\mu)\\&-M_Z\bar{u}_Z\left(-\frac{g_2}{2}u^-\phi^+-\frac{g_2}{2}u^+\phi^-+\frac{g}{2c_W}u_Z\eta\right)\\&-\bar{u}_A\,\partial^\mu\,\partial_\mu u_A+\mathrm{i}e\bar{u}_A\,\partial^\mu(u^+W^-_\mu-u^-W^+_\mu).\end{aligned} \quad (3.3.23)$$

Thus, the GSW theory in this gauge ($\xi_a=1$) is fully defined [Sa 80].

In putting the fermions into pairs as described at the beginning of this section the electric charges of the light, the medium heavy and also the heavy quarks (if the $\frac{2}{3}e$ charged t-quark exists) are always compensated by the e, μ, and τ lepton charges. In this way, the theory formulated by Eqns. (3.3.16), (3.3.21), and (3.3.23) is free from γ_5 anomalies and can therefore be renormalized (see Section 2.4.3(f)). The arrangement of $[(\nu_e, e), (u, d')]$ etc. into 'families' plays an important part in combining strong and electroweak interaction into a unified gauge theory (Section 3.4).

3.3.3 Predictions of the GSW theory

(a) As discussed in Section 1.3, the *GSW theory approaches the Fermi model for the limit of low energies* ($E \ll M_W, M_Z$) *and order* g_1^2, g_2^2. Experiments carried out at low energies and in which accuracy does not permit tests on higher order terms such as radiative corrections, do not discriminate between electroweak interaction gauge theory and the phenomenological Fermi model. However, the numerous experiments carried out so far on neutrino–lepton scattering, neutrino–nucleon scattering, polarized electron–nucleon scattering and parity violation in atomic physics confirm that the low energy limit of the GSW theory—the Fermi model—does agree with nature within the range of measuring accuracy.

This section deals with those predictions of the GSW theory which are typical of its gauge theory character. The predictions for the gauge boson masses must be repeated as a first step:

$$M_W \approx 78 \text{ GeV}, \quad M_Z \approx 89 \text{ GeV}.$$

Radiative corrections [Ve 80, Si 80, Ca 80] can change this by several GeV. Then further effects of the W and Z bosons based on the W^+W^-Z and $W^+W^-\gamma$ coupling which is typical of gauge theories and on effects of Higgs particles are investigated. These effects can be seen either direct at sufficiently high energies $E \gtrsim M_Z, M_W$ or in a more subtle form as radiative corrections at low energies. The latter seems to be a more complicated alternative from both an experimental and a theoretical point of view, hence the high energy aspect will be considered in the main.

(b) The GSW theory predicts energy-dependent interference phenomena between the one-photon exchange and the Z-exchange for *electron–positron annihilation into a fermion–antifermion pair.*

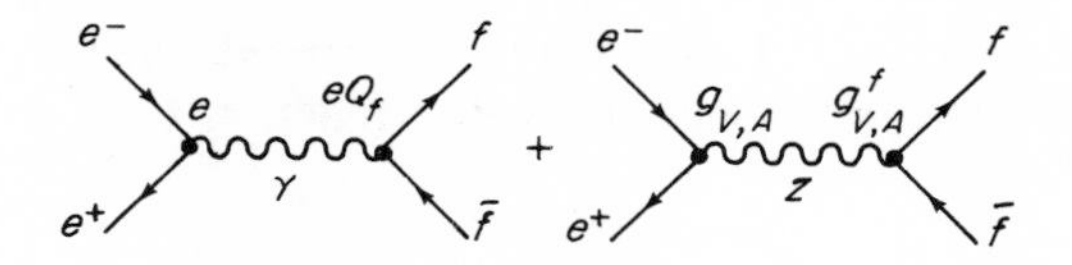

According to Eqn. (3.3.16), the couplings g_V, g_A, g_V^f, g_A^f with $e = g_2 \sin\theta_W$ are given as:

$$g_V = \frac{-1}{2s_W c_W}\left(\tfrac{1}{2} - 2s_W^2\right), \quad g_A = \frac{1}{4s_W c_W},$$

$$g_V^f = \frac{1}{2s_W c_W}\left(I_{W,f}^{(0)} - 2Q_f s_W^2\right), \quad g_A^f = \frac{-1}{2s_W c_W} I_{W,f}^{(0)}.$$

Hence the differential cross-section for the annihilation of polarized electrons (P_L^-, $P_T^- \equiv$ longitudinal or transversal polarization degree) and positrons

(P_L^+, P_T^+) can be calculated as:

$$\begin{aligned}
&\frac{4s}{\alpha^2}\frac{d\sigma}{d\Omega} \\
&\quad = (1+\cos^2\vartheta)(1+P_L^+P_L^-)[Q_f^2 - Q_f D(s) 2g_V g_V^f + D^2(s)(g_V^2 + g_A^2)((g_V^f)^2 + (g_A^f)^2)] \\
&\quad\quad + \sin^2\vartheta\cos 2\varphi P_T^+ P_T^- [Q_f^2 - Q_f D(s) 2 g_V g_V^f + D^2(s)(g_V^2 - g_A^2)((g_V^f)^2 + (g_A^f)^2)] \\
&\quad\quad + 2\cos\vartheta(1+P_L^+P_L^-)[-Q_f D(s) 2 g_A g_A^f + D^2(s) 2 g_V g_A \cdot 2 g_V^f g_A^f] \\
&\quad\quad + (1+\cos^2\vartheta)(P_L^+ + P_L^-)[-Q_f D(s) 2 g_A g_A^f + D^2(s)\cdot 2 g_V g_A((g_V^f)^2 + (g_A^f)^2)] \\
&\quad\quad + 2\cos\vartheta(P_L^+ + P_L^-)[-Q_f D(s) 2 g_V g_V^f + D^2(s)\cdot 2 g_V^f g_A^f (g_V^2 + g_A^2)] \qquad (3.3.24)
\end{aligned}$$

with $$D(s) = \frac{s}{s - M_Z^2} \qquad M_Z = (37.3\ \text{GeV})/\sin\theta_W\cos\theta_W.$$

This presupposes a correction to the first term and a forward–backward asymmetry for unpolarized beams:

$$A(\vartheta) = \frac{d\sigma(\vartheta) - d\sigma(\pi-\vartheta)}{d\sigma(\vartheta) + d\sigma(\pi+\vartheta)} = \frac{2\cos\vartheta}{1+\cos^2\vartheta} D(s)\frac{2g_A g_A^f}{Q_f} + O(D^2(s)), \quad (3.3.25)$$

which measures the product of the axial vector coupling constant of the electron and the final state fermion. Fig. 3.2 shows the behaviour of $D(s)$ for $\sin^2\theta_W = 0.25$ and for the Fermi model $D_{\text{Fermi}}(s) = -s(\sqrt{2})G_F \sin^2\theta_W \cos^2\theta_W/\pi\alpha$. The product of vector and axial vector coupling constants can be measured for longitudinal polarized rays thus supplying direct evidence of parity violating effects. This asymmetry was determined at PETRA to -10.4 ± 1.4 whereas theoretically a value -8.6 ± 0.2 is expected [Br 82N, Ba 82N, Be 82N, Da 82N].

(c) The *existence of the Z boson* can be confirmed using e^+e^- storage rings with high energy. The cross-section for the reaction $e^+e^- \to Z \to$ final state

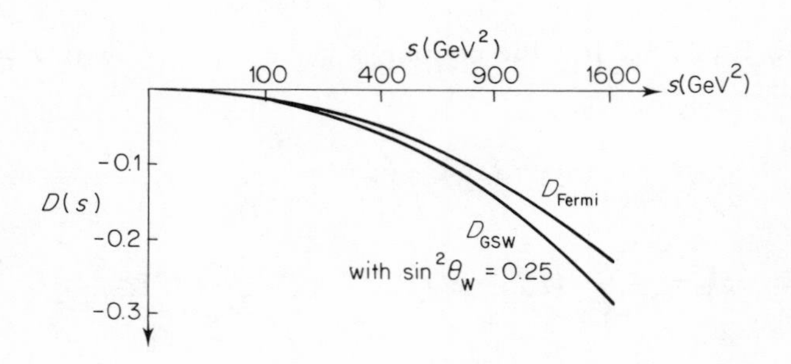

Figure 3.2 Comparison of the functions $D(s)$ in the Fermi model and in GSW theory

(F) is of the Breit–Wigner form around the Z mass:

$$\sigma(e^+e^- \to Z \to F) = 12\pi \frac{\Gamma(Z \to e^+e^-)\Gamma(Z \to F)}{(s - M_Z^2)^2 + (\Gamma_Z)^2(M_Z)^2}. \tag{3.3.26}$$

The leptonic decays of gauge bosons can be calculated direct in the GSW theory. With the values for the Z mass given above, the following can be written [Bj 76]:

$$\Gamma(Z \to f\bar{f}) = \alpha \frac{M_Z}{3}((g_V^f)^2 + (g_A^f)^2), \quad \Gamma(Z \to \mu^+\mu^-) = 69\ \text{MeV},$$
$$\Gamma(W \to f\bar{f}) = \frac{G_F}{\sqrt{2}} \frac{M_W^3}{6\pi}, \quad \Gamma(W^+ \to e^+\nu_e) = 165\ \text{MeV} \tag{3.3.27}$$

The total width for three lepton and three quark pairs can be estimated as:

$$\Gamma_Z \approx \sum_{ff} \Gamma_{f\bar{f}} \approx 2.2\ \text{GeV},$$
$$\Gamma_{W^+} \approx \Gamma(W^+ \to e^+\nu_e)(\underset{e\nu}{1} + \underset{\mu\nu}{1} + \underset{\tau\nu}{1} + \underset{u\bar{d}}{3} + \underset{c\bar{s}}{3} + \underset{t\bar{b}}{3}) \approx 2.0\ \text{GeV}. \tag{3.3.28}$$

In view of $\Gamma_Z \approx 2.2$ GeV, the Breit–Wigner curve is only very slightly widened by the energy distribution of the electron and positron beams. Hence, a cross-section of:

$$\sigma(\sqrt{s} = M_Z) = \frac{12\pi}{(M_Z)^2} B(Z \to e^+e^-)B(Z \to F) \tag{3.3.29}$$

can be expected at the peak. The e^+e^- branching ratio is about 3%; $B_F \approx 80\%$ for hadrons in the final state. Thus, with a luminosity of $L = 10^{31}\ \text{cm}^{-2}\ \text{s}^{-1}$, one Z per second is expected. Comparison with the one-photon exchange result for the μ pair production cross-section gives:

$$R_Z = \frac{\sigma(\sqrt{s} = M_Z)}{\sigma(e^+e^- \to \mu^+\mu^-)} = \frac{9B(Z \to e^+e^-)B(Z \to F)}{\alpha^2} \approx 10^4. \tag{3.3.30}$$

Fig. 3.3 shows the total cross-section expected from the GSW theory, in units of $\sigma(e^+e^- \xrightarrow[1\gamma]{} \mu^+\mu^-)$ as a function of energy. Calculations are carried out for three lepton and three quark pairs. The drastic increase at the resonance caused by the production of Z vector bosons can be clearly seen. It should be sufficient to allow a detailed study of Z decays and thus provide a test for the GSW theory.

(d) The charged vector bosons $W^\pm$ can be produced in pairs by e^+e^- storage rings of sufficiently high energy. (The cross-section for the reaction $e^+e^- \to \bar{\nu}_e e^- W^+$ is two orders of magnitude smaller than for pair production [Ll 79].) If the existence of W bosons with the mass (3.3.19) could be

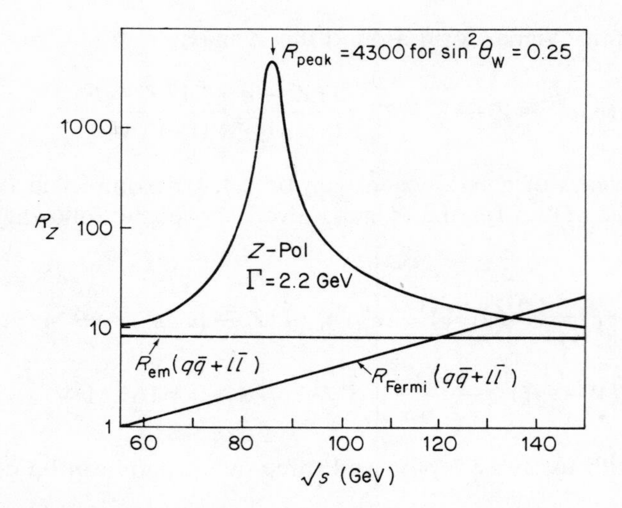

Figure 3.3 The total e^+e^- cross-section in units of σ ($e^+e^- \xrightarrow{\gamma} \mu^+\mu^-$) according to GSW theory, Fermi model, and the 1-photon exchange model

detected, this would provide further proof of the GSW theory. At the same time, the non-commutative character of the GSW theory would be established, as the existence of W^+W^-Z gauge coupling has drastic effects on the size of the cross-section. This reflects the general discussion in Section 3.1, the main theme of which was that the cross-sections of Yang–Mills theories do not increase at high energies. Calculation of the corresponding Feynman graphs:

i.e. of one-photon exchange, neutrino exchange, and Z exchange gives the cross-section shown in Fig. 3.4. The broken curve which increases with energy is the result of a calculation without exchange of a Z with W^+W^-Z gauge coupling.

(e) This discussion has shown how the building stones of the GSW theory can be found in e^+e^- storage ring experiments, and how their properties can be tested. The Higgs sector features as the final point in this section [El 76]. *Higgs particles* are needed to give mass to the gauge bosons without destroying the renormalizability of the theory. The neutral, scalar Higgs

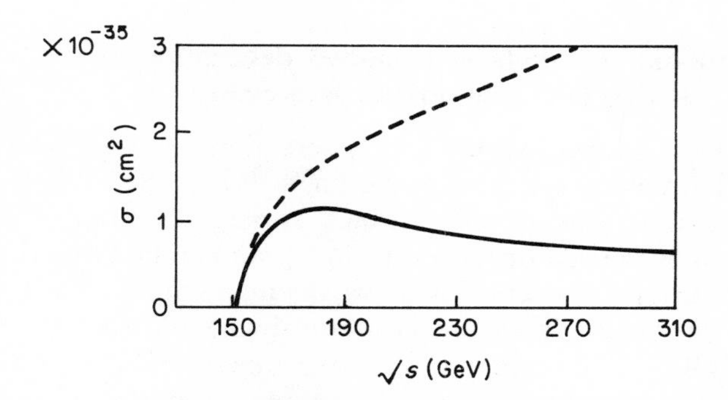

Figure 3.4 Cross-section for the production of W^+W^- pairs in e^+e^- annihilation in the Fermi model (– – –) and in GSW theory (——)

particle H with its couplings, forms the part of this mechanism which can in fact be checked. However, as Section 3.3.2 showed, one parameter in the Higgs sector is unknown, so that the mass and strength of the self-interaction of the Higgs particle cannot be given explicitly. The only possibility is to determine limits for the Higgs mass. The upper limit, a unitarity bound, is found at $M_H < 10^3$ GeV [Le 77] and the lower one, which is calculated from a vacuum stability condition, is $M_H > 3.7$ GeV [We 76b]. If the assumption that the potential which gives the H mass is produced perturbatively by the other fields is accepted, then a typical result is $M_H \approx 10$ GeV [Co 73a, El 79b]. The coupling of H to the fermions is proportional to their mass. Thus light fermions play virtually no direct part in Higgs particle reactions. Significant coupling constant values are not achieved until heavy quarks or leptons are present. For this reason, the following mechanism for detecting H's in storage rings seems promising [Wi 77]:

$$e^+e^- \rightarrow \Upsilon_t \rightarrow \gamma + H \quad \text{if } M_H < M_{\Upsilon_t}.$$

in which Υ_t is as heavy a vector meson as possible (the $t\bar{t}$ state, for example). The monochromatic γ line confirms the presence of H and the fact that it is preferably coupled with heavy fermions.

Because of its small number of indeterminate parameters, the standard model gives relatively precise information on the properties of W and Z bosons. However, for various reasons, models with larger gauge groups have also been used for studying the electroweak interaction [De 77a, Sh 78, Da 79, De 79a, Ba 80]. These have several Z bosons, the lightest of which generally has a smaller mass than that of the standard model. Experiments on e^+e^- storage rings with energies up to 200 GeV and polarized beams can be used to check whether the electroweak interaction has the character of a non-Abelian gauge theory with spontaneous symmetry breaking and which model is actually realized [Ho 80].

3.4 Outlook: Attempts at a unified description of the strong and electroweak interactions

The principle of local gauge invariance of interactions together with spontaneous symmetry breaking concepts is the basis of a unified theory of the electroweak interaction. The question which springs to mind is whether these principles can be carried further to form a grand unified theory (GUT) covering strong, electromagnetic, and weak interactions all together. Speculative attempts in this area are currently being pursued from many different angles [GUT]. The simplest of these models [Ge 74] which has yielded important results from relatively simple premises, is described concluding this whole discussion on gauge theories.

(a) The symmetry group $SU(3)_C \times SU(2)_W \times U(1)$ emerges from a formal composition of the gauge theories of strong, electromagnetic, and weak interactions. The smallest simple Lie group which incorporates this group is $SU(5)$. The 24 infinitesimal generators T_a of $SU(5)$ in the fundamental representation contain the Gell-Mann matrices (1.2.4) of the $SU(3)_C$ group and the Pauli matrices (1.2.3) of the weak isospin group $SU(2)_W$:

$$SU(3)_C\colon\quad T_a = \left(\begin{array}{c|c} \frac{\lambda_a}{2} & 0 \\ \hline 0 & 0 \end{array}\right), \quad a = 1, \ldots, 8 \qquad SU(2)_W\colon\quad T_{8+i} = \left(\begin{array}{c|c} 0 & 0 \\ \hline 0 & \frac{\tau_i}{2} \end{array}\right), \quad i = 1, 2, 3 \tag{3.4.1}$$

The other generators are formed from a diagonal matrix for the weak hypercharge $T_{12} = \sqrt{(\frac{3}{5})} \cdot \frac{1}{2}\, Y_W$ and 12 matrices which produce transformations between $SU(3)_C$ and $SU(2)_W$:

$$T_{12} = \sqrt{(\tfrac{3}{5})} \cdot \tfrac{1}{2}\, Y_W = \frac{1}{\sqrt{(15)}} \left(\begin{array}{ccc|cc} -2 & 0 & 0 & & \\ 0 & -2 & 0 & & 0 \\ 0 & 0 & -2 & & \\ \hline & 0 & & 3 & 0 \\ & & & 0 & 3 \end{array}\right),$$

$$T_{13} = \frac{1}{2} \left(\begin{array}{ccc|cc} & & & 0 & 0 \\ & 0 & & 0 & 0 \\ & & & 1 & 0 \\ \hline 0 & 0 & 1 & & \\ 0 & 0 & 0 & & 0 \end{array}\right), \qquad T_{14} = \frac{1}{2} \left(\begin{array}{ccc|cc} & & & 0 & 0 \\ & 0 & & 0 & 0 \\ & & & -i & 0 \\ \hline 0 & 0 & i & & \\ 0 & 0 & 0 & & 0 \end{array}\right), \quad \text{etc.} \tag{3.4.2}$$

The normalization $Tr(T_a T_b) = \delta_{ab}/2$ corresponds to that in Eqns. (1.2.3)–(1.2.6). The local gauge invariance for this group is mediated by 24 gauge fields which contain 12 additional vector fields X, Y, besides the QCD gluons and the electroweak interaction gauge bosons W^a_μ, B_μ. They can be summarized as follows according to the representation of the generators T_a

(Eqn. (2.2.4)):

$$\left(\begin{array}{c|c} \text{gluons} & X, Y \\ \hline \bar{X} & \\ \bar{Y} & W^a \end{array}\right)_{B}. \tag{3.4.3}$$

As they link the groups $SU(3)_C$ and $SU(2)_W$, the new bosons (X, Y) and their antiparticles $(\bar{X}, \bar{Y})$, have quantum numbers from both groups. They form isospin$_W$–doublet–colour triplets.

As the strong and electroweak interactions can obviously not be described by an $SU(5)$-invariant, universal coupling at low energy, the $SU(5)$ symmetry must be broken. Like in the GSW theory, a Higgs mechanism is assumed to be responsible for these different coupling strengths. In this instance, a large mass of X and Y bosons causes the $SU(5)$ interaction to split up. This is described in more detail below.

(b) Increasing the size of the symmetry group usually results in larger particle multiplets. In the $SU(5)$ gauge theory, the generations $(e, \nu_e; u, d)$, $(\mu, \nu_\mu; c, s)$, $(\tau, \nu_\tau; t, b)$ introduced earlier (Section 1.3.2(c)) can be regarded as symmetry multiplets [Ge 74]. The fermions must be split up into left- and right-handed components, as in the GSW theory:

$$\begin{pmatrix} \nu_e & u_r & u_g & u_b \\ e^- & d_r & d_g & d_b \end{pmatrix}_L, \quad (e^-; u_r, u_g, u_b; d_r, d_g, d_b)_R$$

$$\begin{pmatrix} \nu_\mu & c_r & c_g & c_b \\ \mu^- & s_r & s_g & s_b \end{pmatrix}_L, \quad (\mu^-; c_r, c_g, c_b; s_r, s_g, s_b)_R$$

$$\begin{pmatrix} \nu_\tau & t_r & t_g & t_b \\ \tau^- & b_r & b_g & b_b \end{pmatrix}_L, \quad (\tau^-; t_r, t_g, t_b; b_r, b_g, b_b)_R.$$

Each *generation* is written as a left-handed *multiplet*, by replacing the right-handed particle fields by left-handed antiparticle fields and vice versa: $\tilde{f} = Cf_R = \gamma_0\gamma_2(1+\gamma_5)/2f^*$. Using this notation, a generation becomes the direct sum of a $(\bar{5})$ and a (10) representation of $SU(5) \supset SU(3)_C \times SU(2)_W \times U(1)$:

$$(\bar{5}): \begin{pmatrix} \tilde{d}_r \\ \tilde{d}_g \\ \tilde{d}_b \\ \hline e^- \\ \nu_e \end{pmatrix} \sim \begin{pmatrix} (\bar{3}, 0, \frac{1}{3}) \\ \hline (1, \frac{1}{2}, -\frac{1}{2}) \end{pmatrix},$$

$$(10): \left(\begin{array}{ccc|cc} 0 & \tilde{u}_b & -\tilde{u}_g & u_r & d_r \\ -\tilde{u}_b & 0 & \tilde{u}_r & u_g & d_g \\ \tilde{u}_g & -\tilde{u}_r & 0 & u_b & d_b \\ \hline -u_r & -u_g & -u_b & 0 & \tilde{e} \\ -d_r & -d_g & -d_b & -\tilde{e} & 0 \end{array}\right) \sim \left(\begin{array}{c|c} (\bar{3}, 0, -\frac{2}{3}) & (3, \frac{1}{2}, \frac{1}{6}) \\ \hline (3, \frac{1}{2}, \frac{1}{6}) & (1, 0, 1) \end{array}\right), \tag{3.4.4}$$

in which the complication caused by the Kobayashi–Maskawa mixture (Eqn. (1.3.20)) is suppressed.

The results of this classification are as follows:

(1) The charge operators in these multiplets are described by the T_a's of the fundamental representation (Eqns. (3.4.1), (3.4.2)):

$$(T_a\psi)^{(\bar{5})} = -T_a^*\psi^{(\bar{5})}, \quad (T_a\psi)^{(10)} = T_a\psi^{(10)} + \psi^{(10)}T_a^t. \tag{3.4.5}$$

This, in fact, results in the correct $SU(3)_C \times SU(2)_W \times U(1)$ quantum numbers (colour, I_W, $\frac{1}{2}Y_W$), as described in Eqn. (3.4.4).

(2) The multiplets in Eqn. (3.4.4) describe leptons and hadrons on the same level. In this sense, the $SU(5)$ does represent the idea of a lepton–hadron symmetry.

(3) $SU(5)$ symmetry explains the equality of the electron and proton electric charge $|Q_e| = |Q_p|$ [Ge 74]. Experimentally

$$(|Q_e| - |Q_P|)/|Q_e| < 10^{-21}.$$

The electric charge operator is $Q = I_W^{(0)} + \frac{1}{2}Y_W = T_{11} + T_{12}\sqrt{\frac{5}{3}}$ (Eqns. (3.3.2) and (3.4.1), (3.4.2)). From $\mathrm{Tr}\, Q = 0$ it follows for $(\bar{5})$ $Q_e + 3Q_{\bar{d}} = 0$, i.e. $Q_d = \frac{1}{3}Q_e$. As u and d are isospin doublets in the representation (10), $Q_u = Q_d + 1 = -\frac{2}{3}Q_e$ applies and $Q_P = 2Q_u + Q_d = -Q_e$ results for the proton.

One unattractive feature of the classification (3.4.4) is that the particles of a generation are described by a reducible representation. The $SU(5)$ symmetry can be extended to $SO(10)$ [Fr 75, Ge 75]. In this case an additional heavy neutral particle is introduced for each generation yielding an irreducible 16-dimensional representation. The condition of anomaly freedom as described in Eqn. (2.4.64), which is satisfied for the individual generations, also finds a logical explanation in this model [Fr 75].

(c) For the limit of unbroken $SU(5)$ symmetry, strong, electromagnetic, and weak interactions result as a minimal, gauge-invariant coupling of charged fields with gauge fields. If the term $g_5 T_a A_\mu^a(x)$ is written explicitly in the covariant derivatives:

$$g_5 T_a A_\mu^a = g_5\left(\sum_{a\in SU(3)_C} T_a A_\mu^a + \sum_{a\in SU(2)_W} T_a W^a + \sqrt{\tfrac{3}{5}}\frac{Y_W}{2} B_\mu + \ldots\right) \tag{3.4.6}$$

and this is then compared with Eqn. (2.2.39) for $g_3 = -g_s$ in the $SU(3)_C$ group and with Eqn. (3.3.3) for g_2, g_1 in $SU(2)_W \times U(1)$, then

$$g_5 = g_3 = g_2 = g_1\sqrt{\tfrac{5}{3}}. \tag{3.4.7}$$

is obtained. This results in

$$\sin^2\theta_W = \frac{g_1^2}{g_1^2 + g_2^2} = \frac{3}{8}. \tag{3.4.8}$$

for the Weinberg angle according to Eqn. (3.3.14).

The symmetry relations (3.4.7) as part of a spontaneously broken gauge theory apply only for the running coupling constants g_n, $n=1, 2, 3, 5$ and thus for the running Weinberg angle at energies which are high in comparison with the masses generated by the Higgs mechanism. A comparison with experimental values at low energies can be made by continuing the coupling constants $g_n = \sqrt{(4\pi\alpha_n)}$ to lower energies with help of the solutions of the renormalization group equations (2.5.20) and (2.5.22):

$$\alpha_n^{-1}(Q^2) = \alpha_n^{-1}(\mathcal{M}^2) - \frac{4N_G - 11C_2(SU(n))}{12\pi} \log \frac{Q^2}{\mathcal{M}^2} + \ldots \qquad (3.4.9)$$

[Ge 74a, Bu 78a].

The renormalization group equations are based on the assumption that the masses of the theory are negligible. However, they also apply for effective theories and energies which are high in comparison with the masses of the excited degrees of freedom and small compared with the masses of non-excited degrees of freedom [We 80]. The value of the corresponding effective gauge group should then be substituted for C_2 $(SU(n))$ and the effective number of fermion multiplets for N_G. Therefore the following procedure applies for the case in question:

A Higgs multiplet which transforms under $SU(5)$ according to the adjoint representation and has the following non-zero vacuum components [Bu 78a, El 80b]:

$$\langle 0|\, \phi^{(24)}\, |0\rangle = V \left(\begin{array}{ccc|cc} 1 & 0 & 0 & & \\ 0 & 1 & 0 & \multicolumn{2}{c}{0} \\ 0 & 0 & 1 & & \\ \hline & 0 & & -3/2 & 0 \\ & & & 0 & -3/2 \end{array} \right) \sim VY_W \qquad (3.4.10)$$

is taken. $SU(5)$ is spontaneously broken to $SU(3)_C \times SU(2)_W \times U(1)$ by this Higgs structure (see p. 264). The X, Y gauge bosons are given a mass $M_X^2 = M_Y^2 = \frac{25}{8} g_5^2 V^2$ according to Eqn. (3.2.27). The RG equation (3.4.9) with $G = SU(5)$ and $N_G = 3$ for α_5 is valid for energies greater than M_X. Eqns. (3.4.9) apply for $SU(n)$, $n = 1, 2, 3$, and the corresponding colour or weak isospin multiplets for energies which are low in comparison with M_X but high compared with M_Z, M_W. In the threshold regions which require more accurate treatment [Ro 78], the coupling constants g_1, g_2, and g_3 must approach the symmetry relations (3.4.7) in order to satisfy these conditions beyond these regions. This can then be used for a first estimate of the energy of the unification point [Ge 74a], starting from the experimental values for the strong and electroweak coupling constants and then calculating the energy for which the symmetry conditions (3.4.7), according to Eqn. (3.4.9), are satisfied (see Fig. 3.5).

The requirement that α_3, α_2, and $5\alpha_1/3$, i.e. the running coupling constants, should meet at one point provides a result for the value of the

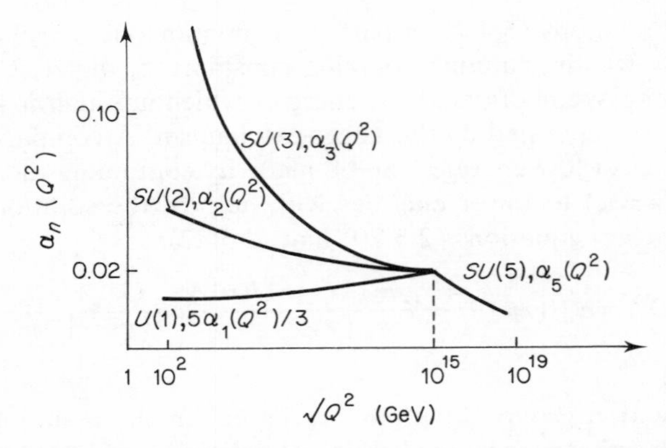

Figure 3.5 Behaviour of running coupling constants for an $SU(5)$ gauge theory broken into $SU(3)_C \times SU(2)_W \times U(1)$ (see Eqn. (3.4.9))

running Weinberg angle at low energies. This gives a unification energy of 10^{16} GeV. A more accurate second-order calculation for the β function in Eqn. (2.5.27), gives the following results [El 80a, Bu 78a]:

$$
\begin{aligned}
M_X &\approx 1.5 \times 10^{15} \Lambda_{\overline{MS}} \approx 6 \times 10^{14}\ \text{GeV}, \\
\sin^2 \theta_W(Q^2 = M_W^2) &\approx 0.209 \pm 0.006 \quad (\text{expt: } 0.228 \pm 0.010), \\
\alpha_5\ (\approx 10^{16}\ \text{GeV}) &\approx 1/40, \quad V \approx 10^{15}\ \text{GeV}.
\end{aligned}
\tag{3.4.11}
$$

This assumes that all fermion multiplets are contained in the three 'low-energy' generations and the result is also based on a more accurate treatment of the gauge boson threshold regions. Both the XY threshold at high energies and the W and Z boson threshold are considered in accordance with the Higgs mechanism of the GSW theory (Section 3.3.2). The Higgs field doublet in the GSW theory can be arranged in a quintuplet $H^{(5)}$ within the $SU(5)$ symmetry [Bu 78a, El 80b], which acquires the vacuum expectation value:

$$
\langle 0|\, H^{(5)}\, |0\rangle = \frac{v}{\sqrt{2}} \begin{pmatrix} 0 \\ 0 \\ 0 \\ 0 \\ 1 \end{pmatrix}, \quad v \approx 250\ \text{GeV}. \tag{3.4.12}
$$

If the fermion masses are produced by an $SU(5)$-invariant Higgs–fermion coupling according to Eqns. (3.3.5) and (3.3.15), this Higgs structure then characterizes fermion masses. Without going into further details, it is clear that this mass determination leaves an $SU(4)$ symmetry and hence the relations:

$$
m_e = m_d, \quad m_\mu = m_s, \quad m_\tau = m_b \tag{3.4.13}
$$

for the $(\bar{5})$-fermions (see Eqn. (3.4.4)) in the symmetry range [Ge 74]. Physical values for effective fermion masses at low energies can be obtained after renormalization with help of the following formula:

$$\frac{m_b(Q^2)}{m_\tau(Q^2)} \approx \left[\frac{\alpha_3(Q^2)}{\alpha_3(M_X^2)}\right]^{4/(11-\frac{2}{3}N_F)} \tag{3.4.14}$$

[Ge 76]. This gives $m_b \approx 5$–5.5 GeV [Ch 77, Bu 78a], and is regarded as a successful result. It is doubtful whether Eqn. (3.4.14) can be used for very low energies; the theoretical result for the mass of the strange quark is $m_s \approx 0.5$ GeV.

The predictions on the Weinberg angle and fermion masses which can be checked experimentally are based on the assumption that the RG equations can be used to give a smooth interpolation between the new 10^{15} GeV energy range and the experimentally accessible one of $\lesssim 10^2$ GeV. This is called the 'Great Desert hypothesis' regarded by many scientists as unnatural [Ha 79, Di 79], as it is based on the premise that no new type of physical phenomena occurs over a very wide intermediate energy range. This unnaturalness becomes obvious when attempting to describe both vacuum expectation values (Eqns. (3.4.10) and (3.4.12)) by a Higgs potential. Taking radiative corrections into account [Co 73a] this is achieved only by introducing very small, but non-vanishing ad hoc parameters [Gi 76, tH 80].

(d) The most spectacular prediction to emerge from a unified theory of the strong, electromagnetic, and weak interactions is *proton decay* [Pa 74]. The universal $SU(5)$ interaction describes a baryon number-violating transition between quarks and leptons based on the following graphs:

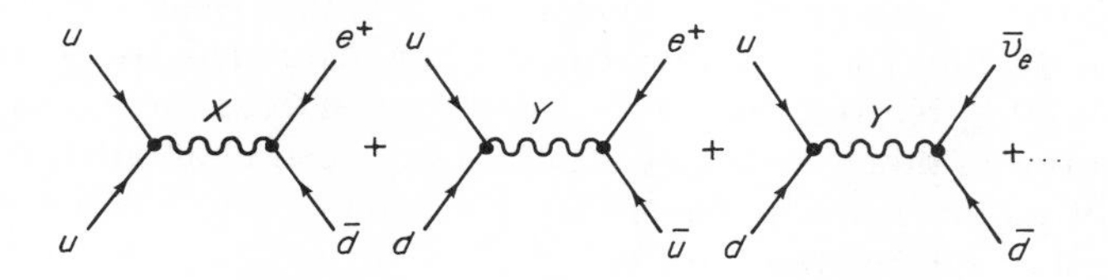

These graphs can be replaced by an effective four-fermion interaction for low-energy processes [We 79, Wi 79b]:

$$\mathscr{L}_{\text{GUT}} = \frac{4G_{\text{GUT}}}{\sqrt{2}} [(\bar{u}_L^C \gamma_\mu u_L)(2\bar{e}_L^+ \gamma^\mu d_L + \bar{e}_R^+ \gamma^\mu d_R) + (\bar{u}_L^C \gamma_\mu d_L)(\bar{\nu}_e \gamma^\mu d_R) + \text{hc}]$$

(colour indices, etc. suppressed). A 'Fermi coupling constant', which determines the proton lifetime, is obtained in the same way as for transition of the GSW theory to the weak interaction Fermi model:

$$\frac{G_{\text{GUT}}}{\sqrt{2}} = \frac{g_5^2}{8M_X^2} = \frac{g_5^2}{8M_Y^2}, \tag{3.4.15}$$

Just like for the μ lifetime in Eqn. (1.3.13), the expression for the proton lifetime takes the following form [El 80c]:

$$\tau_P = \frac{Dg_5^4}{32(G_{GUT})^2 M_P^5} = \frac{D}{M_P}\left(\frac{M_X}{M_P}\right)^4 \approx 8\times 10^{30\pm 2} \text{ years} \qquad (3.4.16)$$

The uncertainty of the factor D is due to the fact that details of the model are unknown. The main decay channels are $P \to e^+\pi^0$, $e^+\rho^0$, $e^+\omega$, $\bar{\nu}_e\pi^+$, $\mu^+K^0, \ldots$ [Ja 79a, Ma 79b]. Experimentally, the lower limit on the proton lifetime is set at 8×10^{30} yr [Pa 82]. Experiments involving a sensitivity which can push this limit up to 10^{33} yr are in preparation.

The overriding importance attached to observation and measurement of proton decay is due mainly to the fact that it gives information on a new energy range in which strong, electromagnetic, and weak interactions can be regarded as a universal interaction. The physical effects of such a discovery are barely conceivable. Even now, aspects of it are being used in attempts to clarify cosmological questions [El 80d], e.g. the number of protons in the cosmos [Sa 67, El 80d]. A certain proximity of the 'grand unified theories' to gravitation and cosmology emerges from the fact that M_X is only a few orders of magnitude below the Planck mass $M_{\text{Planck}} = (\hbar c/\gamma)^{1/2} = 1.2\times 10^{19}\ \text{GeV}/c^2$, $\gamma = 6.67\times 10^{-11}\ \text{m}^3\ \text{kg}^{-1}\ \text{s}^{-2} \equiv$ gravitation constant. This fact has also led to first attempts for the unification of all interactions, including gravitation [El 80e].

Theoretically, violation of the baryon number in proton decay is also satisfactory. Indeed, the replacement of global symmetries by local gauge invariance—possibly broken spontaneously—was one of the most essential aspects in the development of gauge field theories. Although the $SU(5)$ gauge theory still incorporates a global conservation law for the difference between baryon and lepton numbers [Ra 79], this is also replaced by a spontaneously broken local invariance when extended to $SO(10)$ gauge theory [Ma 80a]. No further details can be given here on the extension of gauge theories to other symmetry groups introduced in an attempt to clarify some unsolved problems in elementary particle physics. The literature gives a great deal of information on this.

This short review of a unified interaction theory concludes our discussion on gauge theories. It shows once again how broad the span of theoretical considerations on quantized gauge field theories actually is. It results in a universal conception of the structure of matter and its interactions. One of the most exciting aspects of modern day physics is further theoretical research into these and other highly speculative ideas and their subsequent confirmation or rejection by means of experiment.

Literature

[EP] Elementary particle physics

S. Gasiorowicz, *Elementary Particle Physics* (New York, London, Sydney 1966).

E. Lohrmann, *Hochenergiephysik* (Stuttgart 1978).

R. Omnès, *Introduction to Particle Physics* (London 1971).

M. L. Perl, *High Energy Hadron Physics* (New York 1974).

H. M. Pilkuhn, *Relativistic Particle Physics* (Berlin 1979).

H. Rollnik, *Teilchenphysik* 1/2 (Mannheim 1971).

[GT] Gauge field theory

E. Abers, B. W. Lee, *Phys. Rep.* **9** (1973) 1.

W. Drechsler, M. E. Mayer, *Fibre Bundle Techniques in Gauge Theories, Lecture Notes in Physics* 67 (Berlin, Heidelberg, New York 1977).

R. Jackiw, *Rev. Mod. Phys.* **52** (1980) 661.

J. C. Taylor, *Gauge Theories of Weak Interactions* (Cambridge 1976).

[FM] Fermi model

R. E. Marshak, Riazuddin, C. P. Ryan, *Theory of Weak Interactions in Particle Physics* (New York 1969).

H. F. Schopper, *Weak Interactions and Nuclear Beta Decay* (Amsterdam 1966).

TEPP—Textbook on Elementary Particle Physics: '*Weak Interactions*' (M. K. Gaillard, M. Nikolic, Eds.; Paris 1977).

[FT] Field theory

J. D. Bjorken, S. D. Drell, *Relativistic Quantum Mechanics*; *Relativistic Quantum Fields* (New York 1965).

N. N. Bogoliubov, D. V. Shirkov, *Introduction to the Theory of Quantized Fields* (New York 1979).

L. D. Faddeev, A. A. Slavnov, *Gauge Fields. Introduction to Quantum Theory* (Reading, Mass., 1980).

C. Itzykson, J.-B. Zuber, *Quantum Field Theory* (New York 1980).

P. Roman, *Introduction to Quantum Field Theory* (New York 1969).

S. S. Schweber, *An Introduction to Relativistic Quantum Field Theory* (New York 1964).

[GSW] Glashow–Salam–Weinberg theory

E. Abers, B. W. Lee, *Phys. Rep.* **9,** (1973) 1.

J. C. Taylor, *Gauge Theories of Weak Interactions* (Cambridge 1976).

[GT] Group theory

M. Hamermesh; *Group Theory and its Application to Physical Problems* (Oxford 1964).

D. B. Lichtenberg, *Unitary Symmetry and Elementary Particles* (New York 1978).

H. J. Lipkin, *Anwendung von Lieschen Gruppen in der Physik* (Mannheim 1967).

E. Stiefel, A. Faessler, *Gruppentheoretische Methoden und ihre Anwendung. Eine Einführung mit typischen Beispielen aus Natur- und Ingenieurwissenschaften* (Stuttgart 1979).

B. G. Wyborne, *Classical Groups for Physicists* (New York 1974).

[GUT] Grand unified theories

J. Ellis, 'Grand Unified Theories', Lectures presented at the 21st *Scottish Universities Summer School in Physics, St. Andrews, Scotland* (1980), CERN-TH 2942.

S. Ferrara *et al.*, Eds., *Unification of the Fundamental Particle Interactions* (New York 1980).

F. Wilczek, 'Unification of Fundamental Forces' in *Proceedings of the 1979 International Symposium on Lepton and Photon Interactions at High Energies*, p. 437 (T. B. W. Kirk *et al.*, Eds.; FNAL Batavia, Ill., 1979).

[PM] Parton model

F. E. Close, *An Introduction to Quarks and Partons* (London, New York 1979).

J. Ellis, 'Deep Hadronic Structure' in *Weak and Elektromagnetic Interactions at High Energy, Les Houches* (1976) (R. Balain, C. H. Llewellyn Smith, Eds.; Amsterdam 1977).

R. P. Feynman; *Photon–Hadron Interactions* (Reading, Mass., 1972).

[PQ] Phenomenological applications of QCD

A. J. Buras, *Rev. Mod. Phys.* **52** (1980) 199.

Yu. L. Dokshitzer, D. I. Dyakonov, S. I. Troyan, *Phys. Rep.* **58** (1980) 269.

J. Ellis, 'Status of Perturbative QCD', CERN-TH 2744 (1979) and *1979 International Symposium on Lepton and Photon Interactions at High Energies* (T. B. W. Kirk *et al.*, Eds.; FNAL Batavia, Ill., 1979).

J. Ellis, C. T. Sachrajda, 'QCD and its Applications', CERN-TH 2782 (1979) in *Proceedings of 1979 Cargèse Summer Institute.*

E. Reya, 'Perturbative QCD', *Phys. Rep. 69* (1980) 3.

[QC] Quark confinement

J. M. Drouffe, C. Itzykson; *Phys. Rep.* **38** (1978) 133.

S. Mandelstam; 'Review of Recent Results on QCD and Confinement' in *Proceedings of the 1979 International Symposium on Lepton and Photon Interactions at High Energies*, p. 513 (T. B. W. Kirk *et al.*, Eds.; FNAL Batavia, Ill., 1979).

W. Marciano, H. Pagels, *Phys. Rep.* **36** (1978) 137.

G. 't Hooft *et al.*, Eds.; 'Recent developments in gauge theories', *Proceedings of 1979 Cargèse Summer Institute* (New York, London 1980).

for other literature see:

H. Joos; DESY 76/36 (1976) (published in *Current Induced Reactions, Lecture Notes in Physics* 56, p. 428 (J. G. Körner *et al.*, Eds.; Berlin, Heidelberg, New York 1976)).

[QCD] Quantum chromodynamics

D. J. Gross, in *Methods in Field Theory, Les Houches 1975*, p. 141 (R. Balian, J. Zinn-Justin, Eds.; Amsterdam 1976).

W. Marciano, H. Pagels, *Phys. Rep.* **36** (1978) 137.

H. D. Politzer, *Phys. Rep.* **14** (1974) 129.

[QED] Quantum electrodynamics

A. I. Achieser, W. B. Berestezki, *Quanten-Elektrodynamik* (Frankfurt 1962).

J. D. Bjorken, S. D. Drell, *Relativistic Quantum Mechanics*; *Relativistic Quatum Fields* (New York 1965).

J. M. Jauch, F. Rohrlich, *The Theory of Photons and Electrons. The Relativistic Quantum Field Theory of Charged Particles with Spin one-half* (Berlin 1976).

J. Schwinger, Ed., *Quantum Electrodynamics* (New York 1958).

[RG] Renormalization group
N. N. Bogoliubov, D. V. Shirkov, *Introduction to the Theory of Quantized Fields* (New York 1979).
D. J. Gross, in *Methods in Field Theory, Les Houches 1975* p. 141 (R. Balian, J. Zinn-Justin, Eds.; Amsterdam 1976).
H. D. Politzer, *Phys. Rep.* **14** (1974) 129.
K. Symanzik, in *Particles, Fields and Statistical Mechanics, Lecture Notes in Physics* 32, p. 20 (M. Alexanian, A. Zepeda, Eds.; Berlin 1975).

[RT] Renormalization theory
E. S. Abers, B. W. Lee, *Phys. Rep.* **9** (1973) 1.
C. Becchi, A. Rouet, R. Stora, *Ann. Phys.* **98** (1976) 287.
G. Costa, M. Tonin, *Riv. Nuovo Cim.* **5** (1975) 29.
E. S. Fradkin, I. V. Tyutin, *Riv. Nuovo Cim.* **4** (1974) 1.
J. Zinn-Justin, Bonn 1974: 'Renormalization of gauge theories' in *Lecture Notes in Physics* 37, p. 2 (J. Ehlers *et al.*, Eds.; Berlin 1975).

[Ab 57] A. A. Abrikosov, *Sov. Phys. JETP* **5** (1957) 1174.
[Ab 73] E. Abers, B. W. Lee, *Phys. Rep.* **9** (1973) 1.
[Ab 79] G. S. Abrams *et al.*, *Phys. Rev. Lett.* **44** (1980) 10.
[Ab 80] L. F. Abbott, R. M. Barnett, *Ann. Phys. (NY)* **125** (1980) 276.
[Ad 65] R. Adler, M. Bazin, M. Schiffer, *Introduction to General Relativity* (New York 1965).
[Ad 66] S. L. Adler, *Phys. Rev.* **143** (1966) 1144.
[Ad 68] S. L. Adler, R. F. Dashen, *Current Algebras* (New York 1968).
[Ad 69a] S. L. Adler, *Phys. Rev.* **177** (1969) 2426.
[Ad 69b] S. L. Adler, W. A. Bardeen, *Phys. Rev.* **182** (1969) 1517.
[Ad 70] S. L. Adler, 'Perturbation Theory Anomalies' in *Lectures on Elementary Particles and Quantum Field Theory, 1970 Brandeis University Summer Institute*, Vol. 1, p. 1 (S. Deser *et al.*, Eds.; M.I.T. Cambridge, Mass., 1979).
[Al 56] P. S. Aleksandrov, *Combinatorial Topology* (New York 1956).
[Al 74] G. Altarelli, L. Maiani, *Phys. Lett.* **52B** (1974) 351.
[Al 77] G. Altarelli, G. Parisi, *Nucl. Phys.* B **126** (1977) 298.
[Al 80] A. Ali, E. Pietarinen, G. Kramer, J. Willrodt, *Phys. Lett.* **93B** (1980) 155.
[Am 78] D. Amati, R. Petronzio, G. Veneziano, *Nucl. Phys.* B **140** (1978) 54 and B **146** (1978) 29.
[An 77] D. Antreasyan *et al.*, *Phys. Rev. Lett.* **39** (1977) 906.
[Ap 72] T. Appelquist, H. R. Quinn, *Phys. Lett.* **39B** (1972) 229.
[Ap 75] T. Appelquist, H. D. Politzer, *Phys. Rev.* D **12** (1975) 1404.
[Ap 75a] T. Appelquist, H. D. Politzer, *Phys. Rev. Lett.* **34** (1975) 43.
[Ap 77] T. Appelquist, M. Dine, I. J. Muzinich, *Phys. Lett.* **69B** (1977) 231 and *Phys. Rev.* D **17** (1978) 2074.
[At 78] M. F. Atiyah, N. J. Hitchin, V. G. Drinfeld, Yu. I. Manin, *Phys. Lett.* **65A** (1978) 185.
[Av 72] R. Aviv, A. Zee, *Phys. Rev.* D **5** (1972) 2372.

[Ba 57] J. Bardeen, L. N. Cooper, J. R. Schrieffer, *Phys. Rev.* **106** (1957) 162 and *Phys. Rev.* **108** (1957) 1175.
[Ba 62] M. Baker, S. L. Glashow, *Phys. Rev.* **128** (1962) 2462.
[Ba 69] W. A. Bardeen, *Phys. Rev.* **184** (1969) 1848.
[Ba 70] J. L. Basdevant, B. W. Lee, *Phys. Rev.* D **2** (1970) 1680.
[Ba 72] W. A. Bardeen, R. Gastmanns, B. E. Lautrup, *Nucl. Phys.* B **46** (1972) 319.
[Ba 72a] I. Bars, M. Yoshimura, *Phys. Rev.* D **6** (1972) 374.

[Ba 74] R. Balian, J. M. Drouffe, C. Itzykson, *Phys. Rev.* D **10** (1974) 3376.
[Ba 75] V. Barger, W. F. Long, M. G. Olsson, *Phys. Lett.* **60B** (1975) 89.
[Ba 75a] R. Balian, J. M. Drouffe, C. Itzykson, *Phys. Rev.* D **11** (1975) 1104.
[Ba 77] J. Bailey *et al.* (CERN-Muon-Storage Ring Collaboration), *Phys. Lett.* **67B** (1977) 225.
[Ba 77a] F. Banks, S. Raby, S. Susskind, D. R. Jones, P. N. Scharbach, D. K. Sinclair, *Phys. Rev.* D **15** (1977) 1111.
[Ba 78] M. Bace, *Phys. Lett.* B **78** (1978) 132.
[Ba 78a] W. A. Bardeen, A. J. Buras, D. W. Duke, T. Muta, *Phys. Rev.* D **18** (1978) 3998.
[Ba 79] J. Bailey *et al.* (CERN–Mainz–Daresbury Collaboration), *Nucl. Phys.* B **150** (1979) 1.
[Ba 79a] W. A. Bardeen, A. J. Buras, *Phys. Rev.* D **20** (1979) 166.
[Ba 79b] I. Bars, F. Green, *Phys. Rev.* D **20** (1979) 3311.
[Ba 79c] D. P. Barber *et al.* (MARK-J-Detector), *Phys. Rev. Lett.* **43** (1979) 830.
[Ba 80] V. Barger, W. Y. Keung, E. Ma; *Phys. Rev.* D **22** (1980) 727.
[Be 51] H. A. Bethe, E. E. Salpeter; *Phys. Rev.* **84** (1951) 1232.
[Be 57] H. A. Bethe, E. E. Salpeter, *Quantenmechanik der Ein- und Zweielektronenatome* (Berlin 1957); see also:
L. D. Landau, E. M. Lifshitz, *Relativistische Quantentheorie* (Berlin 1973) and
A. I. Achieser, W. B. Berestezki, *Quanten-Elektrodynamik* (Frankfurt 1962).
[Be 66] F. A. Berezin, *The Method of Second Quantization* (New York 1966).
[Be 69] J. S. Bell, R. Jackiw, *Nuovo Cim.* **51** (1969) 47.
[Be 71] S. M. Berman, J. D. Bjorken, J. Kogut; *Phys. Rev.* D **4** (1971) 3388.
[Be 74] C. Becchi, A. Rouet, R. Stora, *Phys. Lett.* **52B** (1974) 344 and *Commun. Math. Phys.* **42** (1975) 127.
[Be 74a] J. Bernstein, *Rev. Mod. Phys.* **46** (1974) 7.
[Be 75] F. A. Berends, R. Gastmanns, *Phys. Lett.* **55B** (1975) 311.
[Be 75a] A. Belavin, A. Polyakov, A. Schwartz, Y. Tyupkin, *Phys. Lett.* **59B** (1975) 85.
[Be 76] V. B. Berestetzky, *Uspekhi Fiz. Nauk.* **120** (1976) 439.
[Be 77] Ch. Berger *et al.*, *Phys. Lett.* **70B** (1977) 471.
[Be 79] Ch. Berger *et al.* (PLUTO Collaboration), *Phys. Lett.* **81B** (1979) 410 and **78B** (1978) 176.
[Be 79a] B. Berg, M. Lüscher, *Nucl. Phys.* B **160** (1979) 281 and B. Berg, J. Stehr, *Nucl. Phys.* B **175** (1980) 293.
[Be 79b] Ch. Berger *et al.* (PLUTO Collaboration), *Phys. Lett.* **82B** (1979) 449.
[Bj 66] J. D. Bjorken, S. D. Drell, *Relativistic Quantum Mechanics*; *Relativistic Quantum Fields* (New York 1965).
[Bj 69a] J. D. Bjorken, *Phys. Rev.* **179** (1969) 1547.
[Bj 69b] J. D. Bjorken, E. A. Paschos, *Phys. Rev.* **185** (1969) 1975.
[Bj 70] J. D. Bjorken, S. J. Brodsky, *Phys. Rev.* D **1** (1970) 1416.
[Bj 76] J. D. Bjorken, in 'Proceedings of Summer Institute on Particle Physics' SLAC (1976), p. 1 (M. C. Zipf, Ed.; SLAC-Report No. 198, 1976).
[Bl 37] F. Bloch, A. Nordsieck, *Phys. Rev.* **52** (1937) 54; D. R. Yennie, S. C. Frautschi, H. Suura, *Ann. Phys.* **13** (1961) 379; P. P. Kulish, L. D. Faddeev, *Teor. Mat. Fiz.* **4** (1970) 153.
[Bl 63] A. Bludman, A. Klein, *Phys. Rev.* **131** (1963) 2364.
[Bo 65] R. Bott, S. S. Chern, *Acta Math.* **114** (1965) 71.
[Bo 72] C. Bouchiat, J. Iliopoulos, Ph. Meyer, *Phys. Lett.* **38B** (1972) 519.
[Bo 78] P. C. Bosetti *et al.*, *Nucl. Phys.* B **142** (1978) 1.
[Bo 79] A. Bodek *et al.*, *Phys. Rev.* D **20** (1979) 1471.

[Bö 73] M. Böhm, H. Joos, M. Krammer, *Nuovo Cim.* **7A** (1972) 21, *Nucl. Phys,* **B 51** (1973) 397, *Acta Physica Austriaca*, Suppl. XI (1973) 3; CERN TH-1949 (1974).
[Bö 80] M. Böhm, *Z. Phys.* C **3** (1980) 321.
[Br 71] S. J. Brodsky, T. Kinoshita, *Phys. Rev.* D **3** (1971) 356.
[Br 71a] R. A. Brandt, G. Preparata, *Nucl. Phys.* B **27** (1971) 541.
[Br 77] E. Brézin, J. C. Le Guillou, J. Zinn-Justin, *Phys. Rev.* D **15** (1977) 1544, 1558.
[Br 79] L. S. Brown, W. I. Weisberger, *Phys. Rev.* D **20** (1979) 3239.
[Br 79a] R. Brandelik *et al.* (TASSO Collaboration), *Phys. Lett.* **86B** (1979) 243.
[Br 80] R. Brandelik *et al.* (TASSO Collaboration), *Z. Phys.* C **4** (1980) 87.
[Br 80a] R. Brandelik *et al.* (TASSO Collaboration), *Phys. Lett.* **94B** (1980) 437.
[Bu 78] A. J. Buras, K. J. F. Gaemers, *Nucl. Phys.* B **132** (1978) 249.
[Bu 78a] A. J. Buras, J. Ellis, M. K. Gaillard, N. V. Nanopoulos, *Nucl. Phys.* B **135** (1978) 66.
[Bu 80] A. J. Buras, *Rev. Mod. Phys.* **52** (1980) 199.

[Ca 69] C. G. Callan, D. J. Gross, *Phys. Rev. Lett.* **22** (1969) 156.
[Ca 70] C. G. Callan, *Phys. Rev.* D **2** (1970) 1541.
[Ca 71] P. A. Carruthèrs, *Spin and Isospin in Particle Physics* (New York 1971).
[Ca 73] A. Casher, J. Kogut, L. Susskind; *Phys. Rev. Lett.* **31** (1973) 792.
[Ca 74] W. Caswell, *Phys. Rev. Lett.* **33** (1974) 244.
[Ca 76] J. Calmet, S. Narrison, M. Perrottet, E. de Rafael, *Phys. Lett.* **61B** (1976) 283.
[Ca 77] J. Calmet, S. Narrison, M. Perrottet, E. de Rafael, *Rev. Mod. Phys.* **49** (1977) 21.
[Ca 78] C. G. Callan, R. F. Dashen, D. J. Gross, *Phys. Rev.* D **17** (1978) 2717.
[Ca 79] C. G. Callan, R. F. Dashen, D. J. Gross, *Phys. Lett.* **78** (1978) 307 and *Phys. Rev.* D **19** (1979) 1826.
[Ca 80] M. Capdequi-Peyranere, F. M. Renard, M. Talon, *Z. Phys.* C **5** (1980) 337.
[Ce 79] W. Calmaster, R. J. Gonsalves, *Phys. Rev.* D **20** (1979) 1420.
[Ch 46] C. Chevalley, *Theory of Lie Groups* (Princeton, N. J., 1946).
[Ch 72] N. Christ, B. Hasslacher, A. H. Müller, *Phys. Rev.* D **6** (1972) 3543.
[Ch 74] A. Chodos, R. L. Jaffe, K. Johnson, C. B. Thorn, V. F. Weisskopf, *Phys Rev.* D **10** (1974) 2445.
[Ch 77] M. S. Chanowitz, J. Ellis, M. K. Gaillard, *Nucl. Phys.* B **135** (1977) 506.
[Ch 78] N. Christ, E. Weinberg, N. Stanton, *Phys. Rev.* D **18** (1978) 2013.
[Cl 77] D. Cline, 'Charged current weak interactions at high energy' in *Proceedings 1977 International Symposium on Lepton and Photon Interactions at High Energies, Hamburg* (1977), p. 749 (F. Gutbrod, Ed., Hamburg 1977).
[Cl 79] F. E. Close, *An Introduction to Quarks and Partons* (London, New York 1979).
[Co 68] R. Courant, D. Hilbert, *Mathematische Methoden der Physik* I/II (Heidelberg 1968).
[Co 73] E. R. Cohen, B. N. Taylor, *J. Phys. Chem. Ref. Data* **2** (1973) 663.
[Co 73a] S. Coleman, E. Weinberg, *Phys. Rev.* D **7** (1973) 1888.
[Co 74] J. C. Collins, A. J. MacFarlane, *Phys. Rev.* D **10** (1974) 1201.
[Co 75] G. Costa, M. Tonin, *Riv. Nuovo Cim.* **5** (1975) 29.
[Co 75a] F. H. Combley, *Proceedings of the 1975 International Symposium on Lepton and Photon Interactions at High Energies, SLAC (1975)*, p. 913 (T. B. W. Kirk, Ed., Stanford 1975).
[Co 77] S. Coleman, 'Classical lumps and their quantum descendants' in *New*

Phenomena in Subnuclear Physics, p. 297 (A. Zichichi, Ed., New York, London 1977).
[Co 78] E. F. Corrigan, D. B. Fairlie, S. Templeton, P. Goddard, *Nucl. Phys.* B **140** (1978) 31.
[Cr 78] M. Creutz, *Rev. Mod. Phys.* **50** (1978) 561.
[Cr 78a] R. J. Crewther, *Acta Physica Austriaca*, Suppl. IX (1978) 47.
[Cr 79] M. Creutz, *Phys. Rev. Lett.* **43** (1979) 553.
[Cr 80] M. Creutz, *Phys. Rev.* D **21** (1980) 2308 and *Phys. Rev. Lett.* **45** (1980) 313.
[Cu 60] R. E. Cutkosky, *J. Math. Phys.* **1** (1960) 429.

[Da 64] R. F. Dashen, S. C. Frautschi, M. Gell-Mann, Y. Hara, in *The Eightfold Way*, p. 254 (M. Gell-Mann, Y. Ne'eman, Eds., New York, Amsterdam 1964).
[Da 69] R. F. Dashen, *Phys. Rev.* **183** (1969) 1245.
[Da 79] H. D. Dahmen, J. Nölle, L. Schülke, *Nuovo Cim.* **52A** (1979) 573.
[De 61] N. G. De Bruijn, *Asymptotic Methods in Analysis* (Amsterdam 1961).
[De 74] N. Deo, *Graph Theory with Applications to Engineering and Computer Science* (Englewood Cliffs, NY., 1974).
[De 75] A. de Rujula, H. Georgi, S. L. Glashow, *Phys. Rev.* D **12** (1975) 147.
[De 75a] A. de Rujula, S. L. Glashow; *Phys. Rev. Lett.* **34** (1975) 46.
[De 77] A. de Rujula, H. Georgi, H. D. Politzer, *Phys. Rev.* D **15** (1977) 2495.
[De 77a] A. de Rujula, H. Georgi, S. L. Glashow, *Ann. Phys.* **109** (1977) 242.
[De 78] A. de Rujula, J. Ellis, E. G. Floratos, M. K. Gaillard, *Nucl. Phys.* B **138** (1978) 387.
[De 79] J. G. H. de Groot *et al.* (CDHS-Experiment), *Z. Phys.* C **1** (1979) 143.
[De 79a] E. H. de Groot, G. J. Gounaris, D. Schildknecht, *Phys. Lett.* **85B** (1979) 399, *Z. Phys.* C **5** (1980) 127, *Phys. Lett.* **90B** (1980) 470.
[De 79b] J. G. H. de Groot *et al.* (CDHS-Experiment), *Phys. Lett.* **82B** (1979) 292.
[Di 74] J. A. Dixon, J. C. Taylor, *Nucl. Phys.* B **78** (1974) 552.
[Di 79] S. Dimpoulos, L. Susskind, *Nucl. Phys.* B **155** (1979) 237.
[Do 80] V. S. Dotsenko, S. N. Vergeles, *Nucl. Phys.* B **169** (1980) 527.
[Do 80a] Yu. L. Dokshitzer, D. I. Dyakonov, S. I. Troyan, *Phys. Rep.* **58** (1980) 269.
[Dr 70] S. D. Drell, T.-M. Yan, *Phys. Rev. Lett.* **24** (1970) 855.
[Dr 77] W. Drechsler, M. E. Mayer, *Fibre Bundle Techniques in Gauge Theories, Lecture Notes in Physics* **67** (Berlin 1977).
[Dr 78] J. M. Drouffe, C. Itzykson, *Phys. Rep.* **38** (1978) 133.
[Dr 78a] J. M. Drouffe, *Phys. Rev.* D **18** (1978) 1174.
[Dr 80] J. M. Drouffe, J. B. Zuber, *Nucl. Phys.* B **180** [FS2] (1981) 264.
[Dy 49] F. J. Dyson, *Phys. Rev.* **75** (1949) 1736.
A. Salam, *Phys. Rev.* **82** (1951) 217; *Phys. Rev.* **84** (1951) 426.
S. Weinberg, *Phys. Rev.* **118** (1960) 838.
N. N. *Bogoliubov*, D. V. Shirkov, *Introduction to the Theory of Quantized Fields* (New York 1979).
[Dy 79] F. Dydak, 'Neutral currents' in *Proceedings of the International Conference on High-Energy Physics, Geneva (1979)*, Vol. 1, p. 25 (CERN, 1980).

[Ei 22] A. Einstein, *Grundzüge der Relativitätstheorie* (4. Auflage, Braunschweig 1965).
[Ei 75] E. Eichten, K. Gottfried, T. Kinoshita, J. Kogut, K. D. Lane, T.-M. Yan, *Phys. Rev. Lett.* **34** (1975) 369, *Phys. Rev. Lett.* **36** (1976) 500.
E. Eichten, K. Gottfried, *Phys. Lett.* **66B** (1977) 286.

[Ei 78] M. B. Einhorn, B. G. Weeks, *Nucl. Phys.* B **146** (1978) 445.
[Ei 80] E. Eichten, K. Gottfried, T. Kinoshita, K. D. Lane, T.-M. Yan, *Phys. Rev.* D **21** (1980) 203, D **17** (1978) 3050, D **21** (1980) 313 (Err.).
[El 74] S. D. Ellis, M. B. Kislinger, *Phys. Rev.* D **9** (1974) 2027.
[El 76] J. Ellis, M. K. Gaillard, D. V. Nanopoulos, *Nucl. Phys.* B **106** (1976) 292.
[El 76a] J. Ellis, M. K. Gaillard, G. G. Ross, *Nucl. Phys.* B **111** (1976) 253 and B **130** (1977) 516 (Err.).
[El 79] J. Ellis, 'Status of gauge theories' in *Proceedings of Neutrino 79*, Vol. 1, p. 451 (A. Haatuft, C. Jarlskog, Eds., Bergen 1979).
[El 79a] J. Ellis, Status of Perturbative QCD, CERN-TH 2744 (1979) and in *Proceedings of the 1979 International Symposium on Lepton and Photon Interactions at High Energies* (T. B. W. Kirk *et al.*, Eds., FNAL Batavia, *Ill.*, 1979).
[El 79b] J. Ellis, M. K. Gaillard, D. V. Nanopoulos, C. T. Sachrajda, *Phys. Lett.* **83B** (1979) 339.
[El 79c] R. K. Ellis, H. Georgi, M. Machacek, H. D. Politzer, G. G. Ross, *Nucl. Phys.* B **152** (1979) 285.
[El 80] R. K. Ellis, D. A. Ross, A. E. Terrano, *Nucl. Phys.* B **178** (1981) 421.
[El 80a] J. Ellis, 'Grand unified theories', *St. Andrews Proceedings, Gauge Theories and Experiments at High Energies* (K. C. Bowler *et al.*, Eds., (1981)).
[El 80b] J. Ellis, M. K. Gaillard, A. Peterman, C. T. Sachrajda, *Nucl. Phys.* B **164** (1980) 253.
[El 80c] J. Ellis, M. K. Gaillard, D. V. Nanopoulos, S. Rudaz; LAPP-TH-14/CERN-TH 2833 (1980).
[El 80d] J. Ellis, M. K. Gaillard, D. V. Nanopoulos, in *Unification of the Fundamental Particle Interactions*', p. 461 (S. Ferrara *et al.*, Eds., New York (1980) and literature quoted there.
[El 80e] J. Ellis, M. K. Gaillard, B. Zumino, *Phys. Lett.* **94B** (1980) 343.
[EM 80] European Muon Collaboration, 'Measurement of the proton structure function F_2 in muon–hydrogen interactions at 280 and 120 GeV', *Proceedings of the XX International Conference on High Energy Physics 1980* (Madison, U.S.A., 1980).
[En 64] F. Englert, R. Brout, *Phys. Rev. Lett.* **13** (1964) 321.
[Er 55] A. Erdélyi, Ed., *Higher Transcendental Functions*, Vol. 3, Chap. XIX, p. 228 (New York, Toronto, London 1955).

[Fa 67] L. D. Faddeev, V. N. Popov; *Phys. Lett.* **25B** (1967) 29.
[Fa 77] E. Fahri, *Phys. Rev. Lett.* **39** (1977) 1587.
[Fa 80] K. Fabricius, I. Schmitt, G. Schierholz, G. Kramer, *Z. Phys.* **C11** (1982) 315.
[Fe 48] R. P. Feynman, *Rev. Mod. Phys.* **20** (1948) 267.
[Fe 49] R. P. Feynman, *Phys. Rev.* **76** (1949) 769.
[Fe 51] R. P. Feynman, *Phys. Rev.* **84** (1951) 108.
[Fe 65] R. P. Feynman, A. R. Hibbs, *Quantum Mechanics and Path Integrals* (New York 1965).
[Fe 69] R. P. Feynman, *Phys. Rev. Lett.* **23** (1969) 1415.
[Fi 77] R. D. Field, R. P. Feynman, *Phys. Rev.* D **15** (1977) 2590.
[Fi 77a] W. Fischler, *Nucl. Phys.* B **129** (1977) 157.
[Fi 78] R. D. Field, R. P. Feynman, *Nucl. Phys.* B **136** (1978) 1.
[Fl 77] E. G. Floratos, D. A. Ross, C. T. Sachrajda, *Nucl. Phys.* B **129** (1977) 66.
[Fl 79] G. Flügge, *Z. Phys.* C **1** (1979) 121.

[Fl 79a] E. G. Floratos, D. A. Ross, C. T. Sachrajda, *Nucl. Phys.* B **152** (1979) 493.

[Fo 78] G. C. Fox, S. Wolfram, *Phys. Rev. Lett.* **41** (1978) 1581, *Nucl. Phys.* B **149** (1979) 413; *Phys. Lett.* B **82** (1979) 134.

[Fr 53] K. O. Friedrichs, *Mathematical Aspects of the Quantum Theory of Fields* (New York 1953).

[Fr 71] Y. Frishman, *Ann. Phys.* (NY) **66** (1971) 373.

[Fr 72] H. M. Fried, *Functional Methods and Models in Quantum Field Theory* (M.I.T., Cambridge, Mass., London 1972).

[Fr 73] H. Fritzsch, M. Gell-Mann, H. Leutwyler, *Phys. Letters* **47B** (1973) 365.

[Fr 74] E. S. Fradkin, I. V. Tyutin, *Rivista del Nuovo Cim.* **4** (1974) 1.

[Fr 75] H. Fritzsch, P. Minkowski, *Ann. Phys.* **93** (1975) 193.

[Fr 79] E. Fradkin, S. H. Shenker, *Phys. Rev.* D **19** (1979) 3682 and literature quoted there.

[Ga 74] M. K. Gaillard, B. W. Lee, *Phys. Rev. Letters* **33** (1974) 108 and *Phys. Rev.* D **10** (1974) 897.

[Ge 51] M. Gell-Mann, F. Low, *Phys. Rev.* **84** (1951) 350.

[Ge 54] M. Gell-Mann, F. Low, *Phys. Rev.* **95** (1954) 1300.

[Ge 60] M. Gell-Mann, M. Levy, *Nuovo Cim.* **16** (1960) 705.

[Ge 64] M. Gell-Mann, *Phys. Letters* **8** (1964) 214.

[Ge 72] H. Georgi, S. L. Glashow, *Phys. Rev.* D **6** (1972) 429.

[Ge 72a] M. Gell-Mann, 'Quarks' in *Elementary Particle Physics, Multiparticle Aspects, Acta Physica Austriaca*, Suppl. IX (1972), p. 733 (P. Urban, Ed., Wien, New York 1972).

[Ge 74] H. Georgi, S. L. Glashow, *Phys. Rev. Letters* **32** (1974) 438.

[Ge 74a] H. Georgi, H. R. Quinn, S. Weinberg, *Phys. Rev. Letters* **33** (1974) 451.

[Ge 75] H. Georgi, '*Particles and Fields*—1974' (C. E. Carlson, Ed. A.I.P. New York 1975).

[Ge 76] H. Georgi, H. D. Politzer, *Phys. Rev.* D **14** (1976) 1829.

[Ge 76a] J. L. Gervais, A. Jevicki, B. Sakita, *Phys. Reports* **23** (1976) 237 and J. L. Gervais, *Acta Physica Austriaca*, Suppl. XVIII (1977) 385.

[Ge 77] H. Georgi, M. Machacek, *Phys. Rev. Letters* **39** (1977) 1237.

[Ge 78] H. Georgi, H. D. Politzer, *Nucl. Phys.* B **136** (1978) 445.

[Gi 50] V. L. Ginzburg, L. D. Landau, *JETP* (*USSR*) **20** (1950) 1064.

[Gi 64] W. Gilbert, *Phys. Rev. Letters* **12** (1964) 713.

[Gi 76] E. Gildner, *Phys. Rev.* D **14** (1976) 1667.

[Gl 61] S. L. Glashow, *Nucl. Phys.* **22,** (1961) 579.

[Gl 70] S. L. Glashow, J. Iliopoulos, L. Maiani, *Phys. Rev.* D **2** (1970) 1285.

[Gl 76] J. Glimm, A. Jaffe, *Commun. Math. Phys.* **51** (1976) 1.

[Gl 78] M. Glück, E. Reya, *Nucl. Phys.* B **145** (1978) 24.

[Gl 79] M. Glück, E. Reya, *Nucl. Phys.* B **156** (1979) 456.

[Gl 80] J. Glimm, in *Recent Developments in Gauge Theories, Proceedings of 1979 Cargèse Summer Institute*, p. 45 (G. 't Hooft *et al.*, Eds., New York, London 1980).

[Go 58] L. P. Gorkov, *Sov. Phys. JETP* **7** (1958) 505.

[Go 61] J. Goldstone, *Nuovo Cim.* **19** (1961) 154.

[Go 62] J. Goldstone, A. Salam, S. Weinberg, *Phys. Rev.* **127** (1962) 965.

[Go 79] A. González-Arroyo, C. López, F. J. Ynduráin, *Nucl. Phys.* B **153** (1979) 161.

[Gr 64] O. W. Greenberg, *Phys. Rev. Letters* **13** (1964) 598.

[Gr 69] D. J. Gross, C. H. Llewellyn Smith, *Nucl. Phys.* B **14** (1969) 337.

[Gr 71] D. J. Gross, S. B. Treiman, *Phys. Rev.* D **4** (1971) 1059.

[Gr 72] D. J. Gross, R. Jackiw, *Phys. Rev.* D **6** (1972) 477.

[Gr 72a] V. N. Gribov, L. N. Lipatov, *Sov. J. Nucl. Phys.* **15** (1972) 438, 675.
[Gr 73] D. J. Gross, F. Wilczek, *Phys. Rev.* D **8** (1973) 3633 and D **9** (1974) 980.
[Gr 73a] M. Gronau, F. Ravndal, Y. Zarmi, *Nucl. Phys.* B **51** (1973) 611.
[Gr 73b] D. J. Gross, F. Wilczek, *Phys. Rev. Letters* **30** (1973) 1343.
[Gr 74] D. J. Gross, A. Neveu, *Phys. Rev.* D **10** (1974) 3235.
[Gr 76] D. Gromes, I. O. Stamatescu, *Nucl. Phys.* B **112** (1976) 213.
[Gr 78] V. N. Gribov, *Nucl. Phys.* B **139** (1978) 1.
[Gr 79] D. Gromes, *Quarkdynamik und Hadronspektroskopie, Vorlesungen in Maria Laach 1978* (1979).
[Gu 64] G. S. Guralnik, C. R. Hagen, T. W. B. Kibble, *Phys. Rev. Letters* **13** (1964) 585.
[Gü 76] F. Gürsey, P. Ramond, P. Sikivie, *Phys. Letters* **60B** (1976) 177.

[Ha 58] R. Haag, *Phys. Rev.* **112** (1958) 669.
[Ha 62] R. Haag, *Nuovo Cim.* **25** (1962) 287.
[Ha 62a] M. Hamermesh, *Group Theory and its Applications to Physical Problems* (Reading, Mass., 1962).
[Ha 65] M. Y. Han, Y. Nambu, *Phys. Rev.* B **139** (1965) 1006.
[Ha 75] G. Hanson *et al.*, *Phys. Rev. Letters* **35** (1975) 1609.
[Ha 76] H. Harari, 'How many Quarks are there?' in *Storage Ring Physics, Proceedings of the International Colloquium of the C.N.R.S.*, Flaine (France) 1976, p. 461 (J. Tran Thanh Van, Ed., Orsay 1976).
[Ha 77] L. N. Hand, 'Elastic and Inelastic Elektron and Muon Scattering' in *Proceedings 1977 International Symposium on Lepton and Photon Interactions at High Energies*, Hamburg (1977), p. 417 (F. Gutbrod, Ed., Hamburg 1977).
[Ha 78] P. Hasenfratz, *Phys. Reports* **40** (1978) 75.
[Ha 79] H. Harari, *Phys. Letters* **86B** (1979) 83.
[Ha 80] A. E. and P. Hasenfratz, *Nucl. Phys.* B **180** (1981) 341.
[Ha 80a] A. E. and P. Hasenfratz, *Phys. Letters*, **93B** (1980) 165.
[He 28] W. Heisenberg, *Z. Physik* **49** (1928) 619.
[He 62] S. Helgason, *Differential Geometry and Symmetric Spaces* (New York 1962).
[He 67] W. Heisenberg, *Einführung in die einheitliche Feldtheorie der Elementarteilchen* (Stuttgart 1967).
[He 76] W. Heisenberg, *Naturwissenschaften* **63** (1976) 1.
[He 77] S. W. Herb *et al.*, *Phys. Rev. Letters* **39** (1977) 252.
[Hi 64] P. W. Higgs, *Phys. Letters* **12** (1964) 131, *Phys. Rev. Letters* **13** (1964) 508 and *Phys. Rev.* **145** (1966) 1156.
[Hi 80] T. M. Himel *et al.* (MARK II), *Phys. Rev. Letters* **45** (1980) 1146.
[Ho 79] P. Hoyer, P. Osland, H. E. Sander, T. F. Walsh, P. M. Zerwas, *Nucl. Phys.* B **161** (1979) 349.
[Ho 80] W. Hollik, Universität Würzburg Preprint (1980), to be published in *Z. Physik C.*
[Hu 52] C. Hurst, *Proc. Camb. Soc.* **48** (1952) 625.
[Hu 64] K. Huang, *Statistische Mechanik Bd.* 1–3 (Mannheim 1964).

[Il 79] E.-M. Ilgenfritz, D. I. Kazakov, M. Mueller-Preussker, *Phys. Letters* **87B** (1979) 242.
[It 77] C. Itzykson, G. Parisi, J. B. Zuber, *Phys. Rev. Letters* **38** (1977) 306.
[It 80] C. Itzykson, M. Peskin, J. B. Zuber, *Phys. Letters* **95B** (1980) 259.

[Ja 72] R. Jackiw, S. Weinberg, *Phys. Rev.* D **5** (1972) 2396.
[Ja 72a] R. Jackiw, 'Field Theoretic Investigations in Current Algebra' in S. B.

Treiman, R. Jackiw, D. J. Gross, *Lectures on Current Algebra and its Applications* (Princeton 1972).
[Ja 76] R. Jackiw, C. Rebbi, *Phys. Rev. Letters* **37** (1976) 172.
[Ja 77] R. Jackiw, C. Nohl, C. Rebbi, *Phys. Rev.* D **15** (1977) 1642.
[Ja 79] C. Jarlskog, 'Gauge Theories' in *New Phenomena in Lepton Hadron Physics*, p. 1 (D. E. C. Friess, J. Wess, Eds., New York, London 1979).
[Ja 79a] C. Jarlskog, F. J. Yndurain, *Nucl. Phys.* B **149** (1979) 29.
[JA 80] JADE-Collaboration, *Phys. Letters* **91B** (1980) 142.
[Jo 64] G. Jona-Lasinio, *Nuovo Cim.* **34** (1964) 1790.
[Jo 65] R. Jost; *The General Theory of Quantized Fields* (A.M.S., Providence, Rhode Island 1965).
[Jo 70] H. Joos, *Proceedings of the 11th Scottish Universities Summer School in Physics (1970)*, p. 47 (J. Cumming, H. Osborn, Eds., London 1971).
[Jo 74] D. R. T. Jones, *Nucl. Phys.* B **75** (1974) 531.
[Jo 76] S. D. Joglekar, B. W. Lee, *Ann. Phys.* **97** (1976) 160.

[Ka 77] L. D. Kadanoff, *Rev. Mod. Phys.* **49** (1977) 267.
[Ka 78] D. M. Kaplan *et al.*, *Phys. Rev. Letters* **40** (1978) 435.
[Kh 69] I. B. Khriplovich, *Yad. Fiz.* **10** (1969) 409.
[Ki 62] T. Kinoshita, *J. Math. Phys.* **3** (1962) 650.
[Ki 67] T. W. B. Kibble, *Phys. Rev.* **155** (1967) 1554.
[Ki 78] T. Kinoshita, 'Recent Developments of Quantum Electrodynamics' in *Proceedings of the 19th International Conference on High Energy Physics*, Tokyo (1978), p. 571 (S. Homma *et al.*, Eds., Tokyo 1979).
[Kl 61] L. Klein, Ed., *Dispersion Relations and the Abstract Approach to Field Theory* (New York 1961).
[Kl 75] H. Kluberg-Stern, J. B. Zuber, *Phys. Rev.* D **12** (1975) 467, 482, 3159.
[Ko 63] S. Kobayashi, K. Nomizu, *Foundations of Differential Geometry*, vol. I/II (New York, London 1963).
[Ko 69] J. J. J. Kokkedee, *The Quark Model* (New York 1969).
[Ko 73a] M. Kobayashi, K. Maskawa, *Progr. Theor. Phys.* **49** (1973) 652.
[Ko 74] J. Kogut, L. Susskind, *Phys. Rev.* D **9** (1974) 697, 706, 3391.
[Ko 74a] J. Kogut, K. G. Wilson, *Phys. Reports* **12** (1974) 75.
[Ko 75] J. Kogut, L. Susskind, *Phys. Rev.* D **11** (1975) 395.
[Ko 76] J. Kogut, D. K. Sinclair, L. Susskind, *Nucl. Phys.* B **114** (1976) 199.
[Ko 77] J. Kogut, in *Many Degrees of Freedom in Particle Physics*, p. 275 (H. Satz, Ed., New York, London 1978).
[Ko 79] J. Kogut, *Rev. Mod. Phys.* **51** (1979) 659.
[Ko 79a] J. Kogut, R. B. Pearson, J. S. Shigemitsu, *Phys. Rev. Letters* **43** (1979) 484.
[Ko 79b] J. Koller, H. Krasemann, *Phys. Letters* **88B** (1979) 119.
[Ko 80] J. Kogut, R. B. Pearson, J. S. Shigemitsu, *Phys. Lett.* **98B** (1981) 63.
[Kr 79] M. Krammer, H. Krasemann, 'Quarkonia' in *Quarks and Leptons, Acta Physica Austriaca*, Suppl. XXI (1979), p. 259 (P. Urban, Ed., Wien, New York 1979).
[Ku 80] Z. Kunszt, *Phys. Lett.* **99B** (1981) 429.

[La 55] L. D. Landau, I. Ya. Pomeranchuk, *Doklady Akad. Nauk. SSSR* **102** (1955) 489.
[La 69] L. D. Landau, E. M. Lifshitz, *The Classical Theory of Fields* (Oxford 1969 and 1975).
[La 72] B. E. Lautrup, A. Peterman, E. de Rafael, *Phys. Reports* **3** (1972) 193.
[Le 55] H. Lehmann, K. Symanzik, W. Zimmermann, *Nuovo Cim.* **1** (1955) 1425.

[Le 60] T. D. Lee, C. N. Yang, *Phys. Rev. Letters* **4** (1960) 307.
[Le 64] T. D. Lee, M. Nauenberg, *Phys. Rev.* **133** (1964) 1549.
[Le 72] B. W. Lee, J. Zinn-Justin, *Phys. Rev.* D **5** (1972) 3121, 3137, 3155, *Phys. Rev.* D **7** (1973) 1049.
[Le 72a] B. W. Lee, *Phys. Rev.* D **5** (1972) 823.
[Le 72b] B. W. Lee, *Chiral Dynamics* (New York 1972).
[Le 72c] S. Y. Lee, *Phys. Rev.* D **6** (1972) 1701, 1803.
[Le 73] B. W. Lee, *Phys. Letters* **46B** (1973) 214 and *Phys. Rev.* D **9** (1974) 933.
[Le 74] G. Leibbrandt, D. M. Capper, *J. Math. Phys.* **15** (1974) 82, 86.
[Le 74a] B. W. Lee, *Phys. Rev.* D **9** (1974) 933.
[Le 74b] H. Leutwyler, *Nucl. Phys.* B **76** (1974) 413.
[Le 77] B. W. Lee, C. Quigg, B. H. Thacker, *Phys. Rev. Letters* **38** (1977) 883.
[LE 81] LENA-Kollaboration (B. Niczyporuk *et al.*), *Phys. Rev. Letters* **46** (1981) 92.
[Li 75] L. N. Lipatov, *Sov. J. Nucl. Phys.* **20** (1975) 94.
[Li 77] L. N. Lipatov, JETP **72** (1977) 411.
[Li 78] D. B. Lichtenberg, *Unitary Symmetry and Elementary Particles* (New York 1978).
[Ll 70] C. H. Llewellyn Smith, *Nucl. Phys.* B **17** (1970) 277.
[Ll 74] C. H. Llewellyn Smith, 'Phenomenology of Particles at High Energies', *Proceedings of the 14th Scottish Universities Summer School in Physics 1973 at Edinburgh*, p. 459 (R. L. Crawford, J. Jennings, Eds., London 1974).
[Ll 78] C. H. Llewellyn Smith, *Acta Physica Austriaca,*, Suppl XIX (1978) 331.
[Ll 79] C. H. Llewellyn Smith, in *Proc. of the LEP Summer Study*, CERN 79–01 (1979).
[Lo 50] F. London, *Superfluids*, vol. I (New York 1950).
[Lo 78] E. Lohrmann, *Hochenergiephysik* (Stuttgart 1978).
[Lü 77] M. Lüscher, *Commun. Math. Phys.* **54** (1977) 283.
[Lü 80] M. Lüscher, G. Münster, P. Weisz, *Nucl. Phys.* B **180** [FS2] (1981) 1.
[Lü 80a] M. Lüscher, *Phys. Letters* **90B** (1980) 277.
[Lü 80b] M. Lüscher, K. Symanzik, P. Weisz, *Nucl. Phys.* B **173** (1980) 365.
[Lü 80c] M. Lüscher, 'Exact Instanton Gases' in *Recent Developments in Gauge Theories, Proceedings of 1979 Cargèse Summer Institute* (G. 't Hooft, *et al.*, Eds., New York, London 1980).

[Ma 69] R. E. Marshak, Riazuddin, C. P. Ryan, *Theory of Weak Interactions in Particle Physics* (New York 1969).
[Ma 75] W. J. Marciano, A. Sirlin, *Nucl. Phys.* B **88** (1975) 86.
[Ma 78] W. J. Marciano, H. Pagels, *Phys. Reports* **36** (1978) 137.
[Ma 78a] G. Mack, *Phys. Letters* **78B** (1978) 263.
[Ma 78b] G. Mack, V. B. Petkova, *Ann. Phys.* (N.Y.) **125** (1980) 117 and in *Recent Developments in Gauge Theories*, p. 217, op. cit. (see [Ma 80]).
[Ma 79] S. Mandelstam, *Phys. Rev.* D **19** (1979) 2391.
[Ma 79a] Yu. M. Mareenko, A. A. Migdal, *Phys. Letters* **88B** (1979) 135 and **89B** (1980) 437.
[Ma 79b] M. Machacek, *Nucl. Phys.* B **159** (1979) 37.
[Ma 80] G. Mack, DESY 80/03 (1980) and 'Properties of Lattice Gauge Theory Models at Low Temperatures' in *Recent Developments in Gauge Theories, Proceedings of 1979 Cargèse Summer Institute*, p. 217 (G. 't Hooft *et al.*, Eds., New York, London 1980).
[Ma 80a] R. E. Marshak, R. N. Mohapatra, *Phys. Rev. Letters* **44** (1980) 1316.

[Mi 79] P. K. Mitter, 'Geometry of the Space of Gauge Orbits and the Yang-Mills Dynamical System', in *Recent Developments in Gauge Theories, Proceedings of 1979 Cargèse Summer Institute*, p. 265 (G. 't Hooft *et al.*, Eds., New York, London 1980).

[Mo 79] G. Morpurgo, 'Lectures on Quarks' in *Quarks and Leptons, Acta Physica Austriaca*, Suppl. XXI, p. 5 (P. Urban, Ed., Wien, New York 1979).

[Mo 80] K. J. M. Moriarty, DESY 80/67 (1980) (to appear in *J. Phys.*).

[Mü 80] G. Münster, *Phys. Letters* **95B** (1980) 59 and DESY 80/44 (1980) (to appear in *Nucl. Phys. B*).

[Mü 80a] G. Münster, P. Weisz, *Nucl. Phys.* B **180** [FS2] (1981) 330.

[Mü 80b] G. Münster, P. Weisz, *Phys. Letters* **96B** (1980) 119.

[Mü 80c] G. Münster, P. Weisz, *Nucl. Phys.* B **180** [FS2] (1981) 13.

[Mü 80d] G. Münster, *Nucl. Phys.* B **190** [FS3] (1981) 439, B **200** [FS4] (1982) 469.

[Na 61] Y. Nambu, G. Jona-Lasinio, *Phys. Rev.* **122** (1961) 345 and *Phys. Rev.* **124** (1961) 246.

[Na 66] Y. Nambu, *Phys. Letters* **26B** (1966) 626.

[Na 72] O. Nachtmann, *Nucl. Phys.* B **38** (1972) 397.

[Na 73] O. Nachtmann, *Nucl. Phys.* B **63** (1973) 237.

[Na 79] Y. Nambu, *Phys. Letters* **80B** (1979) 372.

[Ni 73] H. Nielsen, P. Olesen, *Nucl. Phys.* B **61** (1973) 45.

[Ni 78] H. Nielsen, *Phys. Letters* **80B** (1978) 133, and H. Nielsen, M. Ninomiya, *Nucl. Phys.* B **169** (1980) 309 and literature quoted there.

[No 78] V. A. Novikov *et al.*, *Phys. Reports* **41** (1978) 1.

[Ol 79] D. Olive, S. Sciuto, R. J. Crewther, *Rivista del Nuovo Cim.* **2** (1979) 8.

[Or 49] A. Ore, I. L. Powell, *Phys. Rev.* **75** (1949) 1696.

[Os 75] K. Osterwalder, R. Schrader, *Commun. Math. Phys.* **42** (1975) 281.

[Os 78] K. Osterwalder, E. Seiler, *Ann. Phys.* (N.Y.) **110** (1978) 440.

[Pa 49] W. Pauli, F. Villars, *Rev. Mod. Phys.* **21** (1949) 434.

[Pa 69] R. D. Parks, Ed., *Superconductivity* (New York 1969).

[Pa 74] J. C. Pati, A. Salam, *Phys. Rev.* D **10** (1974) 275.

[Pa 82] Particle Data Group, *Phys. Lett.* **111B** (1982).

[Pa 80a] H. Pagels, *Phys. Rev.* D **21** (1980) 2336.

[Pa 80b] R. Partridge *et al.* (CRYSTAL BALL), *Phys. Rev. Letters* **45** (1980) 1150.

[Pa 80c] G. Parisi, P. Petronzio, *Phys. Letters* **94B** (1980) 51.

[Pe 72] D. H. Perkins, *Proceedings of the XVIth International Conference on High Energy Physics*, Chicago-Batavia (1972), p. 189 (NAL, Batavia 1972).

[Pe 75] D. H. Perkins, *Proceedings of the 1975 International Symposium on Lepton and Photon Interactions at High Energies*, SLAC (1975), p. 571 (W. T. Kirk, Ed., Stanford 1975).

[PL 79] PLUTO-Collaboration, *Phys. Lettes* **86B** (1979) 418.

[PL 80] PLUTO-Collaboration, *Phys. Letters* **97B** (1980) 459.

[Po 73] H. D. Politzer, *Phys. Rev. Letters* **30** (1973) 1346.

[Po 74] H. D. Politzer, *Phys. Reports* **14** (1974) 129.

[Po 77] H. D. Politzer, *Nucl. Phys.* B **129** (1977) 301.

[Po 77a] A. M. Polyakov, *Nucl. Phys.* B **120** (1977) 429.

[Po 79] A. M. Polyakov, *Phys. Letters* **82B** (1979) 247.

[Ra 79] P. Ramond, CALT-68-709 (1979).

[Re 74] C. Rebbi, *Phys. Reports* **12** (1974) 1.

[Ri 65] R. Rickayzen, *Theory of Superconductivity* (New York 1965).

[Ri 75] E. M. Riordan *et al.*, SLAC-Pub. 1634 (1975), unpublished.
[Ro 67] R. van Royen, V. F. Weisskopf, *Nuovo Cim.* **50** (1967) 617.
[Ro 71a] H. Rollnik, *Teilchenphysik I* (Mannheim 1971).
[Ro 71b] H. Rollnik, *Teilchenphysik II* (Mannheim 1971).
[Ro 78] D. A. Ross, *Nucl. Phys.* B **140** (1978) 1.
[Ru 61] D. Ruelle, *Nuovo Cim.* **19** (1961) 356.
[Ru 62] D. Ruelle, *Helv. Phys. Acta* **35** (1962) 147.

[Sa 67] A. D. Sakharov, *Pis'ma Zh. Eksp. Teor. Fiz.* **5** (1967) 32.
[Sa 68] A. Salam, Proc. 8th NOBEL Symposium, p. 367 (N. Svartholm, Ed., Stockholm 1968).
[Sa 77] G. K. Savvidy, *Phys. Letters* **71B** (1977) 133.
[Sa 78] C. T. Sachrayda, *Phys. Letters* **73B** (1978) 185 and **76B** (1978) 100.
[Sa 80] S. Sakakibara, *Phys. Rev.* D **24** (1981) 1149.
[Sc 49] J. Schwinger, *Phys. Rev.* **76** (1949) 790.
[Sc 51] J. Schwinger, *Proc. Nat. Acad. Sci.* (U.S.A.) **37** (1951) 452.
[Sc 75] J. Scherk, *Rev. Mod. Phys.* **47** (1975) 123.
[Sc 77] A. S. Schwartz, *Phys. Letters* **67B** (1977) 172.
[Sc 79] G. Schierholz, 'e^+e^--Jets' DESY 79/71 (1979), published in *Proceedings of Summer Institute on Particle Physics* (1979), Stanford-SLAC-224 (1979).
[Se 75] R. Sexl, H. Urbantke, *Gravitation and Kosmologie* (Mannheim 1975).
[Se 77] L. M. Sehgal, 'Hadron Production by Leptons' in *Proceedings 1977 International Symposium on Lepton and Photon Interactions at High Energies*, Hamburg (1977), p. 837 (F. Gutbrod, Ed., Hamburg 1977).
[Sh 78] Q. Shafi, Ch. Wetterich, *Phys. Letters* **73B** (1978) 65.
[Si 76] D. Sivers, S. I. Brodsky, R. Blankenbecler, *Phys. Reports* **23** (1976) 1.
[Si 76a] I. Singer, J. A. Thorpe, *Lecture Notes on Elementary Topology and Geometry* (Berlin, Heidelberg, New York 1976).
[Si 78] I. Singer, *Commun. Math. Phys.* **67** (1978) 7.
[Si 80] A. Sirlin, *Phys. Rev.* D **22** (1980) 971.
[Sk 74] A. V. Skorohod, *Integration in Hilbert Space* (Heidelberg, New York 1974).
[Sl 72] A. A. Slavnov, *Theor. and Math. Phys.* **10** (1972) 99.
[Sm 79] A. V. Smilga, M. I. Vysotsky, *Nucl. Phys.* B **150** (1979) 173.
[St 53] E. C. G. Stueckelberg, A. Peterman, *Helvetia Phys. Acta* **26** (1953) 499.
[St 64] R. F. Streater, A. S. Wightman, PCT, *Spin & Statistics, and all that* (New York 1964).
[St 71] H. E. Stanley, *Introduction to Phase Transitions and Critical Phenomena* (Oxford 1971).
[St 76] B. Stech, 'Status of Broken Color Symmetry' in *Current Induced Reactions, Lecture Notes in Physics*, vol. 56, p. 322 (J. G. Körner *et al.*, Eds., Berlin, Heidelberg, New York 1976).
[St 77] G. Sterman, S. Weinberg, *Phys. Rev. Letters* **39** (1977) 1436.
[Su 77] L. Susskind, *Phys. Rev.* D **10** (1977) 3031.
[Su 79] L. Susskind, *Phys. Rev.* D **20** (1979) 2619.
[Sy 54] K. Symanzik, *Z. Naturforschung* **9a** (1954) 809.
[Sy 60] K. Symanzik, *J. Math. Phys.* **1** (1960) 249.
[Sy 66] K. Symanzik, *J. Math. Phys.* **7** (1966) 510.
[Sy 67] K. Symanzik, 'Many-Particle Structure of Green's Functions', *Symposia on Theoretical Physics*, vol. **3** (1967) 121.
[Sy 70] K. Symanzik, *Commun. Math. Phys.* **18** (1970) 227 and *Springer Tracts in Modern Physics* **57** (1971) 222.
[Sy 71] K. Symanzik, *Commun. Math. Phys.* **23** (1971) 49.

[Sy 73] K. Symanzik, *Lett. Nuovo Cim.* **6** (1973) 77.
[Sy 77] K. Symanzik, in *New Developments in Quantum Field Theory and Statistical Dynamics* (M. Lévy, P. Mitter, Eds., New York 1977).
[Sy 79] K. Symanzik, DESY-Preprint 79/76 (1979), published: *Cutoff Dependence in Lattice ϕ_4^4 Theory* in *Recent Developments in Gauge Theories, Proceedings of 1979 Cargèse Summer Institute*, p. 313 (G. 't Hooft *et al.*, Eds., New York, London 1980).

[Ta 57] Y. Takahashi, *Nuovo Cim.* **6** (1957) 370.
[Ta 71] J. C. Taylor, *Nucl. Phys.* B **33** (1971) 436.
[Ta 75] R. E. Taylor, *Proceedings of the 1975 International Symposium on Lepton and Photon Interactions at High Energies*, SLAC (1975), p. 679 (W. T. Kirk, Ed., Stanford 1975).
[Ta 76] J. C. Taylor, *Gauge Theories of Weak Interactions* (Cambridge 1976).
[Ta 78] R. E. Taylor, *Proceedings of the 19th International Conference on High Energy Physics*, Tokyo (1978), p. 285 (S. Homma *et al.*, Eds., Tokyo 1979).
[TA 79] TASSO-Collaboration, *Phys. Letters* **86B** (1979) 243.
[Te 68] C. Teleman, *Grundzüge der Topologie und differenzierbaren Mannigfaltigkeiten* (Berlin 1968).
[Th 53] W. Thirring, *Helv. Phys. Acta* **26** (1953) 33.
[tH 71] G. 't Hooft, *Nucl. Phys.* B **33** (1971) 173 and *Nucl. Phys.* B **35** (1971) 167.
[tH 72] G. 't Hooft, M. Veltman, *Nucl. Phys.* B **50** (1972) 318.
[tH 72a] G. 't Hooft, M. Veltman, *Nucl. Phys.* B **44** (1972) 189, see also:
C. Bollini, J. Gianbiagi, *Nuovo Cim.* **12B** (1972) 20.
J. Ashmore, *Lett. Nuovo Cim.* **4** (1972) 289.
G. M. Cicuta, E. Montaldi, *Lett. Nuovo Cim.* **4** (1972) 329.
P. Butera, G. M. Cicuta, E. Montaldi, *Nuovo Cim.* **19A** (1974) 513.
G. Leibbrandt, *Rev. Mod. Phys.* **47** (1975) 849.
[tH 73] G. 't Hooft, M. Veltman, 'Diagrammar' CERN-Report 73-9 (1973); published in *Louvain 1973, Particle Interactions at very high Energies*, part B, p. 177 (New York 1973).
[tH 73a] G. 't Hooft, *Nucl. Phys.* B **61** (1973) 455.
[tH 74] G. 't Hooft, in *Marseille 1974 Proceedings on Recent Progress in Lagrangian Field Theory and Applications*, p. 58 (Marseille 1975).
[tH 76] G. 't Hooft, *Phys. Rev. Letters* **37** (1976) 8.
[tH 76a] G. 't Hooft, *Phys. Rev.* D **14** (1976) 3432, Err.: *Phys. Rev.* D **18** (1978) 2199.
[tH 78] G. 't Hooft, *Nucl. Phys.* B **138** (1978) 1.
[tH 79] G. 't Hooft, 'Why do we Need Local Gauge Invariance in Theories with Vector Particles? an Introduction' in *Recent Developments in Gauge Theories, Proceedings of 1979 Cargèse Summer Institute*, p. 101 (G. 't Hooft *et al.*, Eds.; New York, London 1980).
[tH 79a] G. 't Hooft, *Nucl. Phys.* B **153** (1979) 141.
[tH 80] G. 't Hooft, 'Naturalness, Chiral Symmetry, and Spontaneous Chiral Symmetry Breaking' in *Recent Developments in Gauge Theories, Proceedings of 1979 Cargèse Summer Institute*, p. 135 (G. 't Hooft *et al.*, Eds.; New York, London 1980).
[Ty 72] I. V. Tyutin, E. S. Fradkin, *Sov. Journ. Nucl. Phys.* **16** (1972) 835.

[Ut 56] R. Utiyama, *Phys. Rev.* **101** (1956) 1597.

[Ve 63] M. Veltman, *Physica* **29** (1963) 186.

[Ve 80] M. Veltman, LAPP Preprint TH-12 (1980).

[Wa 50] J. C. Ward, *Phys. Rev.* **78** (1950) 1824.
[Wa 53] I. Watanabe, *Progr. Theor. Phys.* **4** (1953) 371.
[We 18] H. Weyl, *Raum. Zeit, Materie* (6. unveränderte Auflage, Berlin 1970).
[We 29] H. Weyl, *Z. Physik* **56** (1929) 330.
[We 60] S. Weinberg, *Phys. Rev.* **118** (1960) 838.
[We 67] S. Weinberg, *Phys. Rev. Letters* **19** (1967) 1264.
[We 67a] S. Weinberg, *Phys. Rev. Letters* **18** (1967) 507.
[We 68] S. Weinberg, *Phys. Rev.* **166** (1968) 1568.
[We 71] J. Wess, B. Zumino, *Phys. Letters* **37B** (1971) 95.
[We 72] S. Weinberg, *Phys. Rev. Letters* **27** (1972) 1688.
[We 73] S. Weinberg, *Phys. Rev. Letters* **31** (1973) 494 and *Phys. Rev.* D **8** (1973) 4482.
[We 73a] S. Weinberg, *Phys. Rev.* D **8** (1973) 3497.
[We 75] S. Weinberg, *Phys. Rev.* D **11** (1975) 3594.
[We 76] S. Weinberg, in *Proceedings of the 1976 International School of Subnuclear Physics*, Erice (New York 1978).
[We 76a] S. Weinberg, *Phys. Rev.* D **13** (1976) 974 and *Phys. Rev.* D **19** (1979) 1277.
[We 76b] S. Weinberg, *Phys. Rev. Letters* **36** (1976) 294.
[We 79] S. Weinberg, *Phys. Rev. Letters* **43** (1979) 1566.
[We 80] S. Weinberg, *Phys. Letters* **91B** (1980) 51.
[Wi 39] E. Wigner, *Ann. Math.* **40** (1939) 149.
[Wi 56] A. S. Wightman, *Phys. Rev.* **101** (1956) 860.
[Wi 58] N. Wiener, *Nonlinear Problems in Random Theory* (New York 1958).
[Wi 69] K. G. Wilson, *Phys. Rev.* **179** (1969) 1499.
[Wi 74] K. G. Wilson, *Phys. Rev.* D **10** (1974) 2445.
[Wi 75] K. G. Wilson, *Rev. Mod. Phys.* **47** (1975) 773.
[Wi 77] F. Wilczek, *Phys. Rev. Letters* **39** (1977) 39.
[Wi 77a] K. G. Wilson, in *New Developments in Quantum Field Theory and Statistical Mechanics* (Cargèse 1976) (M. Lévy, P. Mitter, Eds., New York 1977).
[Wi 79] B. H. Wiik, G. Wolf, *Electron–Positron Interactions* (Berlin, Heidelberg, New York 1979).
[Wi 79a] K. Winter, 'Review of Experimental Measurements of Weak Neutral Current Interactions', *Proceedings of the International Symposium on Lepton and Photon Interactions at High Energies*, FNAL, Batavia (1979).
[Wi 79b] F. Wilczek, A. Zee, *Phys. Rev. Letters* **43** (1979) 1571.
[Wi 80] B. Wiik, DESY 80/124 (1980), to appear in *Proceedings of the 1980 Madison-Conference.*
[Wi 80a] K. G. Wilson, 'Monte-Carlo Calculations for the Lattice Gauge Theory' in *Recent Developments in Gauge Theories, Proceedings of 1979 Cargèse Summer Institute*, p. 363 (G. 't Hooft *et al.*, Eds., New York, London 1980).
[Wo 80] G. Wolf, 'Selected Topics on e^+e^--Physics', DESY 80/13 (1980).

[Ya 54] C. N. Yang, R. L. Mills, *Phys. Rev.* **96** (1954) 191.

[Zi 58] W. Zimmermann, *Nuovo Cim.* **X** (1958) 597.
[Zi 59] W. Zimmermann, *Nuovo Cim.* **13** (1959) 503 and *ibidem* **16** (1960) 690.
[Zi 74] J. Zinn-Justin, Bonn 1974: 'Renormalization of Gauge Theories' in *Lecture Notes in Physics*, vol. 37, p. 2 (J. Ehlers *et al.*, Eds., Berlin 1975).
[Zw 64] G. Zweig, CERN-TH 401 and 412 (1964).

Conventions Used

(1) Natural system of units $\hbar = c = 1$.

(2) Space–time coordinates $x^\mu \equiv (x^0, x^1, x^2, x^3) \equiv (t, \mathbf{x})$, similar for other four-vectors:

$$A^\mu \equiv (A^0, \mathbf{A}), \quad \partial^\mu = \left(\frac{\partial}{\partial x_0}, \frac{\partial}{\partial x_1}, \frac{\partial}{\partial x_2}, \frac{\partial}{\partial x_3}\right) = (\partial^0, \boldsymbol{\partial}).$$

(3) Metric: $g^{\mu\nu} = g_{\mu\nu} = \text{diag}(+1, -1, -1, -1)$.

(4) Summation convention also in the form that as far as possible all indices which are not explicitly written down are contracted.

(5) Dirac matrices and spinors according to Table 1.4 on p. 20.

(6) Commutator and propagator functions of free fields according to J. D. Bjorken and S. D. Drell [Bj 66].

(7) Experimental results not explicitly referenced are all taken from the *Review of Particle Properties* table, published by the Particle Data Group [Pa 82].

Appendix to the Second Edition

The publications on gauge field theories of the strong and electroweak interaction and the corresponding experiments which appeared since the first edition of this book did not introduce major alteration in the ideas on the interaction of elementary particles. Remarkable progress has occurred in the following fields:

(a) In *particle spectroscopy*, these are the discovery [Be 83N] of mesons B with bottom quantum numbers ±1 and masses $M_{(B^{\pm})} = 5270.8 \pm 5$ MeV, $M_{(B^0)} = 5274.2 \pm 4$ MeV. In addition, there are more results on the bottomonium system [Fr 82N]. They have given a confirmation and further refinement of the potential models for the quarkonia systems [Bu 81aN, Ma 82N]. The continuous improvement of the general particle data especially of the τ-lepton is reported in the 82 edition of the *Review of Particle Properties* [Pa 82].

(b) The experimental basis of the *parton model* (Section 1.4) including the interpretation of *jets* (section 2.7.1) has been enlarged further. In addition, to improved experimental distribution functions of quarks in nucleons [Ab 83N, Ei 82N], there are now gluon distributions [Ab 82N], indirect measurements of the structure functions of mesons [Ba 79N, Ba 80N] and first measurements of the photon structure functions [Be 81N, Ba 83N]. Jets resulting from parton reactions [Wo 82N] have been observed not only in the standard e^+e^--experiments [Ha 82N, Cr 82N] but also in deep inelastic lepton scattering [Au 81N], in hadronic interactions [Ba 82N, Ak 82N] and photon–photon scattering [Br 81N, Ba 81N, Cr 82N]. The angular distribution of the jet-axes in the parton reactions $e^+e^- \to q\bar{q}$ or $e^+e^- \to q\bar{q}g$ allow the determination of the quark spin [Ha 82aN, Cr 82N] and the gluon spin [Br 80N]. On the other hand, the expected relationship between the charge structure of the partons and the charge distributions in the jets still escapes verification. Hints on 4-jet events in e^+e^--reactions as a result of multiple gluon-bremsstrahlung have been seen [Ba 82aN].

The improved theoretical calculation of gluon-bremsstrahlung in second order QCD [Fa 82N] and its comparison with experiments ([Ba 82N], see also [Ab 82aN]) demonstrates the non-Abelian character of the gluon coupling and results in a coupling constant

$$(30\ \text{GeV}) = 0.16 \pm 0.015\ (\text{stat.}) \pm 0.03\ (\text{syst.}).$$

(c) An experimental *test of perturbative QCD* (Section 2.7, see also [Sö 81N]) is now based in almost all cases on the required second order calculations [Ba 81N]. From a critical point of view, one cannot yet state a complete success of the quantitative comparison between theory and experiment [Po 82N].

(d) There are many attempts to calculate the *hadron spectrum in the framework of QCD* (section 2.6). Essentially these are based on the evaluation of the lattice approximation of QCD with help of Monte Carlo calculations. The calculations [G 82N] give for the relation between the string constant and the asymptotic scale parameter in a pure gauge theory $\Lambda_{MOM} = (0.5 \pm 0.1)$ (see Eqn. 2.6.124). A critical comparison of the various glueball masses with experimental data is not yet possible. On the one hand, the experimental candidates ι (1440) [Sc 80N] and θ (1640) [Ed 82N] are not yet clearly identified as glueball states [Be 82N, D0 82N]. On the other hand, the numerical calculations are still subject of critique on the applied methods. This is even more true for the calculations of the meson- and baryon spectrum which were possible by including the quark fields in the lattice calculations [Ha 81N, Fu 82N, We 82N, Ha 82N]. Successful calculations of hadron masses were performed with help of QCD-(ITEP)-sum rules [No 81N, Re 82N].

(e) The main theoretical development in the field of the gauge theory of the electroweak interaction (sections 3.3 and 3.4) consisted in attempts to embedding this in more general theories like 'grand unified theories' super-symmetries or a subquark theory (section 3.4). Such theories are beyond the scope of this book. Within the Glashow–Salam–Weinberg theory mainly calculations of radiative corrections have been performed [Be 82N]. Remarkable experiments are the observation of the electroweak interference in $e^+e^- \to \mu^+\mu^-$ at Petra [Ba 82N, Be 82N, Br 82N, Da 82N] and in electron nucleon scattering at SLAC [Pr 79N]. The operation of the $P\bar{P}$-collider at CERN [St 81N] has opened the possibility to discover the intermediate W- and Z-bosons. First experimental results indicate the existence of these particles with masses which agree with the predictions of the Glashow–Salam–Weinberg model [Ar 83N].

Appendix to the Literature

[Ab 82N] H. Abramowicz *et al.*, *Z. Physik C* **12** (1982) 289.

[Ab 82aN] H. Abramowicz *et al.*, *Z. Physik C* **13** (1982) 199.

[Ab 83N] H. Abramowicz *et al.*, *Z. Physik C* (to appear).

[Ak 82N] T. Akesson *et al.*, *Phys. Letters* **118B** (1982) 185, 193.

[Ar 83N] G. Arnison *et al.* (UA1 Coll.), *Phys. Lett.* **122B** (1983) 103; G. Banner *et al.* (UA2 Coll.), *Phys. Lett.* **122B** (1983) 476; G. Arnison *et al.* (UA1 Coll.), *CERN-EP*/83-73 (1983).

[Au 81N] J. J. Aubert *et al.*, *Phys. Letters* **100B** (1981) 433.
[Au 82N] J. J. Aubert *et al.*, *Phys. Letters* **114B** (1982) 291.
[Ba 79N] R. Barate *et al.*, *Phys. Rev. Letters* **43** (1979) 1541.
[Ba 80N] J. Badier *et al.*, *Phys. Letters* **93B** (1980) 354.
[Ba 81N] W. Bartel *et al.*, *Phys. Letters* **107B** (1981) 163.
[Ba 82N] W. Bartel *et al.*, *Phys. Letters* **108B** (1982) 140.
[Ba 82aN] W. Bartel *et al.*, *Phys. Letters* **115B** (1982) 338.
[Ba 82bN] W. Bartel *et al.*, *Phys. Letters* **119B** (1982) 239.
[Ba 82cN] M. Banner *et al.*, *Phys. Letters* **118B** (1982) 203.
[Ba 83N] W. Bartel *et al.*, *Phys. Letters* **121B** (1983) 203.
[Be 81N] Ch. Berger *et al.*, *Phys. Letters* **107B** (1981) 168.
[Be 82N] M. A. Bég, A. Sirlin, *Phys. Reports* **88** (1982) 1.
[Be 82aN] B. Berg *et al.*, *Phys. Letters* **113B** (1982) 65, **114B** (1982) 324 and *Annals Phys.* **142** (1982) 185.
[Be 82bN] H. J. Behrend, *Phys. Letters* **114B** (1982) 282.
[Be 83N] S. Behrends *et al.*, 'Observation of Exclusive Decay Modes of b-flavoured Meson', *Phys. Rev. Letters* to appear).
[Bl 82N] E. D. Bloom; *Paris-Conference 82*,* p. C3-407.
[Br 80N] R. Brandelik *et al.*, *Phys. Letters* **94B** (1980) 437.
[Br 81N] R. Brandelik *et al.*, *Phys. Letters* **107B** (1981) 290.
[Br 82N] R. Brandelik *et al.*, *Phys. Letters* **117B** (1982) 365.
[Bu 81N] A. J. Burgas; 'A Tour of Perturbative QCD' in *Proceedings 1981 International Symposium on Lepton and Photon Interactions at High Energies*', Bonn (1981), p. 636 (W. Pfeil, Ed.; Bonn 1981).
[Bu 81aN] W. Buchmüller, S. H. Tye, *Phys. Rev.* **D24** (1981) 132.
[Ca 82N] L. Caneshi *et al.*, *Nucl. Phys.* **B200** [FS4] (1982) 409.
[Co 82N] B. Cox, *Paris-Conference 82*,* p. C3-140.
[Cr 82N] L. Criegee, G. Knies, *Phys. Reports* **83** (1982) 152.
[Cr 82aN] M. Creutz, K. J. M. Moriarty, *Phys. Rev.* **D26** (1982) 2166.
[Da 82N] M. Davier, *Paris-Conference 82*,* p. C3-471.
[Do 82N] J. F. Donoghue, *Paris-Conference 82*,* p. C3-89.
[Dr 82N] J. M. Drouffe *et al.*, *Phys. Letters* **115B** (1982) 301.
[Ed 82N] C. Edwards *et al.*, *Phys. Rev. Letters* **48** (1982) 458.
[Ei 82N] F. Eisele; *Paris-Conference 82*,* S. C3-337.
[Fa 82N] K. Fabricius *et al.*, *Z. Physik C* **11** (1982) 315.
[Fr 82N] P. Franzini, J. Lee-Franzini, *Phys. Reports* **81** (1982) 239.
[Fu 82N] F. Fucito *et al.*, *Nucl. Phys. B* **210** [FS 6] (1982) 407.
[Ga 82N] J. Gasser, H. Leutwyler, *Phys. Reports* **87** (1982) 77.
[Ha 81N] H. Hamber, G. Parisi, *Phys. Rev. Letters* **47** (1981) 1792.
[Ha 82N] A. Hasenfratz *et al.*, *Phys. Letters* **110B** (1982) 282.
[Ha 82aN] G. Hanson *et al.*, *Phys. Rev.* **D26** (1982) 991.
[He 80N] F. W. Hehl, J. Nitsch, P. v. d. Heyde; 'Gravitation and the Poincaré Gauge Field Theory with Quadratic Lagrangian' in *General Relativity and Gravitation*, vol. 1, p. 329 (A. Held, Ed.; New York 1980).
[Is 82N] K. Ishikawa *et al.*, *Phys. Letters* **116B** (1982) 429, 120B (1983) 387.
[Kr 82N] G. Kramer, *13. Spring Symposium on High Energy Physics*, Bad Schandau, DESY-Preprint, DESY 82/29 (1982).
[Ma 82N] A. Martin, *Paris-Conference 82*,* p. C3-96.
[No 81N] V. A. Novikov *et al.*, *Nucl. Phys. B* **191** (1981) 301.
[Po 82N] H. D. Politzer, *Paris-Conference 82*,* p. C3-659.

* The complete reference of 'Paris Conference 82' is the following: *Proceedings of the 21st International Conference on High-Energy Physics*, Paris (1982) (P. Petiau, M. Porneuf, Eds.; *J. Phys.* (France), No. 12, t. 43, 1982).

[Pr 79N] C. Prescott *et al.*, *Phys. Letters* **77B** (1978) 347, 84B (1979) 524.
[Re 82N] L. J. Reinders *et al.*, *Nucl. Phys. B* **196** (1982) 125.
[Sc 80N] D. L. Scharre *et al.*, *Phys. Letters* **97B** (1980) 329.
[Sö 81N] P. Söding, G. Wolf, *Ann. Rev. Nucl. Sci.* **31** (1981) 231.
[St 81N] The Staff of the CERN Proton-Antiproton Project, *Phys. Letters* **107B** (1981) 306.
[We 82N] D. H. Weingarten, *Phys. Letters* **109B** (1982) 57.
[Wo 82N] G. Wolf; *Paris-Conference 82*,* C3-525.

Index